A New Focus on
Groundwater–Seawater Interactions

IAHS AISH IAPSO

A New Focus on Groundwater–Seawater Interactions

Edited by

WARD SANFORD
National Center, US Geological Survey
Reston, Virginia, USA

CHRISTIAN LANGEVIN
Florida Integrated Science Center, US Geological Survey
Fort Lauderdale, Florida, USA

MAURIZIO POLEMIO
Consiglio Nazionale delle Ricerche,
Istituto di Ricerca per la Protezione Idrogeologica, Bari, Italy

PAVEL POVINEC
Mathematics, Physics and Informatics, Comenius University
Bratislava, Slovakia

Proceedings of an international symposium, HS1001, convened jointly by the IAHS International Commission on Groundwater and the International Association for the Physical Sciences of the Oceans (IAPSO) during IUGG2007, the XXIV General Assembly of the International Union of Geodesy and Geophysics at Perugia, Italy, July 2007.

IAHS Publication 312
in the IAHS Series of Proceedings and Reports

Published by the International Association of Hydrological Sciences 2007

IAHS Publication 312
ISBN 978-1-901502-04-6

British Library Cataloguing-in-Publication Data.
A catalogue record for this book is available from the British Library.

The papers included in this volume have been reviewed and some were extensively revised by the Editors, in collaboration with the authors, prior to publication.

IAHS is indebted to the employers of the Editors for the invaluable support and services provided that enabled them to carry out their task effectively and efficiently.

Publications in the series of Proceedings and Reports are available from:
IAHS Press, Centre for Ecology and Hydrology, Wallingford, Oxfordshire OX10 8BB, UK
tel.: +44 1491 692442; fax: +44 1491 692448; e-mail: jilly@iahs.demon.co.uk

Printed in The Netherlands by Krips BV, Meppel.

Preface

Water and chemical fluxes across the sea bottom provide an important linkage between terrestrial and marine environments. From the marine perspective, these water fluxes, commonly referred to as submarine groundwater discharge (SGD), may contain elevated nutrient concentrations or high levels of other potentially harmful contaminants. Terrestrially derived SGD can also be an important source of freshwater for estuarine ecosystems that require relatively low salinities. For these reasons, the past decade has shown a rapid increase in the level of interest from estuary and marine scientists toward a better understanding of SGD. From the terrestrial perspective, SGD has also been a topic of interest to those studying saltwater intrusion and management of coastal aquifers. Saltwater intrusion studies commonly employ some form of a water balance method, whether through numerical modelling or volumetric calculations, to explain intrusion patterns and develop predictions and management plans. In developing a water balance for a coastal aquifer, estimates for all of the key components, including SGD, are synthesized. Although the motivation may be different depending on whether one works from the marine or terrestrial perspective, both groups have a common goal of obtaining accurate SGD estimates.

Unlike rivers and streams, where discharges to the ocean can be observed and quantified using proven techniques, estimates of SGD are much more difficult to obtain. Not only are there issues with making accurate field measurements in difficult environments where wind, waves and storms pose logistical concerns, but upscaling field measurements to other spatial and temporal scales significantly increases the uncertainties associated with those estimates. Consequently, a diverse suite of methods and approaches has evolved to characterize, quantify, and better understand the exchange of water and chemicals across the sea floor. For example, marine scientists have developed sophisticated strategies for using the chemical isotopes of radium to estimate SGD. From the terrestrial side, the same variable-density flow models used to simulate saltwater intrusion are also capable of providing quantitative estimates of SGD. Both groups rely on physical methods using head measurements, seepage meters or electromagnetic geophysical techniques to obtain qualitative and quantitative SGD estimates. As this diverse suite of techniques continues to improve, so does our ability to quantify the interaction between seawater and groundwater.

To unify the broad interest in the subject of SGD and groundwater–seawater interactions, a symposium entitled: *A New Focus on Groundwater–Seawater Interactions* was held during the 24th General Assembly of the International Union of Geodesy and Geophysics (IUGG) in Perugia, Italy, from 2–13 July 2007. The symposium was organized by the International Association of Hydrological Sciences (IAHS) International Commission on Groundwater (ICGW) and by the International Association for the Physical Sciences of the Oceans (IAPSO). Members from both the marine and terrestrial groups elected to participate in the exchange. This IAHS Redbook contains 38 peer-reviewed papers on one or more aspects of water and chemical fluxes across the sea floor. The papers have been organized into three general categories: physical, chemical, and modelling.

The editors are grateful to the symposium participants for their scientific contributions, which together form an impressive volume on the topic of groundwater–seawater interactions. We also thank symposium participants for their prompt submission of manuscripts and adherence to a tight publication schedule. Cate Gardner and her colleagues at IAHS Press are graciously thanked for their tireless effort in preparing the papers for publication. Lastly, the editors thank Mary Hill (President of ICGW) for suggesting this topic for the IUGG meeting in Perugia and Pierre Hubert (Secretary General of IAHS) for coordinating the symposium details.

Editor-in-Chief

Ward Sanford
National Center, US Geological Survey
Reston, Virginia, USA

Co-Editors

Christian Langevin
Florida Integrated Science Center
US Geological Survey
Fort Lauderdale, Florida, USA

Maurizio Polemio
Consiglio Nazionale delle Ricerche,
Istituto di Ricerca per la Protezione Idrogeologica
Bari, Italy

Pavel Povinec
Mathematics, Physics and Informatics,
Comenius University
Bratislava, Slovakia

Contents

3 CHEMICAL APPROACHES

1 OVERVIEWS

BACKGROUND AND SUMMARY

A new focus on groundwater–seawater interactions

CHRISTIAN LANGEVIN[1], WARD SANFORD[2], MAURIZIO POLEMIO[3] & PAVEL POVINEC[4]

1 *Florida Integrated Science Center, US Geological Survey, Fort Lauderdale, Florida, USA*
langevin@usgs.gov

2 *National Center, US Geological Survey, Reston, Virginia, USA*

3 *Consiglio Nazionale delle Ricerche, Istituto di Ricerca per la Protezione Idrogeologica, Bari, Italy*

4 *Mathematics, Physics, and Informatics, Comenius University, Bratislava, Slovakia*

INTRODUCTION

Water and chemical fluxes across the sea floor provide an important linkage between terrestrial and marine environments. Oceanographers recognize that these fluxes may act as a source of nutrients or other harmful contaminants to marine systems (e.g. Johannes, 1980; Valiela *et al*., 1990). These fluxes may also act as a beneficial source of freshwater for coastal marine estuaries that require relatively low salinities. Hydrologists and hydrogeologists recognize that fluxes across the sea floor comprise an important part of the water balance for coastal aquifers. Most fresh groundwater discharge to the ocean is derived from terrestrial aquifer recharge. Management of coastal aquifers requires careful estimates of recharge and other hydrological components, such as groundwater discharge. These estimates are commonly combined into a comprehensive water budget to evaluate how much groundwater might be available for municipal uses and whether saltwater intrusion may be a potential concern. Excessive groundwater withdrawals can cause saltwater intrusion by intercepting the seaward flux of freshwater that prevents saltwater from intruding a coastal aquifer. Quantitative estimates of fresh groundwater discharge toward the coast can provide a basis for determining safe withdrawal rates. Oceanographers, marine scientists, and those studying and managing saltwater intrusion in coastal aquifers, share a common goal of quantification and understanding of groundwater and seawater interactions.

Submarine groundwater discharge, or SGD, has become a popular term in the literature for describing the flux of water across the sea floor. Burnett *et al*. (2003) specifically define SGD as the discharge of aquifer porewater across the sea floor and into the ocean. They define flow in the opposite direction as submarine groundwater recharge (SGR). SGR is the recharging flux of seawater into the aquifer. The presence of SGR does not necessarily indicate saltwater intrusion, which occurs when saline water moves into parts of the aquifer previously occupied by freshwater. Figure 1 shows a conceptual model of the types of flow patterns that are expected to exist in

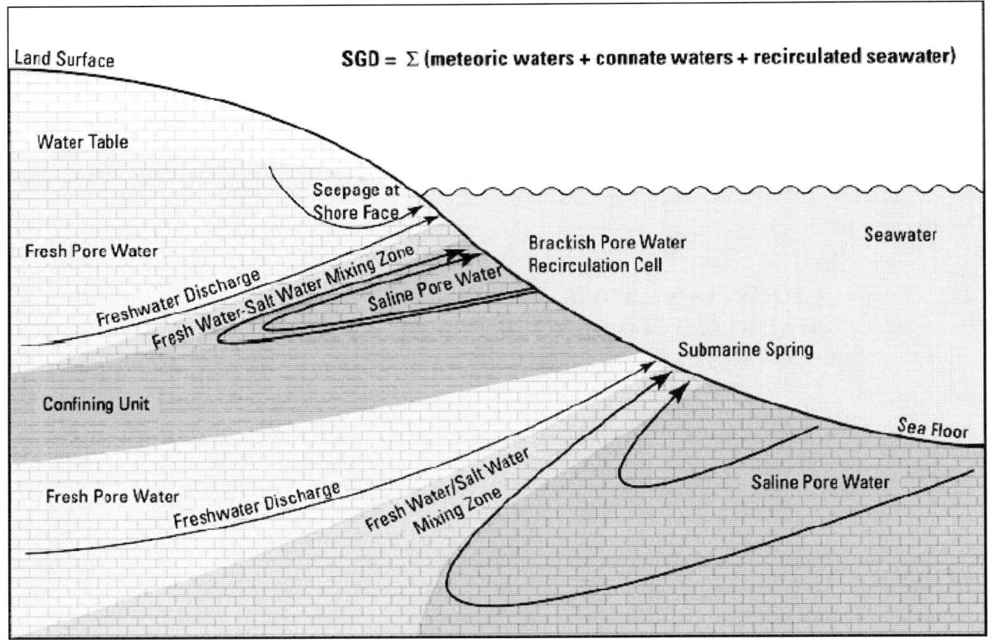

Fig. 1 Schematic showing various components of submarine groundwater discharge (SGD), from Swarzenski *et al.* (2004).

many coastal aquifers at the terrestrial–marine boundary. Through a detailed field study, Kohout (1960) showed that as fresh groundwater flows seaward, it meets and mixes with saline groundwater before discharging into the ocean. Because of this mixing and subsequent discharge to the ocean, seawater is drawn into the aquifer to replace the saline groundwater that discharged to the ocean. Seawater drawn into the aquifer is sometimes referred to as recirculated seawater. Michael *et al.* (2005) suggest that the seasonality of terrestrial recharge may also act as a mechanism for recirculating seawater through a coastal aquifer. As the mixing zone moves landward and seaward in response to seasonal fluctuations in recharge, seawater may be drawn in and flushed out of the aquifer over the course of a year. SGD can also occur at the bottom of the open ocean, even in the absence of a terrestrial connection. Waves, tides, and ocean currents can create hydraulic gradients that pump seawater across the sea floor.

Interest in groundwater–seawater interactions continues to receive a significant amount of attention in the literature. Saltwater intrusion, freshwater deliveries to marine estuaries, and nutrient loading are persistent problems of global importance. The problems are difficult to address, however, because of the elusive nature of SGD. Fortunately, the science is advancing. The journal *Biogeochemistry*, dedicated the entire November 2003 issue to the subject of SGD as did the journal *Ground Water*, in the December 2004 issue. Zektser & Dzhamalov (2007) released a comprehensive review on the subject of SGD and groundwater–seawater interactions in a new book: *Submarine Groundwater*. In their book, they suggest that SGD and related studies should be categorized into a new field called "Marine Hydrogeology". The concentrated efforts of numerous researchers from a wide range of disciplines have led to substantial advancements in characterizing SGD, but there is still more work to be done.

This IAHS volume contains 38 papers presented at the symposium on groundwater–seawater interactions. This introductory paper briefly introduces the topic by describing and characterizing the approaches used to measure SGD, and presents a brief summary of each paper. The concept of integrating the efforts of those studying SGD from the marine perspective and those working from the terrestrial side is promoted here as a "new focus on groundwater–seawater interactions".

APPROACHES FOR MEASURING SUBMARINE GROUNDWATER DISCHARGE

Unlike rivers and streams where discharges to the ocean can be observed and quantified using proven techniques, estimates of SGD are much more difficult to obtain. Not only are there issues with making accurate field measurements in difficult environments where wind, waves and storms pose logistical concerns, but upscaling field measurements to other spatial and temporal scales significantly increases the uncertainties associated with those estimates. SGD can occur as widespread diffuse flow across the sea floor, or as channelized flow through fractures, breaches in low-permeability marine sediments, karst conduits, or through other preferential flow zones. As one might expect, the presence of heterogeneities can complicate upscaling of even the best point measurements of SGD. To overcome some of these inherent complications with measuring SGD, a diverse suite of methods and approaches have evolved to characterize, quantify, and better understand the exchange of water and chemicals across the sea floor. Following a classification scheme by Burnett *et al.* (2003), these approaches can be loosely grouped into categories that are based on physical measurements, chemical tracers, and modelling.

Physical measurements of SGD are typically made using seepage meters; however, Darcy-based calculations using field data or qualitative estimates based on results from geophysical surveys are also included here in the category of physical measurements. These approaches based on physical measurements have continued to improve as viable methods for assessing SGD. For example, the Lee-type seepage meter (Lee, 1977), which is based on measurements of SGD that has been funnelled into a plastic bag, has served as a model for the development of more sophisticated meters. Today's (2007) seepage meters are capable of digitally logging seepage rates at frequent intervals and use high-precision thermal, electromagnetic, and ultrasonic technologies to measure fluxes.

Naturally occurring chemical tracers such as the isotopes of radium or radon can be used to provide indirect measurements of SGD rates. Moore (1996, 1999, 2003) describes the use of four radium isotopes with different decay rates that can be used to quantify SGD rates. Burnett & Dulaiova (2003) demonstrate the use of radon to quantify SGD. The radium and radon methods are based on the general principle that groundwater is enriched with respect to these isotopes in comparison to surface waters, and can therefore be used as a quantitative indicator of SGD. An advantage of these chemical tracers is that they can provide an integrated average rate of SGD over very large areas.

Variable-density groundwater flow and solute transport models are often used to simulate saltwater intrusion, but these same models can also be used to provide quantitative estimates of SGD. Numerical modelling is a sophisticated way of distributing aquifer recharge, in space and time, to the various outflow boundaries. Use of numerical models is appealing because a simulation can provide spatially and temporally detailed estimates of SGD rates. A well-calibrated numerical model is a defensible tool for estimating SGD within an entire study area by interpolating and extrapolating field measurements in space and time. Numerical models can also be used to predict future SGD rates or rates for other hydrological conditions (e.g. predevelopment), provided a reasonable set of hydrological stresses can be assigned for those conditions. The number of reported studies using numerical models specifically for SGD continues to increase (e.g. Langevin, 2001; Smith & Nield, 2003; Smith & Zawadski, 2003).

OVERVIEW OF THE PAPERS IN THIS VOLUME

This volume contains 38 papers on one or more aspects of groundwater–seawater interactions. The papers are grouped into three categories (physical approaches, chemical approaches, and mathematical modelling approaches) based on the dominant approach used in the study. An exception to this is the three overview papers that contain information from more than one approach. These overview papers are summarized in the next paragraph. Physical approaches are those that use hydrogeological analysis of monitoring well data, seepage meters, or geophysical measurements. This volume presents eight papers in this group. Chemical approaches can be either terrestrial- or marine-based and involve analysis of marine or coastal aquifer water quality. Chemical and isotopic tracer methods, such as the use of radon and radium for estimating SGD also fall into this group. This volume presents 14 papers in this group. The last group consists of 13 papers and contains those that report the use of a mathematical model to characterize groundwater–seawater interactions. Numerical modelling with finite-difference or finite-element models is the most common approach in this group.

Three of the papers provide overviews of groundwater–seawater interactions. The paper by Zektser & Dzhamalov provides an overview of the general topic of marine hydrogeology. Water fluxes between the terrestrial and marine systems cannot be directly measured, and thus, SGD may be one of the largest sources of uncertainty in the water balance for the ocean. This new field of marine hydrogeology focuses on the subsurface water exchange between land and sea. Kontar identifies the hydrological and water quality contamination problems resulting from the December 2004 tsunami in the Indian Ocean. He also describes a study by the International Commission on Groundwater–Seawater Interaction to evaluate the processes of groundwater–seawater interactions in areas affected by the tsunami. Bratton provides an overview on the importance of shallow confining units in controlling patterns of SGD.

Physical Approaches: hydrogeological analysis, seepage meters, and geophysical methods

Freshwater resources in coastal and island aquifers continue to be of paramount importance. The paper by Bonacci & Gabric provides an overview on the status of eight brackish springs along the coast of the Croatian Adriatic Sea. They describe technical concerns with protecting these springs from saltwater intrusion. Majumdar & Das used electrical resistivity soundings and water quality measurements to characterize the chemistry and extent of fresh groundwater resources on Sagar Island, the largest island in the Ganga Delta of West Bengal, India.

Japanese researchers are actively involved with studies of SGD. Ishitobi *et al.* developed a method for using temperature measurements from a fibre-optic cable to estimate SGD rates. Results from Lee-type and heat-flow seepage meters were used to calibrate temperature measurements from the optic cable. The total SGD was estimated to be 37% of the river discharge in Osaka Bay, Japan. Miyaoka quantified the strong seasonal variability of groundwater–seawater interactions in a coastal aquifer in Japan using measurements from two sets of nested monitoring wells and electrical resistivity surveys. Shimada *et al.* conducted a comprehensive hydrological study of a small mountainous basin in the Uto Peninsula of Kumamoto, Japan. Using a variety of physical, chemical, and modelling techniques, they characterized hydrological conditions and verified the presence of SGD. Taniguchi *et al.* used heat-type automated seepage meters and electrical resistivity methods to evaluate spatial and temporal SGD patterns along the coast of Yatsushiro, Japan. Their results are presented in the context of a collection of global estimates, which indicate that SGD may be as much as 7% of the total discharge from rivers.

Sophisticated new geophysical methods are gaining popularity in studies of groundwater–seawater interactions. Ozorovich & Kontar describe the use of the MARSES TEM geophysical instrument for coastal groundwater–seawater applications. MARSES TEM is a small portable instrument, based on the transient electromagnetic method (TEM), and has been used to characterize spatial and temporal variations in the freshwater–saltwater interface at depths of up to 300 m. Swarzenski *et al.* demonstrated the use of improved, multi-channel, electrical resistivity geophysical techniques to characterize subsurface salinity patterns. A stationary resistivity profile, extending 20 m onshore and about 110 m offshore in Tampa Bay, Florida, clearly showed the presence of a shallow freshwater tongue. The tongue could be seen to depths of about 8 m and extended out beneath the bay.

Chemical Approaches: terrestrial water quality analysis, marine chemistry, and nutrients

Chemical tracer methods for quantifying SGD have been rapidly improving. Burnett *et al.* provide a discussion on some of the uncertainties that remain with using ^{222}Rn as a quantitative tracer of SGD. Difficulties in using this approach can arise when trying to assign reasonable estimates of end-member radon concentrations and atmospheric and mixing losses. This paper describes approaches for handing these difficulties and quantifying the uncertainties. Povinec describes the novel application of an underwater gamma-ray spectrometer in studies of groundwater–seawater interactions to measure ^{222}Rn activity. Tests off the coast of Sicily, Italy, and southeastern Brazil indicate that the method is capable of monitoring spatial and temporal variations of SGD. Weinstein *et al.* used seepage meters and radon concentrations to quantify SGD in Dor Bay along the southern Carmel Coast of Israel. Anomalous decreases in discharge at some seepage meters were attributed to localized seawater recharge events. Moore characterized the spatial pattern of ^{226}Ra along the southern Atlantic coasts of North and South Carolina, Georgia, and Florida, USA. Radium enrichments measured in offshore seawater samples show clear patterns of spatial and seasonal variability of regional-scale SGD. Peterson *et al.* developed a lumped-parameter box model based on water, radon, and salinity fluxes to calculate rates of SGD off the coast of Hawaii, USA. Estimated discharges from the box model show clear tidal variations with overall average rates that compare well with independent estimates.

Several papers in this volume focus on estimating nutrient fluxes and identifying their sources. Onodera *et al.* used a network of piezometers and water quality measurements to characterize nutrient transport dynamics within a tidal flat along the southern coast of Ikuchijima Island, Japan. Yasumoto *et al.* quantify nutrient inputs to Ariake Bay, Kyushu Island, Japan. Lee-type seepage meters were installed, and chemical analyses were performed to measure nutrients, major ions, and other chemical species. Results suggest that SGD may act as a significant source of nutrients to Ariake Bay. Umezawa *et al.* used a suite of physical and chemical methods to characterize SGD and associated nutrient transport in the Gulf of Thailand, just offshore from Bangkok, Thailand. Their results indicate that upward water and chemical fluxes measured in the bay could be attributed to recirculation of the overlying water.

Several authors used chemical methods to address unique research issues related to groundwater–seawater interactions. An analysis of metal chemistry and transport in the Mirim-Patos Lagoon system in southeastern Brazil was conducted by Windom *et al.* Their results suggest that SGD may act as a sink and source for metals in coastal regions. Mahara *et al.* characterized the chemical and isotopic nature of various water types collected from a coal mine in Japan that extends 8.5 km offshore and 700 m below sea level. They conclude that pore water trapped in Cretaceous formations beneath the sea floor has been isolated for over 2 million years. Using chlorine isotopic ratios, Tokunaga *et al.* concluded that diffusion is the dominant process for chloride transport beneath the sea floor in Yatsushiro Bay, Japan.

Water quality analyses are commonly employed in coastal aquifers as part of saltwater intrusion studies. For example, Bocanegra *et al.* used chemical indicators to compare and contrast the anthropogenic impacts at coastal aquifers in Argentina and Italy. Gattacceca *et al.* combined an isotopic analysis and electrical conductivity measurements from 12 monitoring wells to identify causes of saltwater intrusion in the semiconfined aquifer of the southern Venice lagoon system, Italy. Chulli *et al.* collated existing data for the coastal Gabes aquifer system in southern Tunisia to determine the causes of salinity increases.

Modelling Approaches: numerical, analytical, and transfer functions

Application of unique mathematical models or use of mathematical models to address unique problems has been performed by several authors. Cardenas & Wilson developed a numerical model to quantify the effects of turbulent sea currents and bedform topography on SGD. A current flowing over an irregular sea bottom can create areas of low and high pressure, which drive downwelling or upwelling through the sediments. The area and effective penetration depth of the SGD is shown to be a function of the ocean current Reynolds number. Ambient, or regional, groundwater flow can reduce the current-driven exchange. Attanayake & Sholley used an analytical flow net model to estimate hydraulic gradients at a proposed low-level nuclear waste disposal site near an island in the Taiwan Straits. Their modelling predicted that relatively stagnant conditions could be expected beneath the sea floor at the proposed disposal site. Van der Velde *et al.* used transfer function theory to predict the change in groundwater salinity in response to rainfall and dilution. An analysis using data from the Kingdom of Tonga, located in the southern Pacific Ocean, suggests that most lag times, which result from a combination of percolation times and buoyancy stabilization, are less than 2000 days (about 5.5 years).

Numerical models that include the effects of density variations can be used to provide insight into coastal hydrological processes. Fratesi *et al.* used a three-dimensional numerical model to quantify the effect of an offshore sinkhole on SGD patterns. Depending on the location of the sinkhole with respect to the freshwater–saltwater interface, as much as 20% of the terrestrially derived recharge may be routed through the sinkhole and into the ocean. Swain & Wolfert used results from an integrated surface water and groundwater model of the coastal Everglades, Florida, to show that the freshwater–saltwater interface beneath coastal wetlands is quite different from the classical interface. Rather than being located at the coastline, the top of the interface is found to be tens of kilometres inland. Dausman *et al.* used a generalized numerical model of variable density flow and solute and heat transport for southern Florida. The model was used to identify the hydrogeological conditions under which various remote detection methods (airborne electromagnetic geophysical surveys or infrared surveys) might identify SGD areas. Lee *et al.* used a numerical model to evaluate the effects of heavy rainfall events on SGD. Results show that heavy rainfall events and their effects on SGD could play an important role in delivering dissolved nutrients and other contaminants to the ocean.

Freshwater resources in coastal aquifers continue to receive widespread attention, and evaluating susceptibility to saltwater intrusion is commonly performed with a numerical model. Zhenghua *et al.* used a numerical model to characterize saltwater intrusion patterns on Xiamen Island, China, and to predict future intrusion patterns in response to the construction of a 15-km water diversion project. Model results suggest that intrusion patterns reach equilibrium after about one year of operation. Park *et al.* developed a three-dimensional numerical model for a coastal aquifer along the western coastline of Korea. They show the importance of explicitly including aquifer heterogeneity and anisotropy in saltwater intrusion models used to predict the effects of groundwater pumping. Ranjan *et al.* developed a numerical model of the Walawe River Basin, Sri Lanka. Analyses with the model suggest it is capable of predicting the effects of countermeasures designed to prevent saltwater intrusion. Sherif & Kacimov developed a numerical model of the coastal aquifer of Wadi Ham to quantify the effects of artificial recharge on saltwater intrusion. Results from the model suggest that a recharge pond could significantly improve water quality patterns in the aquifer and cause the freshwater–saltwater interface to move more than a kilometre seaward.

In addition to simulating saltwater intrusion, variable-density numerical models can be used to predict nutrient and contaminant transport to seas and estuaries. Sanford

& Pope used a numerical model to predict nitrogen loading to the Chesapeake Bay from the Delmarva Peninsula, USA. Predictions from the model suggest that nitrogen loads from SGD will continue to increase unless future surface application rates are drastically reduced. Furthermore, there appears to be a lag time of up to a decade or more before reductions in surface application rates will reduce nitrogen loads to the bay. La Licata *et al.* developed a numerical model of a coastal refinery in Italy to evaluate the chemical migration patterns to the sea. The effects of tides on chemical transport through the coastal aquifer were shown to be important under certain conditions.

SUMMARY

In summary, the papers in this volume present research by those working from the marine and the terrestrial sides of issues related to SGD and groundwater–seawater interactions. The first part of this paper provides an introduction and background information on the subject of SGD and groundwater–seawater interactions. The second part of this paper provides an overview of the 38 symposium papers and places them in context according to the methods used to quantify SGD. The papers presented in this volume describe important contributions to the literature and document a variety of investigative approaches applied over a range of conditions at locations across the globe.

REFERENCES

Burnett, W. C. & Dulaiova, H. (2003) Estimating the dynamics of groundwater input into the coastal zone via continuous radon-222 measurements. *J. Environ. Radioact.* **69**(1-2), 21–35.

Burnett, W. C., Bokuniewicz, H., Huettel, M., Moore, W. S. & Taniguchi, M. (2003) Groundwater and pore water inputs to the coastal zone. *Biogeochemistry* **66**, 3–33.

Johannes, R. E. (1980) The ecological significance of the submarine discharge of ground water. *Marine Ecol. Prog. Ser.* **3**, 365–373.

Kohout, F. A. (1960) Cyclic flow of saltwater in the Biscayne aquifer of southeastern Florida. *J. Geophys. Res.* **65**, 2133–2141.

Langevin, C. D. (2003) Simulation of submarine ground water discharge to a marine estuary: Biscayne Bay, Florida. *Ground Water* **41**, **6**, 758–771.

Lee, D. R. (1977) A device for measuring seepage flux in lakes and estuaries. *Limnol. Oceanogr.* **22**, 140–147.

Michael, H. A., Mulligan, A. E. & Harvey, C. F. (2005) Seasonal oscillations in water exchange between aquifers and the coastal ocean. *Nature* **436**(7054), 1145–1148.

Moore, W. S. (1996) Large groundwater inputs to coastal environments revealed by ^{226}Ra enrichments. *Nature* **380**, 612–614.

Moore, W. S. (1999) The subterranean estuary: a reaction zone of ground water and sea water. *Marine Chem.* **65**, 111–126.

Moore, W. S. (2003) Sources and fluxes of submarine groundwater discharge delineated by radium isotopes. *Biogeochemistry* **66**, 75–93.

Smith, A. J. & Nield, S. P. (2003) Groundwater discharge from the superficial aquifer into Cockburn Sound Western Australia: estimation by inshore water balance. *Biogeochemistry* **66**(1-2), 125–144.

Smith, L. & Zawadzki, W. (2003) A hydrogeologic model of submarine groundwater discharge: Florida intercomparison experiment. *Biogeochemistry* **66**(1-2), 95–110.

Swarzenski, P. W., Bratton, J. F. & Crusius, J. (2004) Submarine ground-water discharge and its role in coastal processes and ecosystems. *US Geological Survey Open-File Report 2004-1226*.

Valiela, I., Costa, J., Foreman, K., Teal, J. M., Howes, B. L. & Aubrey, D. G. (1990) Transport of groundwater-borne nutrients from watersheds and their effects on coastal waters. *Biogeochemistry* **10**, 177–197.

Zektser, I. S. & Everett, L. G. (2007) *Submarine Groundwater*. CRC Press, Taylor & Francis Group, Boca Raton, Florida, USA.

Regional assessment of groundwater discharge into seas: present-day concepts and methods

IGOR S. ZEKTSER & ROALD G. DZHAMALOV
Water Problems Institute, Russian Academy of Sciences, 3 Gubkina Street, 119991 Moscow, Russia
zektser@aqua.laser.ru; dzhamal@aqua.laser.ru

Abstract Studies of groundwater discharge into the seas and oceans are part of a complex hydrological–hydrogeological problem of underground water exchange between land and sea. Submarine discharge into seas and oceans is the least studied element of the present and prospective water and salt balance of the seas. Primarily, this is because groundwater inflow is the only water balance component that cannot be measured, and data needed for a well-grounded calculation of a water balance underground component are often missing. Therefore, it is important to determine this directly by hydrogeological methods. These methods permit areas of submarine groundwater discharge to be singled out and quantitatively characterized and, in some cases, make it possible to calculate the value of groundwater discharge causing these anomalies. The results of estimating the groundwater discharge to some seas and major lakes are considered.

Key words groundwater discharge; water balance; subsurface water exchange; subaqueous groundwater

The theory of subsurface water exchange between land and sea is closely connected with the general theory on groundwater flow, and began its development as a branch of hydrogeology during the mid-20th century. Concrete investigations of submarine groundwater discharge and, in particular, the intrusion of seawater into coastal areas have been carried out much earlier, for example, during the exploration for groundwater in coastal areas. However, the study of water exchange between land and sea at regional and global scales began relatively recently in conjunction with the need to obtain a reliable assessment of the role of groundwater in the water and salt balances of particular seas, and in global water circulation.

Investigations of the hydrogeological cycle, water balance and water resources of particular regions, sea basins, continents, and the Earth as a whole, have been hampered due to the absence of sufficient quantitative data on the subsurface water exchange between the land and the sea. The problems occurring in recent decades in inland seas and lakes have also brought to the fore the task of direct measurement and quantitative assessment of the role of submarine groundwater discharge in the water and salt balances of these water bodies. The poor knowledge of submarine water and chemical discharge is partly due to the difficulty in measuring these elements of the water balance directly. Until recently they were determined by calculating the discrepancy in the water balance equation. As a result, the final values included all the errors arising from the measurements of the rest of its terms. The prediction of changes in the water, salt and hydrobiological regimes of some water bodies, and the analysis of measures for their maintenance and protection requires a comprehensive study of the water exchange between the land and the sea.

The achievements of marine geology in obtaining geophysical data and unique deep-drilling data have allowed the analysis of this information in terms of the distribution and conditions of occurrence of submarine waters, their features, specific circulation, and verifying the parameters of their interaction with rocks, the sea and groundwaters of the land, as well as their influence upon biota. All this substantiates the need to investigate the role of submarine groundwater, not only in the water balance of the seas and oceans, but also in the geological processes occurring at their floors. In the other words, an independent branch is being formed in general hydrogeology – *marine hydrogeology*. This paper presents the scientific fundamentals of a large section of marine hydrogeology, dealing with the study of subsurface water exchange between land and sea. Special attention is paid to regional assessment and the identification of regularities in submarine groundwater discharge to seas and oceans, because these processes are manifested everywhere and can exert a significant influence upon the water and salt balances of individual water bodies or their parts. The intrusion of seawaters into coastal shores is also an element of subsurface water exchange between land and sea, but it has only a limited character and is activated under the presence of human activity.

The present-day achievements of marine geology and geophysics make it possible to perform hydrogeological zoning of the bottom of the seas and oceans, and to distinguish hydrogeological structures having or not having analogues on the continents. Of special interest in this respect are the most studied structures in the shelf areas of seas and oceans where geo-structural, hydrodynamic and hydrochemical features of water exchange between land and sea are very clearly manifested (Korotkov *et al.*, 1980; Zektser *et al.*, 1984). The factual data on the distribution and the features of groundwater migration at different depths have allowed the quantitative assessment of groundwater flows in the covers of the Earth's crust, and of their role in different geological–hydrogeological processes (Kononov, 1983; Zverev, 1993; Dzhamalov *et al.*, 1999; Shvartsev, 1999).

As mentioned above, a new scientific branch has been formed at the intersection between two allied sciences – *marine hydrogeology*. It has its own subject and objectives, aims and investigations. A system of the basic concepts and terms of this new science and its related definitions are given below. In the future, the proposed conceptual-terminological base will be improved in accordance with development of the science itself. Some concepts were defined using particular terminology used for the study of groundwater flow on the land (Zektser *et al.*, 1984).

The general concepts and terms of marine hydrogeology are as follows:

- *subaqueous* or submarine *groundwater* is the water enclosed in rocks composing the bottoms of large lakes, seas and oceans;
- *submarine groundwater flow* is the groundwater movement in rocks under the bottom of lakes, seas and oceans, occurring as the result of the general water circulation and geodynamic processes in the Earth's crust;
- *submarine ionic or chemical discharge* is the transfer of salts and chemical elements, dissolved in submarine groundwater, to lakes, seas and oceans.

The general term "*submarine groundwater flow*" is defined as the water exchange between rocks, composing the bottom (floor), and the marine basin. *Submarine groundwater discharge* implies an influx (discharge) of groundwater, generated on land, directly to the sea. The reverse process is the *penetration (intrusion) of seawaters* into the shores and aquifers of the land under the influence of different natural and artificial factors. All these processes in combination define the term "*subsurface water exchange between land and sea".*

The quantitative characteristics of the subsurface water exchange between the land and sea can be represented by, besides absolute values, such specific indices as *modulus* and the *linear flow rate* of groundwater discharge to the sea, as well a modulus of submarine groundwater discharge. The *modulus of groundwater discharge to the sea* is understood as losses of groundwater flow to the sea from a drainage area of 1 km^2, the discharge from which is directed directly to sea. The *modulus of submarine groundwater discharge* is defined by the characteristics of groundwater flow rate from an area of 1 km^2 of aquifer discharge on the sea bottom. The *linear groundwater discharge to the sea* is defined as the losses of groundwater flow per one width unit of its front or the shoreline of the sea. The linear discharge can also characterize seawater intrusion into the shore.

By analogy with groundwater discharge to sea, it is reasonable to introduce specific characteristics for the submarine chemical discharge. The term *chemical discharge* insufficiently reflects the essence of the physical-chemical processes of the transfer of dissolved salts with groundwater, but it is already used in literature and it is unfeasible to replace it. In this case, the amount of dissolved salts or particular chemical elements (compounds) transferred with groundwater directly to sea from 1 km^2 of drainage area, will be termed as a *modulus*, and the amount from a width unit of groundwater flow front or from 1 km of the shoreline will be defined as *linear losses of submarine chemical discharge.*

Groundwater is discharged to lakes, seas and oceans in the form of:

- juvenile waters during the degassing of the Earth's mantle;
- sedimentation waters at the expense of their expulsion during lithogenesis of marine sediments;
- subsurface component of the total river discharge (river low water runoff);
- direct groundwater discharge to seas not involving the river network.

The juvenile water discharge represents a "subsurface" component of the water balance of seas and oceans. The quantitative assessment of the volume of juvenile waters discharged to the seas is presently rather difficult. As reported by Timofeev *et al.* (1988), the amount of juvenile waters at the present-day phase of the Earth's evolution usually does not exceed 5% of the total hydrothermal discharge from volcanic areas. The isotopic composition of inert gases indicates that major volatile substances, including hydrogen, were degassed at the initial phase of the Earth's evolution (Verkhovsky *et al.*, 1985). However, the complete degassing of the mantle has not yet happened and the current emission of the mantle hydrogen and methane through rift zones confirms this (Kononov, 1983). In the current geological epoch the juvenile gas- and water-containing fluids are associated with rift zones; the mid-oceanic ridges play a leading role. According to Vinogradov (1967), the annual amount of juvenile water, contributed from volcanoes, hot springs and deep-seated faults, does not exceed 0.5–1.0 km^3 (an extremely small value for the current balance of the World's Oceans).

The methodical procedure of calculating groundwater discharge drained from rivers, which in the water balance equations is included in total river runoff, is presently well worked out. It should be taken into account that approximately one third of the water discharge from rivers flowing into the seas is generated at the expense of groundwater drained from the zone of intensive water exchange.

The waters in the underground part of the hydrosphere can be subdivided into: (1) *groundwater of the land* that has an unsaturated zone in the upper part of the hydrogeological cross-section and is linked with the atmosphere, and (2) *subaqueous groundwater* under the bottom of large lakes and seas and hydraulically linked with the waters of these water bodies. The most widely distributed type of underground water is subaqueous water, i.e. submarine waters occurring under the bottom of seas and oceans, the hydrodynamics and hydrochemistry of which are largely determined by their interconnection with seawaters. The submarine waters are subdivided into *infiltration waters* that are generated on land at the expense of precipitation and surface runoff, *sedimentation waters* that are formed directly within the marine area due to accumulation of sediments and their subsequent diagenesis and katagenesis, and *juvenile waters* that are connected with the degassing of the mantle.

The infiltration waters discharging from the land are mainly distributed in the shelf zone. Their current filtration recharge per unit time is much higher than the amount of recharge from sedimentation waters during sedimentary processes. Therefore, when conditions are favourable, the tongues of the confined filtration waters can intrude far into the sea area, reaching the continental slope and displacing the sedimentation waters. The results from drilling of the sea bottom have provided vivid evidence of the intrusion of fresh or low-mineralized groundwater formed on land (initial reports of the Deep Sea Drilling Project). The constant interaction between infiltration water, sedimentation water, and directly with seawater leads to the gradual equalization of the chemical composition and mineralization of submarine groundwater of different origin, owing to the convective–diffusive processes and physical-chemical reactions. It is possible that at certain depths a transient zone exists, the spatial characteristics of which are, to a great degree, determined by hydrodynamic and physical-chemical gradients of the counter-moving submarine groundwater, and, depending on the geofiltrative properties of the water-bearing rocks, the groundwater of different genesis may have a by-layer occurrence.

The specific features of the interaction between unconfined filtration water flow and seawaters are considered below. Theoretically, if the filtration properties are homogenous, the shallow groundwater would be entirely discharged in the area of the sea and groundwater line (coastal zone). When a sufficiently thick shallow aquifer has a close hydraulic linkage with seawater, usually in the coastal zone, a counter seawater flow is formed, which restricts the shallow groundwater current and causes it to wedge out and, hence, the discharge has a greater velocity. The velocity of the discharging shallow groundwater in this zone can be several times greater than the average velocity of the filtration flow above the shoreline or the tidal zone. Such occurrences are true only for ideal homogenous aquifers. In nature, the shallow aquifers are usually heterogeneous, whether for lithological composition and/or for filtration qualities. Closer to the shoreline, the filtration properties of such sedimentary confined/unconfined aquifers are, as a rule, decreased due to the presence of clayey materials and increased water pressure head. The entire submarine discharge of infiltration flow occurs not only in the narrow coastal or tidal zone, but also at a distance of several kilometres offshore. At the same time, the results of the direct measurements show that a considerable part of the total shallow groundwater filtration flow is discharged in the coastal zone with a width of several hundreds of metres.

The shallow groundwater of the land, like confined groundwater, transforms its composition and mineralization during submarine discharge because of mixing and physical-chemical reactions with seawaters. However, due to the agricultural and industrial development of many coastal areas, the composition of natural (including shallow groundwater) waters is subject to significant changes. The nitrate concentrations in shallow groundwater during submarine discharge in some coastal areas of the USA, Australia, islands of Jamaica and Guam vary from 20–80 mg/L to 120–380 mg/L. Such high concentrations of biogenic elements in the submarine waters cause the anomalous growth of biomass in the seawater, the appearance of specific species of microorganisms and algae, which, in turn, can serve as indicators of submarine discharge of groundwater of a certain composition. In the other words, the intensive human activity in the coastal drainage areas causes direct changes in shallow groundwater composition and exerts a significant influence upon the environmental state of seawaters due to the transfer of biogenes, pesticides, heavy metals and other toxic substances.

Submarine discharge of confined groundwater takes place through submarine springs usually associated with tectonic disturbances and the areas with developing fissured and karst rocks, as well as through seepage via low-permeability covers of aquifers and sea-bottom sediments.

It is important to analyse the principal possibility of a hydrodynamic interconnection between hydrogeological structures of the land and adjacent parts of oceans. It is acknowledged that the shelf and continental slope are parts of the land, which do not continue on the ocean bottom. A continental slope usually represents a tectonic flexure, within which the thickness of sedimentary rocks sharply reduces until the full wedging out and disappearance of some beds (Lisitsyn, 1978). In other words, the hydrogeological structures of continents and oceanic depressions are dissociated tectonically, and the boundary between them runs along the continental slope. In this case, the submarine infiltrating waters, formed on the land, are wedged out mainly within the shelf and continental slope. The submarine groundwater discharge in the shelf and continental slope is considerably favoured by the existing submarine canyons, i.e. deeply entrenched erosional valleys stripping not only Quaternary, but also more ancient rocks.

The interaction between the infiltration and sedimentation waters in the shelf has not been fully studied until now. Some researchers consider that elision processes, i.e. the pressing-out of water from clayey sediments during the compaction and diagenesis of the latter, play a leading role in all the phases of the formation of artesian basins. However, the modelling of the gravitational consolidation of clayey rocks has shown that with real initial parameters the pressure distribution in clayey strata is controlled by the conditions at their upper and lower boundaries (Dyunin, 2000). The determining influence upon the pore pressure distribution is exerted by the permeability of clays and the velocity of their sedimentation. In the actual conditions of the seas and oceans, the sedimentation velocity of clayey rocks seldom exceeds 10^{-3} m/year, and their permeability is usually very small and can increase with depth due to the formation of macro- and meso-jointing during lithogenesis. In these cases, according to the data from the model, the pore pressure distribution in the clayey strata actually has a linear character and depends on a ratio of the pressures at the upper and lower boundaries of clays. The hydrostatic pressure of the clayey strata in the upper cross-section of marine sediments increases with depth. The pressure is almost always higher at the base than at the strata roof. Therefore, the submarine sedimentation waters, pressed out during clay consolidation, move predominantly from underneath upward into the source sea basin.

It should be noted that the upper part of the cross-section of the bottom sediments is mainly composed of sedimentary clays with anomalously high moisture and porosity (to 80%) and a low density (1.2–1.3 g/cm^3). The resultant changes in their physical properties with depth are chiefly caused by compressive (gravitational) consolidation. The most intensive reduction of the porosity and moisture of clayey rocks take place within the first tens and hundreds of metres and gradually the effect diminishes with greater depth. Thus, within the first 100 m of depth, the clays lose more pore water than within the next 100 m, and at a depth of 300–500 m, the loss amounts to 70–80% of the initial moisture volume. Below that, the clay consolidation rates progressively slow down. Thus, a decrease in the porosity and moisture of clays with depth occurs in relation to an exponential dependence on the rate at which the major pore sedimentation waters are pressed out during the initial steps of sediment consolidation and then discharged to the source basin. In general, the compression regime of rock consolidation depends on the lithological composition, and on the ratio of pressure and temperature, which determines the physical-chemical and mineralogical processes of diagenesis and lithification of the sediment (Dzhamalov *et al.*, 1999).

The vertical filtration of sedimentation waters provides their constant linkage with seawaters in this case. The water exchange between compacting sediments and the sea basin is accompanied by salt exchange. The invariable composition of the seawaters during the recent geological epochs and salt exchange between the sedimentation and seawaters may cause the mineralization and salt composition of these waters to be similar. This is confirmed by the drilling experiences from the ship "Glomar-Challenger", when the boreholes, drilled into the ocean bottom are stripped at depths to several hundreds of metres from the bottom surface, the waters have a total salt concentration and a content of basic components almost similar to those in the modern seawater (initial reports of the Deep Sea Drilling Project).

One specific feature of submarine sedimentation waters is the invariability of their chemical composition. In submarine groundwater of the infiltration type one can observe a change in the chemical composition and mineralization, both in area and in section. However, in the submarine waters of the sedimentation type in the upper part of the geological cross-section, the regional hydrochemical zonality is actually absent.

The data from sea-bottom drilling indicates that the similarity of the chemical compositions of submarine and seawaters is most plainly apparent in areas with slow sedimentation rates. In other deeper parts of the oceans, a substantial change is often observed in the concentrations of some chemical elements in the submarine sedimentation waters. Namely, the Ca concentration usually increases, whereas concentrations of Mg and K decrease with depth. According to McDuff & Gieskes (1976), the breakdown of basalts in the deep parts of the oceanic depressions and the breakdown, though to a lesser degree, of dispersed volcanic material and carbonates in the sedimentary rocks are the main reasons for the increase of dissolved Ca and the decrease of Mg in the submarine waters.

Of greatest interest, in the hydrogeochemical respect, are the areas with modern submarine volcanism in the zones of the median oceanic ridges, where the major juvenile submarine waters are probably discharged. The magma in these regions is effused onto the ocean bottom or close to it, causing a powerful heat flux. The young oceanic crust is highly fissured because of the processes of cooling, compaction and extension. The seawater saturates the fissured zone and cools the magmatic body, solidifying it. The hydrothermal solutions of the rift zones are usually enriched with CO_2, ^3He, H_2, metals and other components. The temperature of these solutions can approach the magmatic solidification point (980°C), but close to the bottom surface it is usually around 10–30°C. At higher temperatures, the water has low values of density and viscosity, causing active convection and an increased ability to penetrate into rocks, to dissolve and leach water-bearing rocks. The thermal gas- and water-containing fluids, enriched with different chemical elements, serve as the basic source of polymetals. The isotopic compositions of carbon and helium in the hydrothermal solutions often indicate the age of these elements. The fluids of the oceanic rift zones represent a mixture of juvenile and, chiefly, seawater; due to the contact with the magma, the temperature in the latter increases and chemical composition changes.

The conceptual model of the water exchange between land and sea under consideration is based on the analysis of up-to-date information, and makes it possible to formulate the following conclusions that can be used as the methodical basis for creating regional mathematical models of the formation, distribution and migration of submarine groundwater:

- Submarine waters are subdivided into infiltration, sedimentation and juvenile waters. The infiltration waters reside only within shelf areas and are entirely discharged to the continental slope. The sedimentation waters are generated everywhere on the bottom of seas and oceans, but prevail within the floor of the World Ocean. The juvenile waters are mainly found in the zone of mid-oceanic ridges and their role in the current water balance of the World Ocean is not significant.
- The predominant type of submarine groundwater discharge is seepage from below. The constant interaction of sedimentation and seawaters due to the vertical water exchange makes their mineralization and salt composition similar (if there are no additional salt sources).
- According to the conditions of the formation and distribution of different types of submarine waters, the bottom of the oceans is subdivided into shelf, floor and mid-oceanic ridges, which differ from each other by peculiar hydrodynamic and hydrochemical regimes of interaction of submarine waters with groundwater of the land, rocks and seawaters.
- Submarine waters serve as the basic agent and medium of the migration of chemical elements in the Earth's crust. Due to the higher mineralization of groundwater compared with surface water, the influence of submarine chemical

discharge upon the salt balance of seas (especially of inland ones) can be significant. The migration of chemical elements with submarine waters is most boldly distinguished in the zone of mid-oceanic ridges, where, due to the convection of seawater, the active leaching of young basalts and thermal enrichment by gas- and water-containing fluids with different components, takes place.

REFERENCES

Dyunin, V. vI. (2000) *Hydrodynamics of Deep Layers of Oil- and Gas-bearing Basins*. Nauchny Mir, Moscow, Russia.

Dzhamalov, R. G. & Safronova, T. I. (1999) Influence of submarine sedimentation waterts on water and salt balance of oceans. *Water Resour.* **26**(6), 722–730.

Kononov, V.vI. (1983) *Geochemistry of Thermal Water of Areas of the Modern Volcanism*. Nauka Publ., Moscow, Russia.

Korotkov, A. I., Pavlov, A. I. & Yurovskiy, Yu. G. (1980) *Hydrogeology of the Shelf Areas*. Nedra, Leningrad, Russia.

Lisitsyn, A. P. (1978) *Processes of Ocean Sedimentation*. Nauka, Moscow, Russia.

McDuff, R. E. & Gieskes, J. M. (1976) Calcium and magnesium profiles in DSDP interstitial water: diffusion or reaction? *Earth Planet Sci. Lett.* **33**, 1–10.

Timofeyev, P. P., Kholodov, V. N. & Zverev, V. P. (1988) Hydrosphere and evolution of the Earth. *Izv. An. SSSR. Geology Series* **6**, 3–10.

Shvartsev, S. L. (1999) *Hydrogeochemistry of the Hypergenesis Zone*. Nedra, Moscow, Russia.

Verkhovsky, A. B., Yurgina, E. K. & Shukolyukov, Yu. A. (1985) *Degassing of the Earth and Geotectonics*. Nauka, Moscow, Russia.

Vinogradov, A. P. (1967) *Introduction to Geochemistry of the Ocean*. Nauka, Moscow, Russia.

Zektser, I. S., Dzhamalov, R. G. & Meskheteli, A. V. (1984) *Subsurface Water Exchange Between Land and Sea*. Hydrometizdat, Leningrad, Russia.

Zverev, V. P. (1993) *Hydrogeochemistry of Sedimentary Processes*. Nauka, Moscow, Russia.

Groundwater–seawater interactions in tsunami affected areas: solutions and applications

EVGENY A. KONTAR

International Commission on Groundwater–Seawater Interactions (CGSI), Moosstrasse 25, CH-3113 Rubigen, Switzerland and *P. P. Shirshov Institute of Oceanology, Russian Academy of Sciences, 36, Nakhimovsky Prospekt, Moscow 117851, Russia*

cgsi@bluewin.ch; ekontar@ocean.fsu.edu

Abstract The December 2004 tsunami in the Indian Ocean caused a disaster affecting thousands of kilometres of coastal zone in SE Asia. Many coastal wetlands were affected in the short term by the large inflow of salt seawater and littoral sediment deposited during the tsunami, and in the longer-term by changes in their hydrogeology caused by changes to coastlines and damage to sea-defences. Many water quality and associated problems were generated by the tsunami. The tsunami has created an accelerating process of salt-water intrusion and freshwater contamination in the affected regions that now require drastic remediation measures. According to the International Commission on Groundwater–Seawater Interaction (CGSI) these measures have to be economically feasible, environmentally sound and socially acceptable. We report here some results of preparation of the CGSI EU FP7 project related to the study of the processes of groundwater–seawater interactions in tsunami affected areas.

Key words coastal zone; submarine groundwater discharge; salt-water intrusion; tsunami; groundwater–seawater interactions

INTRODUCTION

On 26 December 2004, devastating tsunami waves caused a terrible humanitarian disaster affecting thousands of kilometres of the coastal belt in SE Asia. It is likely that many coastal wetlands were affected by the large inflow of salt seawater and littoral sediments that were deposited during the tsunami, with longer-term effects that include changes in the local hydrogeology caused by changes to coastlines and damage to sea-defences (Gupta, 2005). Serious problems pertaining to salinity changes were encountered in coastal south India during this tsunami (Fig. 1) and the consequent loss of fertility of agricultural land has been reported in requests for remedial measures to revitalize economic growth in these regions. Many water quality and associated problems generated by the tsunami and influencing coastal environments, are related to past and on-going contamination of terrestrial groundwaters, because those groundwaters are now seeping out along the shorelines affected by the tsunami. For example, chronic inputs of fertilizers and sewage on the land surface over several decades has resulted in higher groundwater nitrogen concentrations which, because of slow yet persistent discharge along the coast, eventually result in coastal marine eutrophication (Alagarswami, 1993; Krishnamoorthy, 1996; Sonak, 2000; Ramesh, 2001). Such inputs may thus contribute to the increased occurrence of coastal hypoxia, detrimental algal blooms, and associated ecosystem consequences. Tsunamis, in addition to increasing the magnitude of salt-water intrusion, can significantly affect these nutrient pollution problems.

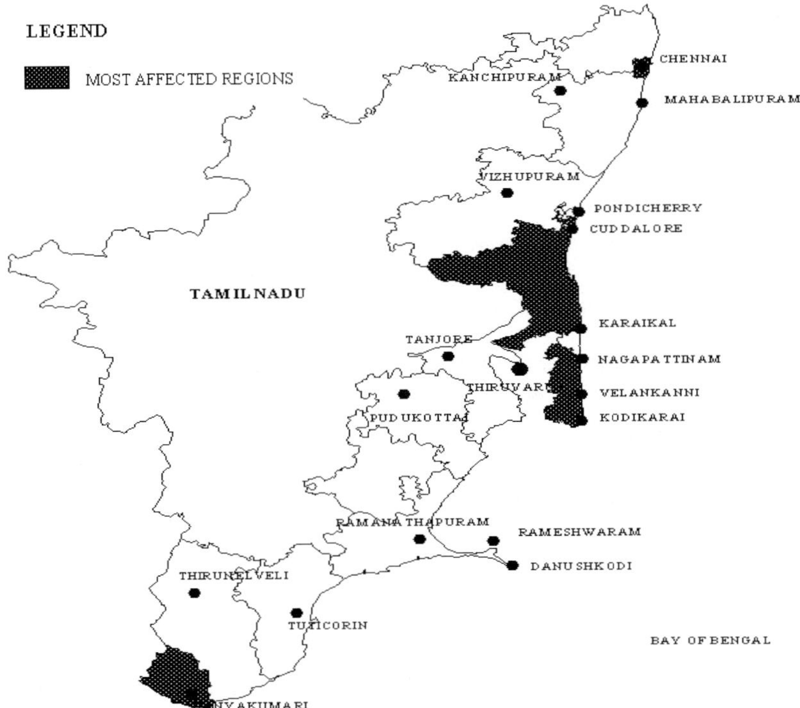

Fig. 1 Coastal areas most affected by the December 2004 tsunami in Tamil Nadu, India.

The International Commission on Groundwater-Seawater Interaction (CGSI) is working on preparation on of the EU FP7 integrated project related to the study of processes of groundwater–seawater interaction in tsunami affected areas.

RESEARCH DIRECTION

The coastal zone of the Indian Ocean is an intensively populated region where the December 2004 tsunami waves tragically altered land forms and land-use practice, causing perturbations and impacts on the river–sediment–soil–groundwater system. Saltwater intrusion caused severe modifications of the hydraulic regime of coastal aquifers, the position of the salt–freshwater interface, the chemical composition of surface and subsurface waters, and adversely affected the fauna and flora. The studies on the impact of tsunami-wave-induced changes to coastal zone soils and sediments are being focused on interfaces and transition zones characterized by steep gradients. Shifts in these boundary lines and transition zones are sensitive indicators for ongoing processes and trends. Specifically, the Research Direction will investigate the tsunami-related shifts of the subsurface salt–freshwater interface, phase switches between dissolved and free gas (e.g. methane), relocation of the oxic–anoxic boundary in soils and sediments, and changes of the spatial distribution of key organisms sensitive to changes in salinity, nutrients, oxygen and methane concentrations. The Research

Direction is building on the "state of the art" by combining hydrological, chemical, and biological expertise on sampling and modelling of regions characterized by steep gradients generated by the tsunami. This allows an assessment of interrelations between different forcing factors, such as hydraulic head, re-mineralization kinetics, and subsurface fluid flow.

The spectre of groundwater contamination looms over industrialized, suburban and rural coastal areas. The sources of groundwater contamination in tsunami affected countries are many, and the contaminants numerous. The disposal of domestic wastewater is accomplished in many areas through the use of septic tanks and drainage canals. The tsunami has accelerated the process of contamination in these affected regions, which now require drastic remediation measures.

These measures have to be economically feasible, environmentally sound and socially acceptable. In this project, a basic estimate will be conducted to assess the economic, environmental and social impacts of the tsunami and the feasibility of remediation measures suggested to overcome the damage. Interventions to secure livelihood through innovative solutions will consider gender as a part of the social analysis.

The scientific and technological objectives can only be covered by a well coordinated interaction of data mining (Zektser, 2000), expert knowledge (Kontar *et al.*, 2002; Lobkovsky *et al.*, 2003), well focused field work (Bokuniewicz *et al.*, 2004; Burnett *et al.*, 2002, 2003; Taniguchi *et al.,* 2006), and modelling (Kontar *et al.*, 2004, 2006). Field work is scheduled after intense compilation of data contributed by the partners of the Research Direction and data mining of former EU Projects (e.g. ELOISE and other research frameworks), LOICZ and CGSI reports. Data acquisition in the tsunami-effected transition zone between land and sea is requiring innovation, so cost and time efficient techniques are being applied by the partners. This includes shallow seismic, geoelectric and radar systems for characterization of the subsurface aquifer system (Kontar & Ozorovich, 2006). Investigations within former EU projects revealed that research is generally not limited to the availability of basic parameters such as salinity, hydraulic heads or major chemical components, so the Research Direction includes a strong expertise in trace gases, isotopes, and radio-nuclides suitable for dating of coastal groundwater, detection of subsurface freshwater discharge, and to derive kinetics of re-mineralization processes (Valiela *et al.*, 2002).

Data mining and field work will be compiled for spatial modelling by Geo-Information System (GIS). This allows computation of spatial patterns (Batelaan & De Smedt, 2001). Difference maps, visualizing changes between present and former maps, provide essential information on shifts in ecology and chemical transition zones in the soil–sediment–water system affected by the tsunami. Furthermore, GIS will be applied to deliver essential data for the assessment of hydrological budgets derived by modelling, e.g. by MODFLOW (Valstar, 2001). For numerical modelling of transport and reaction processes in coastal aquifers and sub-seafloor, well established algorithms derived by CGSI partners will be used. The model results will be validated by field measurements. The impact of scenarios associated with tsunami, global change, and smaller-scale perturbations will be evaluated by prognostic modelling.

The soil–groundwater–sediment system function will be investigated in detail for selected coastal areas affected by the tsunami using data acquisition, data mining and selected field work, as well as spatial analysis by GIS, and numerical modelling. The present status of target areas that represent a large section of Indian Ocean coastal environments will be derived and upscaled via spatial and numerical process-oriented modelling. In cooperation with CGSI, the numeric modelling of soil–water–sediment functioning in tsunami affected coastal areas will be implemented in an integrated coastal zone soil–water numerical model.

During the project, special groundwater information will be collected and used. For example pre-tsunami groundwater quality data is available (a few months before and one month before) in the affected areas (selected locations through out the affected areas). There is also post-tsunami data for one month, two months and four months for groundwater quality throughout the affected areas in Tamil Nadu, India. There is also data for groundwater level changes (including fluctuation maps) throughout the region, before and after tsunami. There are statistics about the number of hand pumps, bore wells, how people rejuvenated their open wells after the tsunami, statistics of new wells dug after the tsunami, emergency water supply for tsunami impacted areas, impact of water scarcity after the tsunami on the village population, etc. There is some data on the seawater intrusion in the coastal areas over the past decade (coastal Tamil Nadu region), and on the subsurface groundwater discharge in selected locations of tsunami affected regions (see Fig. 1). This data will supplement these studies.

The project will be concentrated on these lines because there are already some sufficient data and experience.

IMPLEMENTATION PLAN

This Research Direction combines intensive exploitation and compilation of existing data sources complemented by land and marine field research in India, geo-statistical analysis and numerical modelling, to achieve an integrated view of the impact of perturbations on groundwater quality and ecology, and shifts of the salt–freshwater interface in the areas affected by tsunami. For several tsunami affected key regions, land and marine soils and sediments, hydrology, biogeochemistry of groundwater and coastal waters, and ecological indices will be integrally addressed from existing data sources and by targeted field research, respectively.

Target zones, considered as representative of the tsunami affected areas, were identified as: (1) karstic coasts off North Sentinel, Nicobar; (2) soft, glacial sediments characteristic for delta system of Adyar River at the confluence with the Bay of Bengal; and (3) the coastal lagoon system in Pichavaram Mangrove, Tamil Nadu. The karstic environments at Nicobar have been experiencing strongly increasing coastal urbanization, and, thus, undergo predominantly anthropogenic perturbations. The delta system of the Adyar River coast is predominantly of glacial origin with diffusive subterranean and submarine flow. Its coastal lagoon system in Pichavaram Mangrove, Tamil Nadu, is situated within a humid temperate climate within the target regions that are affected by the tsunami, sea level rise, and an increasing number of storm events.

Historical data/Data mining and compilation

Within the project's initial phase, existing data and published pre-investigations are used to gather information needed to characterize the coastal soil–water continuum at the selected tsunami-effected regions. A database management system (DBMS) will be coupled to GIS and topographic and bathymetric charts, remote sensing data, thematic maps of the coastal zone (land and sea), and attribute data like biological and geochemical measurements that will be made available for modelling purposes. The DBMS will be used as a base from which to proceed with field work. The format and structure of data will be harmonized with all partners in order to have compatible data sets for integrative modelling.

Assessment of aquifer contamination

The three-dimensional study of salinization of waters and soils along the east and west coasts of India, with particular reference to the coasts of Tamil Nadu and the Car – Nicobar Islands caused by the tsunami of 26 December 2004, will be investigated by geophysical (resistivity and self-potential), geochemical and hydrological methods, with the objective of developing mitigation scenarios. In some cases, the tsunami-caused salinization may have been superimposed on anthropogenic salinization caused by excessive exploitation of groundwater. Within this project, the localization and quantification of submarine groundwater discharge, the efflux of dissolved constituents and trace gases from sub-seafloor aquifers, and the shifts in the subsurface freshwater–seawater interface towards the shore will be investigated by geochemical analyses of dissolved and particulate components (including stable and radio-isotopes, trace gases, and xenobiotics). The importance of sub-seafloor pathways will be estimated by CGSI studying interfacial transport and reaction processes with techniques especially appropriate for the individual regions. These activities include natural tracer studies (^{222}Rn, ^{226}Ra), physical parameters measurements, major ion and nutrient measurements, and sediment pore water sampling. They are all planned to be performed hand-in-hand with bathymetric and shallow seismic surveys. Special emphasis is focused on the interfacial exchange of carbon, nitrogen and phosphorous compounds. Their processes will be observed throughout all research directions. For consideration of shifts in the landside freshwater–saltwater interface, advanced radar systems and geoelectric techniques will be applied, along with seafloor photography.

On land as well as in the marine environment, several natural (including tsunami) and man-made perturbations are affecting the occurrence and fate of organisms. In the marine environment, taxa such as polychaetes, bivalve and bacteria, and plants such as seaweed, are closely associated with salinity or nutrient level supplied by ambient waters. In coastal margin sediments, it is well known that such organisms and consortia are indicative of seepage of fluids through sediments. In the sediments of the Bay of Bengal there is a strong indication that the occurrence or disappearance of polychaetes is linked to freshwater seepage from the land to the marine sediments. For coastal areas characterized by glacial sediments and a karstic/rocky seabed, the impact of the freshwater cycle on the benthic community will be studied. This suits two objectives: (1) to use organisms as an indicator for sub-seafloor freshwater discharge, and (2) to study the impact of perturbations on benthic communities in the coastal zone by analysing remains in deeper sediment strata. The study will use existing data sets, will investigate selected sites in the different target areas, and will conduct field experiments at the seafloor. Tsunami-generated changes in salinity, nutrient supply, etc. on the benthic community will be simulated. This allows simulation of perturbations, such as a sea level rise that affects the hydraulic head and therefore the admixture of freshwater at seep sites, the influence of dissolved nutrients, or changes in turbidity, caused e.g. by an increase in storm intensity and frequency. These aspects are directly linked to the changes of the physical, chemical and biological properties of soils/sediments generated by tsunami.

The hydrological budget for the different target areas will be derived by using models well established in civil engineering (e.g. MODFLOW). This allows comparison of model results derived for the different target areas and transfer of the modelling concept to other coastal regions impacted by tsunami. Based on the model calibrated with respect to the field data, sensitivity analyses of the impact of tsunami sea level rise and the natural and man-made perturbations mentioned above on the hydrological cycle, will be conducted. By this means, the response times of the three distinct environmental settings to perturbations can be studied and compared. This is a step towards a detailed understanding of water–soil system functioning and as a tool for the integrated management of coastal soil–water systems under the influence of tsunami.

The modelling efforts can be divided into three branches: (1) modelling of the water cycle in coastal areas considering groundwater renewal and submarine groundwater discharge, (2) shifts of the freshwater–saltwater interface, and (3) transport reaction modelling.

Shifts of the freshwater–saltwater interface due to over pumping or changing hydraulic heads (as a consequence of tsunami or sea level rise or intense storms that transport seawater into the groundwater recharge areas) are severe risks for coastal aquifers, as seawater can spoil water supply wells. Within this project, data sets obtained by chemical analysis and geoelectric techniques will provide detailed information about the density and concentration distribution in coastal aquifers. Such data sets will be used to support a calibration of numerical models suitable to resolve sharp density gradients and to predict the reaction of an area under tsunami and anthropogenic forcing. Shifts of the oxic–anoxic interface will also be considered in detail as a common issue throughout the Research Direction. Scenarios of enhanced erosion of soils and transfer of DOM to coastal waters via surface or subsurface pathways can be considered in detail. This component proceeds from the end of the initial data mining through to the field work components.

Tracing sewage contamination

Sewage is a generic term for the faecal waste from animals, although it is usually applied to human derived materials. This highlights one of the principal problems with assessing sewage inputs: is it human derived or from other, predominantly agricultural, sources? Notwithstanding this issue, the term sewage used here will predominantly apply to human wastes. Unlike most contamination, sewage is not a single compound, element, or even a class of compounds. Rather it is primarily a mixture of organic and inorganic components, along with intact biological entities (bacteria and viruses); together, this makes a very complex mixture. This mixture changes from region to region, diurnally and with distance from source due to partitioning between solid and solution phases. Some of the more water-soluble components will be moved with the liquid phase, while others are principally associated with the solid or particulate phase and may only move small distances. Therefore, different tracers may exist for each environment and at a range of distances from potential sources. Chemicals that are intrinsic to sewage, such as the stanols and sterols associated with human faecal matter, additives like detergents, microbiological communities present in wastewaters and effects caused by sewage to communities will be examined.

Sampling will take place in key locations and on key resources. These include drinking water, groundwaters (especially those used for potable waters), agricultural land and crops, bathing beaches and urban (semi-urban) environments where inhalation or ingestion pathways are important. Principal chemicals to be studied will be those derived from human faeces, such as the stanols and sterols, while key bacterial components will also be investigated. The often-used bacterial markers for sewage (faecal coliforms, streptococci, Intestinal Enterococci, etc.) are not appropriate in some locations. These organisms are not pathogens in themselves but are used as markers for other pathogenic ones. However, many of the bacteria are susceptible to UV irradiation and saline waters, and die off fairly quickly (<<24 hours in many cases). The differential die-off may lead to a false sense of security as negative *E. coli* results do not necessarily mean no bacteria are present. Therefore, a combination microbiological approach will be used.

Agricultural solutions for use of coastal contaminated aquifers

A systematic study pertaining to salinity problems encountered in coastal south India during tsunami, and the consequent loss of fertility of agricultural land, including remedial measures to revitalize economic growth in the region, will be conducted.

The study suggests that the tsunami affected area will be unproductive due to: (a) deposition of littoral sediments containing heavy minerals on the surface; and (b) contamination of soil and groundwater aquifers due to sea water intrusion. To revive the fertility of the land, the following measures are suggested: (1) Removal of littoral deposits – the littoral deposits containing heavy minerals should process through weathering and change into clay minerals or soil. Moisture, temperature climatic condition, salinity-stressed vegetation and time are the main factors that affect weathering (artificial weathering). Although the scraping of the inert littoral sediments from the agricultural field is a permanent solution, it is not economically viable due to the large size and extent of the affected area. (2) Desalinization of soil and groundwater – the surface water and groundwater of the tsunami-effected area are affected by seawater intrusion into the aquifer. To prevent salt-water contamination, over pumping and unnecessary mining of groundwater should cease. Earth-filled barriers may be constructed along the coast to limit the salt-water intrusion. Further, earthen bunds should be provided in flat areas so that rainwater gets trapped and can percolate through the soil. The amount of salt and harmful elements leached from the soil profile depends on the quantity of water that passes through the soil. For leaching to be effective, salt rich leachates must be discharged out of the area. The desalinization of the land can be accomplished by: (1) providing subsurface drainage; (2) surface water management; and (3) growing salt resistant crops. An ideal package of cultural practices besides other soil management and agronomic practices can ensure a good crop stand vis-à-vis good yield. In saline soils, germination of seeds is a very serious problem. Poor germination and crop stand can be counter-balanced by heavy pre-sowing irrigation, increased seed rate or plant population, treatment of seeds/seedlings for inducing tolerance to salinity, good seed bed preparation, optimum time and proper method of sowing/planting, appropriate crop geometry and direction of sowing, inter-tillage to ensure good aeration and control of weeds, etc. Removal of salts through leaching from the root-zone is an important aspect in the reclamation and management of salt-affected lands and in maintaining the long-term productivity of irrigated lands. As reclamation is meant for growing crops, leaching should be in conformity with the tolerance limits of the crops to be grown. The cost will be high if salts are leached beyond a threshold value, as no additional benefits will accrue, because yield stabilizes at the threshold value. Site-specific suitable agricultural solutions for use of the contaminated aquifer can be adapted to revitalizing agricultural land and to promote economic growth in the tsunami affected areas.

Conventional agricultural techniques to combat salinization processes due to the contamination of the aquifers can be characterized by four practices: (1) Root zone desalination by soil leaching – two options occur – when there is an impermeable layer, salts will be concentrated above this layer; on the other hand, when there is no impermeable layer, aquifer contamination may be observed. (2) Use of subsurface trickle irrigation – economical of water, and therefore less additional salts; however, the problem of groundwater contamination due to natural rain or artificial leaching remains. (3) Enhanced fertilization increases the tolerance to salinity (however, sensitivity also increases), but the contamination will be increased by other hazardous chemicals such as nitrate. (4) Use of salt tolerant species – this technique will be very useful to the plants, but does not solve the problem of soil or groundwater contamination.

Agricultural environmentally safe and clean techniques to combat the salinization process due to the contamination of the aquifers can be characterized by: (1) use of salt removing species; (2) use of drought tolerant crops species; (3) reduction of salt application by deficit irrigation; (4) re-use of minimal levels of saline water enough to obtain a good visual appearance GVA of the landscape. The study of these new techniques is being carried out. The research work will be focused on the combination of conventional agricultural techniques and environmentally safe and clean techniques in order to increase the best economic and environmental solutions to control salinity and maintain the sustainability of the agricultural areas and landscape.

Costing solutions and remediation/economic, environment and social impacts of tsunami and the feasibility of remedial actions (innovative solutions)

The spectre of groundwater contamination looms over industrialized, suburban and rural areas (Zektser, 2000). The sources of groundwater contamination in tsunami affected countries are many and the contaminants numerous. The disposal of domestic waste-water is accomplished in many areas through the use of septic tanks and drainage canals. The tsunami accelerated these processes of contamination in affected regions, and they now require drastic remediation measures. These measures have to be economically feasible, environmentally sound and socially acceptable. In this project we estimate the economic, environment and social impacts of tsunami and the feasibility of remediation measures suggested in previous projects to overcome the tsunami damage.

Interventions to secure livelihood through innovative solutions will consider the involvement of gender as a part of the social analysis. Most of the affected people are generally poor, and are also more vulnerable to environmental risk and disasters, and therefore the remedial solutions should address the issue of poverty alleviation. The effectiveness of the solutions to alleviate poverty in the affected areas will be investigated. The main consideration of this project is to ensure the proposed remedial actions are appropriate for the local institutional, environmental, social and economic conditions. The analyses will be carried out through field surveys and standard environmental/economic/econometric analyses.

REFERENCES

Alagarswami, K. (1993) *Sustainable Management and Development of Coastal Aquaculture* (ed. by M. S. Swaminathan & R. Ramesh). MSSRF, Chennai, India.

Batelaan, O. & De Smedt, F. (2001) WetSpass: a flexible, GIS based, distributed recharge methodology for regional groundwater modelling. In: *Impact of Human Activity on Groundwater Dynamics* (ed. by H. Gehrels, N. E. Peters, E. Hoehn, K. Jensen, C. Leibundgut, J. Griffoen, B. Webb & W. J. Zaadnoordijk), 11–17. IAHS Publ. 269. IAHS Press, Wallingford, UK.

Bokuniewicz, H., Kontar, E., Rodrigues, M. & Klein, D. A. (2004) Submarine Groundwater Discharge (SGD) Patterns through a fractured rock: a case study in the Ubatuba coastal area, Brazil. *AAS Revista, Asociacion Argentina de Sedimentologia* **11**(1), 9–16.

Burnett, W. C., Chanton, J., Christoff, J., Kontar, E. A., Krupa, S., Lambert, M., Moore, W., O'Rourke, D., Paulsen, R., Smith, C., Smith, L. & Taniguchi, M. (2002) Assessing methodologies for measuring groundwater discharge to the ocean. *EOS* **83**(11), 117, 122–123.

Burnett, W. C., Chanton, J. P. & Kontar, E. A. (eds) (2003) *Submarine Groundwater Discharge. Biogeochemistry* **66**. Kluwer Academic Publishers, The Netherlands.

Gupta, H. (2005) Mega-tsunami of the 26th December 2004: Indian initiative for early warning system and mitigation of oceanogenic hazards. *Episodes* **28**(1), 2–5.

Kontar, E. A. & Ozorovich, Y. R. (2006) Geo-electromagnetic survey of fresh/salt water interface in the coastal southeastern Sicily. *Continental Shelf Research* **26**(7), 843–851.

Kontar, E. A., Salokhiddinov, A. T. & Khakimov, Z. M. (2004) Assessment of groundwater–seawater interactions in the Aral Sea Basin. In: *Isotopes in Environmental Studies* (IAEA Aquatic Forum, Monaco), 196–201. IAEA.

Kontar, E. A., Burnett, W. C. & Povinec, P. P. (2002) Submarine groundwater discharge and its influence on hydrological trends in the Mediterranean Sea. In: *Tracking Long Term Hydrological Change in the Mediterranean* Sea, 109–114. MEL-IAEA, Monaco.

Kontar, E. A., Lobkovsky, L. I. & Korotenko, K. A. (2006) On risk assessment and sustainable industrial development of shelf zones. In: *Recent Geodynamics, Georisk and Sustainable Development in the Black Sea to Caspian Sea Region* (ed. by A. T. Ismail-Zadeh), 84–94. American Institute of Physics, vol. 825. Melville, New York, USA.

Krishnamoorthy, R. (1996) Remote sensing of mangrove forests in Tamil Nadu Coast, India. PhD Thesis, Anna University, Chennai, India.

Lobkovsky L. I., Kontar., E. A., Garagash I. A. & Ozorovich, Y. R. (2003) Monitors and methods for investigation of submarine landslides, seawater intrusion and contaminated groundwater discharge as coastal hazards. In: *NATO Risk Science and Sustainability: Science for Reduction of Risk and Sustainable Development of Society* (ed. by T. Beer & A. Ismail-Zadeh), 191–207. Kluwer Academic Publishers, Dordrecht, The Netherlands.

Ramesh, R. (2001) Effect of land-use change on groundwater quality in a coastal habitat of south India. In: *Impact of Human Activity on Groundwater Dynamics* (ed. by H. Gehrels, N. E. Peters, E. Hoehn, K. Jensen, C. Leibundgut, J. Griffoen, B. Webb & W. J. Zaadnoordijk), 161–166. IAHS Publ. 269. IAHS Press, Wallingford, UK.

Sonak, S. (2000) Aquaculture in India. In: *Proc. On Land Use, Land Cover Changes and Modeling in Coastal Areas* (ed. by S. Ramachandran), 20–32. Anna University, Chennai, India.

Taniguchi, M., Burnett, W., Duevalova, H., Kontar, E., Povinec, P. & Moore, W. (2006) Submarine groundwater discharge measured by seepage meters in Sicilian coastal waters. *Continental Shelf Research* **26**(7), 835–842.

Valiela, I., Bowen, J. L. & Kroeger, K. D. (2002) Assessment of models for estimation of land-derived nitrogen loads to shallow estuaries. *Appl. Geochem.* **17**, 935–953.

Valstar, J. (2001) Inverse modelling of groundwater flow and groundwater mass transport. In: *Impact of Human Activity on Groundwater Dynamics* (ed. by H. Gehrels, N. E. Peters, E. Hoehn, K. Jensen, C. Leibundgut, J. Griffoen, B. Webb & W. J. Zaadnoordijk), 239–245. IAHS Publ. 269. IAHS Press, Wallingford, UK.

Zektser, I. S. (2000) *Groundwater and the Environment: Applications for the Global Community*. Lewis Publishers, Boca Raton, Florida, USA.

The importance of shallow confining units to submarine groundwater flow

JOHN F. BRATTON
US Geological Survey, 384 Woods Hole Rd, Woods Hole, Massachusetts 02556, USA
jbratton@usgs.gov

Abstract In addition to variable density flow, the lateral and vertical heterogeneity of submarine sediments creates important controls on coastal aquifer systems. Submarine confining units produce semi-confined offshore aquifers that are recharged on shore. These low-permeability deposits are usually either late Pleistocene to Holocene in age, or date to the period of the last interglacial highstand. Extensive confining units consisting of peat form in tropical mangrove swamps, and in salt marshes and freshwater marshes and swamps at mid-latitudes. At higher latitudes, fine-grained glaciomarine sediments are widespread. The net effect of these shallow confining units is that groundwater from land often flows farther offshore before discharging than would normally be expected. In many settings, the presence of such confining units is critical to determining how and where pollutants from land will be discharged into coastal waters. Alternatively, these confining units may also protect fresh groundwater supplies from saltwater intrusion into coastal wells.

Key words Atlantic; coastal aquifer; confining unit; glaciomarine; groundwater; mangrove; peat; salt marsh; saltwater intrusion; nutrients

INTRODUCTION

As the phenomenon of submarine groundwater discharge has been investigated more intensively over the last decade, several critical data gaps and technological obstacles have emerged (Taniguchi *et al.*, 2002; Slomp & Van Capellen, 2004; Burnett *et al.*, 2006; Gallardo & Marui, 2006). Among these barriers to answering research questions and solving coastal management issues has been the problem of poorly documented submarine stratigraphy and hydrogeology. Many modelling or field studies that initially assumed simple "sandbox" geology at shoreline sites, as suggested by the surficial appearance of many beaches (Fig. 1(a)), had to be modified as field results became available (Fig. 1(b)–(d)) Seepage meter measurements and subsurface sampling have commonly indicated the influence of unexpected lower permeability sediments at depths of less than a metre to several metres beneath the beach and seafloor. Often these investigations have treated the fine-grained units as curious anomalies that complicated experiments which were intended to be more straight-forward. Rarely, however, have investigators attempted to synthesize such phenomena into a coherent typological framework.

For land-based hydrological investigations, stratigraphy of unconsolidated units is normally determined by soil borings. Piezometers or monitoring wells are installed to measure heads, determine flow directions, and collect water-quality samples. In shoreline and submarine studies, these types of investigations become more difficult due to problems associated with accessing the areas with drilling rigs, and installing

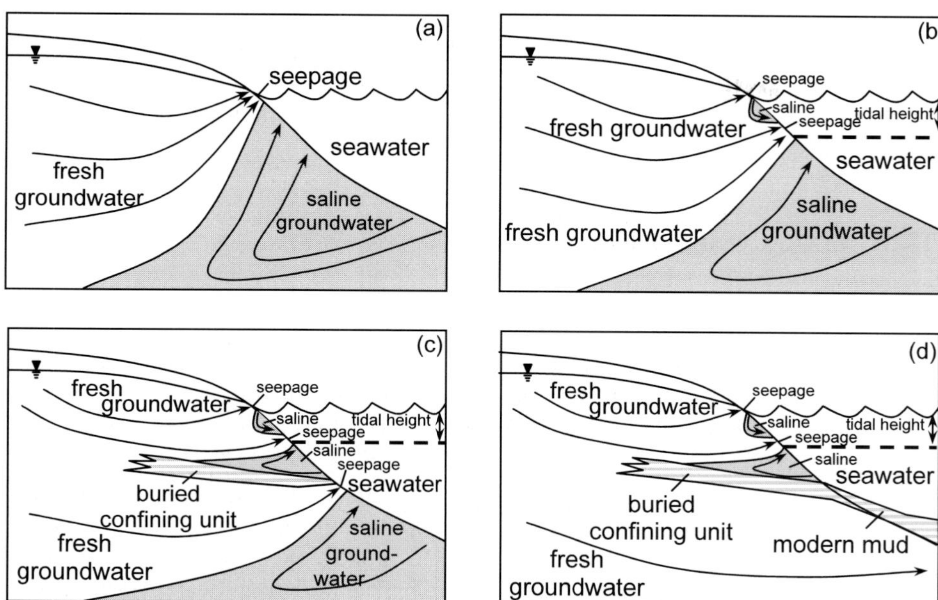

Fig. 1 Schematic diagrams showing: (a) the geometry of a typical fresh–saline interface in a coastal groundwater system, and the circulation within each portion of the aquifer; (b) how the addition of tides can create a shallow saline circulation cell in groundwater beneath the intertidal zone, and two discharge zones, as described in the text; (c) how the presence of a shallow submarine confining unit can create an additional seepage zone and saline circulation cell; and (d) how a more extensive submarine confining unit, formed by the combination of a buried unit and/or fine-grained sediments that are being actively deposited, can create a long (up to kilometres) submarine flow path for fresh groundwater.

wells and piezometers that can withstand the effects of waves, tides, and moving ice. Between 2000 and 2003, six intercomparison experiments were performed using combinations of seepage meters, piezometers, models, geophysical instruments, and other techniques (Burnett *et al.*, 2006). In several of these studies, including sites in Australia, Florida, and New York, discharge patterns were influenced by the presence of shallow confining units. At the Florida site, an anomalous band of discharge was observed between 80 and 100 m offshore (Smith & Zawadzki, 2003) and was attributed to leakage through a "window" in a buried confining unit. Seepage meters deployed during other experiments in Waquoit Bay, Massachusetts (Michael *et al.*, 2005) showed a 15-m wide band of unusually high discharge parallel to the shore, but offset about 15 to 20 m offshore, that may have been influenced by a confining unit. Submarine flow of fresh or brackish water controlled by shallow confining units has also been documented beneath coastal bays in Delaware and Maryland (Bratton *et al.*, 2004; Krantz *et al.*, 2004; Manheim *et al.*, 2004), and beneath the Neuse River Estuary in North Carolina (Cross *et al.*, 2006) and Nauset Marsh in Massachusetts (Bratton *et al.*, 2006).

This paper synthesizes available information on the origins and significance of submarine confining units, with the goal of providing a foundational conceptual model for considering their role in the solution of current scientific and societal problems.

The focus is on examples from the Atlantic coast of the USA. The discussion, however, broadens the application of these examples to other similar coastal settings. The conclusions particularly apply to other passive margin coastlines with active sedimentation, such as those found around much of the Atlantic, Mediterranean, and Arctic basins, as well as the coasts of Australia, eastern Africa, and India.

ORIGINS OF CONFINING UNITS

The ways in which submarine confining units are formed vary both temporally and areally. The process that indirectly produces most of the conditions necessary for formation of such deposits is global climate change. Climate-driven cycles of ice sheet growth and retreat cause vertical sea-level fluctuations and lateral shoreline migrations. As described by Walther's law (Middleton, 1973), lateral movement of adjacent high-energy and low-energy environments, such as sandy beaches and muddy lagoons, results in the stacking of sediment units with different permeabilities. Sands and gravels from high-energy environments act as conduits for groundwater flow. Fine-grained sediments, which are generally deposited in sheltered embayments or depressions during sea-level stillstands and highstands, are barriers to groundwater flow.

The most extensive coastal confining units in many modern systems were deposited between glaciations, particularly during the last interglacial (Marine Isotope Stage [MIS] 5, especially 5e [i.e. Eemian, Ipswichian, or Sangamon age]). Deposits of this age are found throughout tropical and mid-latitude coastal areas, except where they have been eroded away during intervening glacial periods with corresponding low sea levels and subaerial exposure of highstand deposits. In many high-latitude coastal areas such deposits were removed by glacial scour during the last ice age.

There are many regionally extensive deposits of fine-grained sediments from interglacial highstands, or from relative highstand deposits that have been subsequently uplifted. These formations commonly form shallow onshore confining units that extend offshore (Fig. 1(c)). Examples include shallow marine or back-barrier deposits found around the shore of Chesapeake Bay, the Virginia and North Carolina coasts, the northern Gulf of Mexico, and the southern Baltic Sea and North Sea of northern Europe.

The second major category of modern submarine confining unit is deposits formed during the late Pleistocene or Holocene transgression. These units consist of latitudinally distinct biogenic deposits, proglacial deposits, and common estuarine and marine muds (Fig. 1(d)). The biogenic deposits are peats formed at the shoreline, or slightly inland, and derived from mangroves or sawgrass in the tropics, and salt marshes and coastal freshwater marshes and swamps at higher latitudes (Fig. 2). Many such deposits that are no longer actively forming are now submerged below sea level and either exposed on the seafloor, or buried at shallow depths along the coast and offshore.

Mangroves are present along modern tropical and subtropical shorelines up to approximately 28°N and S latitude on western margins of ocean basins, and about 15 to 20°N and S along eastern margins (Fig. 2). In the mid-1990s the total surface area covered by mangroves was 181 000 km^2, or about 8% of the global coastline and 25% of tropical coastlines (Spalding *et al.*, 1997). Submerged mangrove peats have been

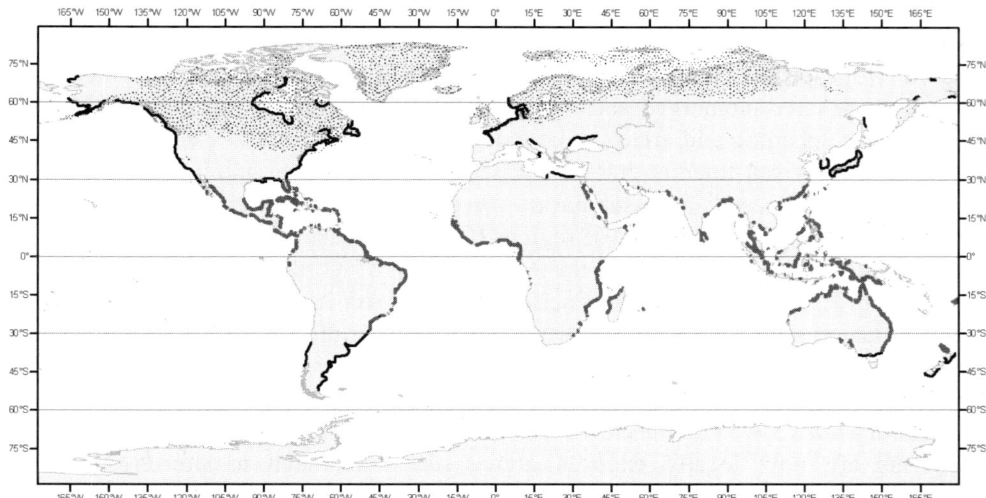

Fig. 2 Map showing global coastlines where mangrove swamps (heavy grey lines) and salt marshes (heavy black lines) are present. The stippled area was covered by ice at the last glacial maximum. Salt marsh and mangrove information is modified from Sumich (1999) and Spalding *et al.* (1997). Glacial limits are from Ray & Adams (2001).

observed around the world at sites including Florida (Robbin, 1984), Belize (Macintyre *et al.*, 2004), and the South China Sea (Hanebuth *et al.*, 2000). Where submerged peats from drowned mangroves either extend close enough to shore to intersect the unconfined seepage face, or are laterally continuous with peats beneath living mangroves, conditions are present for these peats to act as shallow submarine confining units. This is particularly true in areas with very low coastal relief, such as the carbonate platforms of Florida or the Bahamas. The tendency of living mangroves to create brines further complicates the groundwater flow systems in these types of settings (Greenwood *et al.*, 2006). In some tropical and subtropical settings, such as the Florida Everglades, areas inland of mangroves are covered by extensive sawgrass (*Cladium* sp.) marshes (Willard *et al.*, 2006), which also form peats. As sea level rises, mangroves may migrate inland, causing mangrove peat to be deposited over marsh peat (Willard & Holmes, 1997). In Belize, mangrove peat up to 10 m thick has been documented (Macintyre *et al.*, 2004).

Mid- to high-latitude coastal plain shores contain intertidal salt marshes dominated by cordgrass (*Spartina* sp.) and other salt-tolerant vegetation (Fig. 2). The range of these marshes (~25°N and S along western ocean boundaries, ~20°N and S on eastern boundaries due to colder boundary currents) overlaps slightly with that of mangroves, and extends into the subarctic. In the United States, submerged peats have been documented along the Atlantic coast in Maryland, New York, Massachusetts, and Maine, among other locations (Emery *et al.*, 1967; Field *et al.*, 1979, Thieler *et al.*, 1999; Buynevich & FitzGerald, 2002; Bratton *et al.*, 2004, 2007). In low-lying coastal areas inland of salt marshes, freshwater marshes of cattails, bulrushes, phragmites, and sphagnum form peats as well. Forested wetlands or swamps in some areas up to about

38°N or S along the western Atlantic coast also contain cypress and gum trees. At higher latitudes, vegetation in coastal swamps is dominated by red maples, oaks, or Atlantic white cedars. Like salt marshes, these swamps can form low-permeability peats that are later submerged and buried as sea level continues to rise (e.g. Dillon, 1970). In the subarctic and arctic, at latitudes higher than about 55°, coastal wetlands are dominated by salt-tolerant grasses and sedges, with muskeg and tundra vegetation further inland. Coastal and submarine permafrost also becomes hydrologically significant in the subsurface in these regions (Gosink & Baker, 1990).

At latitudes north of 40° along the Atlantic coast of North America and 50° along the coasts of Europe, Asia, and the Pacific coast of North America, large areas of fine-grained glaciomarine and glaciolacustrine deposits from the end of the last glaciation are present on the inner continental shelf (Fig. 2, stippled area). Along coasts that were isostatically depressed due to glacial loading and that subsequently rebounded (Barnhardt *et al.*, 1995; Lambeck, 1995), extensive deposits of these silty and clayey sediments are now locally exposed above sea level, such as the Presumpscot Formation of Maine (Bloom, 1963; Fig. 3, hatched area). Other glacially-derived sediments remain submerged, such as the fine sands, silts, and clays deposited in ephemeral proglacial lakes during the late Pleistocene on the inner continental shelf of southern New England (Fig. 3; Uchupi *et al.*, 2001). In areas where these units reach

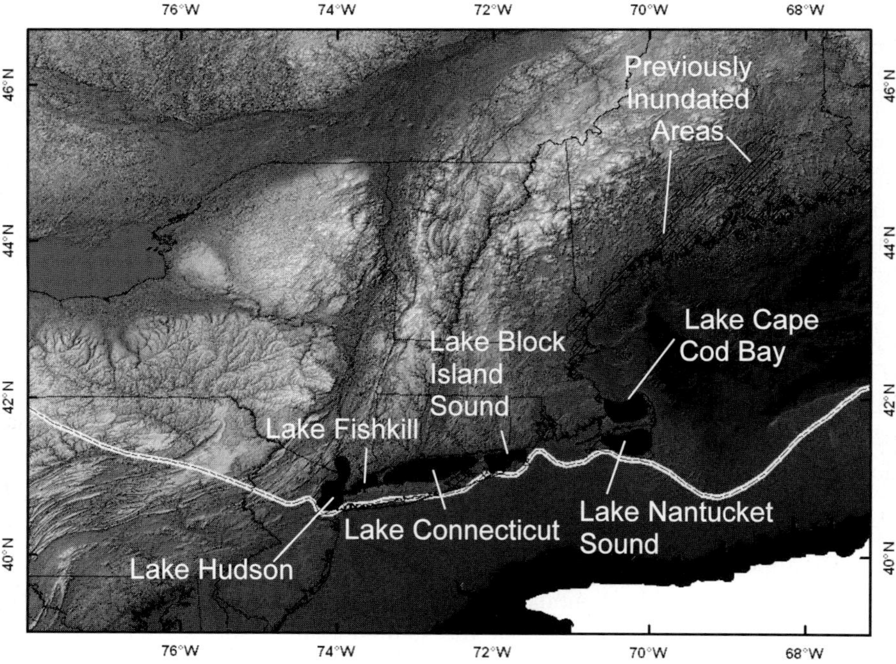

Fig. 3 Shaded relief and bathymetric map of the northeastern United States and Canada showing locations of maximum glacial ice extent during the last Ice Age (dashed white line), proglacial lake beds now mostly submerged (black), and areas of coastal Maine and Massachusetts inundated shortly after deglaciation (hatched area) (modified from Uchupi & Mulligan, 2006).

the nearshore or continue onshore, they serve as very effective confining units. In some locations, fine-grained sediments associated with previous glaciations or marine highstands may also be present beneath late Pleistocene deposits on the inner shelf (Person *et al.*, 2003; Uchupi & Mulligan, 2006).

Estuarine and marine muds, like sediments from previous sea-level highstands such as MIS 5e, are currently being deposited below the wave base and on intertidal mudflats in many coastal areas. This sediment type is especially common in sheltered or deep estuaries, back-barrier lagoons, and near the mouths of large rivers. These surficial confining units can be more than 20 m thick in deep palaeochannels or drowned basins (e.g. Bratton *et al.*, 2003), but are generally much thinner. Thickness often decreases toward the shore where wave action prevents silt and clay from settling out. Although fine grained, the effective permeability of some of these units is increased by bioturbation of infaunal organisms such as bivalves, polychaetes, crabs, and shrimp, some species of which burrow up to 2 m into the seafloor. The uppermost tens of centimetres of estuarine sediments also tend to be only minimally consolidated, with water content values often exceeding 70 percent. Estuarine muds have been observed to overlie buried peats in shallow coastal estuaries (Bratton *et al.*, 2004, 2007; Krantz *et al.*, 2004); and fluvial sand and gravel, beach sand, or oysters (Bratton *et al.*, 2003).

SIGNIFICANCE OF CONFINING UNITS TO SUBMARINE GROUNDWATER FLOW

Historically, coastal and submarine groundwater flow systems have generally been oversimplified. Situations where the groundwater follows a simple Ghyben-Herzberg geometry, with a single wedge of saline groundwater underlying a fresh lens that discharges in a narrow band at the shoreline (Fig. 1(a)), may be more the exception than the rule. Shallow confining units can produce a pattern of two or more discharge zones (Fig. 1(c) and (d)). Permeable shorelines that are subject to significant tides, however, also tend to develop shallow saline flow cells that can mimic the effect of a shallow confining unit by creating two discharge zones (Fig. 1(b)). This was recently observed at a study site in Australia and simulated with a variable-density model (Robinson *et al.*, 2006).

Early variable-density models treated the ocean side of coastal model domains as a simple vertical boundary with hydrostatic saltwater pressure (Dominick *et al.*, 1971). Results of numerous field studies that showed the importance of a more realistic sloping (non-vertical) seepage face subject to tidal pumping are now being incorporated into variable-density models of shoreline aquifer systems (Ataie-Ashtiani *et al.*, 2001). Some modelling studies (e.g. Kooi & Groen, 2001; Bakker, 2006) have also incorporated semi-confined offshore conditions into their analyses; results include migration of the fresh–salt boundary in groundwater farther offshore.

The overall result of these refinements of coastal groundwater flow models is that they are better able to reproduce actual conditions observed in the field. As field measurement techniques have improved, new processes and problems have been identified and models have facilitated more sophisticated quantitative testing of ideas about submarine flow, reaction, and discharge processes. The conceptual framework and instrumental toolkit for approaching submarine geological heterogeneities in something other than case-by-case mode is now starting to emerge. Development of shoreline typologies is likely to be a fruitful approach (Bokuniewicz *et al.*, 2003).

The significance of submarine confining units to submarine groundwater flow and discharge ultimately comes down to questions of scale: How common are these units? How thick are they? How continuous are they? How far onshore and offshore do they extend? How much head is required to establish offshore flow in submarine confined systems? How quickly do these systems respond to changes in recharge? The more prevalent and extensive submarine confining units are, the more they impact coastal aquifers and land–sea interactions. Their primary effect is to broaden or split the zone of contact in the subterranean estuary (Moore, 1999) between land-derived water and ocean water. This, in turn, increases the lengths of submarine flow paths and residence time of water in submarine flow systems. The result is a greater time lag in the response of the groundwater systems to changing surface conditions, and more complicated or substantial submarine biogeochemical reactions.

IMPLICATIONS FOR MANAGEMENT AND SCIENCE

In some estuarine settings, groundwater discharge may be the primary route by which freshwater (and associated nutrients and contaminants) enters the coastal system. Submarine confining units influence discharge patterns in important ways, such as displacing discharge offshore (e.g. seaward of barrier islands rather than into lagoons) or creating conditions conducive to more diffusive rather than advective discharge. In settings where onshore groundwater has been impacted by pollutants, detecting the presence of coastal and submarine confining units is critical to understanding how and where the pollutants will be discharged into coastal waters, and what remedial measures might be appropriate (McCobb & LeBlanc, 2002; Kroeger *et al.*, 2006).

In the particular case of nitrate contamination, the presence of submarine confining units that are enriched in organic matter is likely to increase the amount of nitrate removal from groundwater prior to discharge via denitrification (Capone & Slater, 1990; Bratton *et al.*, 2004, 2007). There are also potential linkages between submarine groundwater discharge and the presence of microbial pathogens at beaches, which would be affected by submarine flow paths (Boehm *et al.*, 2004). Finally, confining units may protect fresh groundwater supplies from saltwater intrusion into onshore wells used for drinking water or irrigation. Such situations would call for the protection of submarine confining units from breaching by dredging and other marine construction (e.g. Foyle *et al.*, 2002).

CONCLUSIONS

It is not surprising that the stratigraphy of coastal aquifers plays a critical role in determining their flow and discharge behaviour. What has been less clear, however, is that offshore stratigraphy may differ significantly from that of nearby onshore areas. Layered sediments with different hydrogeological properties provide many potential pathways for submarine groundwater flow and discharge, as well as for saltwater intrusion. Intensive recent investigations of groundwater systems spanning the shoreline in many settings have shown: (1) that submarine confining units are widespread, and (2) that they can strongly influence patterns of submarine groundwater discharge and associated phenomena such as pollutant discharge, oceanic elemental budgets, microbial contamination of beaches, and saltwater intrusion. Further development of a typological understanding of the eustatic, hydrographic, biogenic, and glaciogenic origins of such confining units will help in predicting where they may be important, both scientifically and societally, in previously unstudied locations.

Acknowledgements Discussions with John Crusius, Kevin Kroeger and Peter Swarzenski were helpful in developing some of the ideas presented in this manuscript. VeeAnn Cross assisted with preparation of figures. Comments on earlier versions of the manuscript by Rob Thieler, Walter Barnhardt and two anonymous reviewers improved it significantly.

REFERENCES

Ataie-Ashtiani, B., Volker, R. E. & Lockington, D. A. (2001) Tidal effects on groundwater dynamics in unconfined aquifers. *Hydrol. Processes* **15**(4), 655–669.

Bakker, M. (2006) How far does a seawater intrusion model need to be extended below the sea bottom? In: *Abstracts, First International Joint Salt Water Intrusion Conference* (1st SWIM-SWICA, Cagliari–Chia Laguna, Italy, September 2006).

Barnhardt, W. A., Gehrels, W. R., Belknap, D. F. & Kelley, J. T. (1995) Late Quaternary relative sea-level change in the western Gulf of Maine; evidence for a migrating glacial forebulge. *Geology* **23**(4), 317–320.

Bloom, A. L. (1963) Late Pleistocene fluctuations of sea level and postglacial crustal rebound in coastal Maine. *Am. J. Sci.* **261**(9), 862–879.

Boehm, A. B., Shellenbarger, G. G. & Paytan, A. (2004) Groundwater discharge: a potential association with fecal indicator bacteria in the surf zone. *Environ. Sci. Technol.* **38**(13), 3558–3566.

Bokuniewicz, H., Buddemeier, R., Maxwell, B. & Smith, C. (2003) The typological approach to submarine groundwater discharge (SGD). *Biogeochem.* **66**(1-2), 145–158.

Bratton, J. F., Colman, S. M., Thieler, E. R. & Seal, R. R., II (2003) Birth of the modern Chesapeake Bay estuary 7,400 to 8,200 years ago and implications for global sea-level rise. *Geo-Mar. Lett.* **22**(4), 188–197.

Bratton, J. F., Böhlke, J. K., Manheim, F. T. & Krantz, D. E. (2004) Ground water beneath coastal bays of the Delmarva Peninsula: ages and nutrients. *Ground Water* **42**(7), 1021–1034.

Bratton, J. F., Colman, J. A., Crusius, J., McCobb, T. D., Massey, A. J., Koopmans, D. J. & Masterson, J. P. (2006) Evidence and implications of a fresh groundwater flow system beneath a shallow coastal estuary in New England. In: *Abstracts, First International Joint Salt Water Intrusion Conference* (1st SWIM-SWICA, Cagliari–Chia Laguna, Italy, September 2006).

Bratton, J. F., Böhlke, J. K., Krantz, D. E. & Tobias, C. R. (2007) Flow of groundwater beneath a back-barrier lagoon: geometry and geochemistry of the subterranean estuary at Chincoteague Bay, Maryland. *Mar. Chem.* (in press).

Burnett, W. C., Aggarwal, P. K., Aureli, A., Bokuniewicz, H., Cable, J. E., Charette, M. A., Kontar, E., Krupa, S., Kulkarni, K. M., Loveless, A., Moore, W. S., Oberdorfer, J. A., Oliveira, J., Ozyurt, N., Povinec, P., Privitera, A. M. G., Rajar, R., Ramessur, R. T., Scholten, J., Stieglitz, T., Taniguchi, M. & Turner, J. V. (2006) Quantifying submarine groundwater discharge in the coastal zone via multiple methods. *Sci. Total Environ.* **367**(2-3), 498–543.

Buynevich, I. V. & FitzGerald, D. M. (2002) Organic-rich facies in paraglacial barrier lithosomes of northern New England: preservation and paleoenvironmental significance. *J. Coast. Res.*, Special Issue **36**, 109–117.

Capone, D. G. & Slater, J. M. (1990) Interannual patterns of water table height and groundwater derived nitrate in nearshore sediments. *Biodegradation* **10**(3), 277–288.

Cross, V. A., Bratton, J. F., Bergeron, E., Meunier, J. K., Crusius, J. & Koopmans, D. (2006) Continuous resistivity profiling data from the upper Neuse River Estuary, North Carolina, 2004-2005. *US Geol. Survey Open-File Report 2005-1306.*

Dillon, W. P. (1970) Submergence effects on a Rhode Island barrier and lagoon and inferences on migration of barriers. *J. Geol.* **78**(1), 94–106.

Dominick, T. F, Wilkins, B. & Roberts, H. (1971) Mathematical model for beach groundwater fluctuations. *Water Resour. Res.* **7**(6), 1626–1635.

Emery, K. O., Wigley, R. L., Bartlett, A. S., Rubin, M. & Barghoorn, E. S. (1967) Freshwater peat on the continental shelf. *Science* **158**(3806), 1301–1307.

Field, M. E., Meisburger, E. P., Stanley, E. A. & Williams, S. J. (1979) Upper Quaternary peat deposits on the Atlantic inner shelf of the US. *Geol. Soc. Am. Bull.* **90**(7), 618–628.

Foyle, A. M., Henry, V. J. & Alexander, C. R. (2002) Mapping the threat of seawater intrusion in a regional coastal aquifer–aquitard system in the southeastern United States. *Environ. Geol.* **43**(1-2), 151–159.

Gallardo, A. H. & Marui, A. (2006) Submarine groundwater discharge: an outlook of recent advances and current knowledge. *Geo-Mar. Lett.* **26**(2), 102–113.

Gosink, J. P. & Baker, G. C. (1990) Salt fingering in subsea permafrost: some stability and energy considerations. *J. Geophys. Res.* **95**(C6), 9575–9583.

Greenwood, W. J., Kruse, S. & Swarzenski, P. (2006) Extending electromagnetic methods to map coastal pore water salinities. *Ground Water* **44**(2), 292–299.

Hanebuth, T., Stattegger, K. & Grootes, P. M. (2000) Rapid flooding of the Sunda Shelf: a late-glacial sea-level record. *Science* **288**(5468), 1033–1035.

Kooi, H. & Groen, J. (2001) Offshore continuation of coastal groundwater systems; predictions using sharp-interface approximations and variable-density flow modeling. *J. Hydrol.* **246**(1-4), 19–35.

Krantz, D. E., Manheim, F. T., Bratton, J. F. & Phelan, D. J. (2004) Hydrogeologic setting and ground water flow beneath a section of Indian River Bay, Delaware. *Ground Water* **42**(7), 1035–1051.

Kroeger, K. D., Cole, M. L., York, J. K. & Valiela, I. (2006) N transport to estuaries in wastewater plumes: modeling and isotopic approaches. *Ground Water* **44**(2), 188–200.

Lambeck, K. (1995) Late Devensian and Holocene shorelines of the British Isles and North Sea from models of glacio-hydro-isostatic rebound. *J. Geol. Soc. Lond.* **152**(3), 437–448.

MacIntyre, I. G., Toscano, M. A., Lighty, R. G. & Bond, G. B. (2004) Holocene history of the mangrove islands of Twin Cays, Belize, Central America. *Atoll Res. Bull.* **510**, 1–16.

Manheim, F. T., Krantz, D. E. & Bratton, J. F. (2004) Studying ground water under Delmarva coastal bays using electrical resistivity. *Ground Water* **42**(7), 1052–1068.

McCobb, T. D. & LeBlanc, D. R. (2002) Detection of fresh ground water and a contaminant plume beneath Red Brook Harbor, Cape Cod, Massachusetts, 2000. *US Geol. Survey Water-Resources Investigations Report 02-4166*.

Michael, H. A., Mulligan, A. E. & Harvey, C. F. (2005) Seasonal oscillations in water exchange between aquifers and the coastal ocean. *Nature* **436**(7054), 1145–1148.

Middleton, G. V. (1973) Johannes Walther's law of the correlation of facies. *Geol. Soc. Am. Bull.* **84**(3), 979–988.

Moore, W. S. (1999) The subterranean estuary: a reaction zone of ground water and sea water. *Mar. Chem.* **65**(1-2), 111–126.

Person, M. A., Dugan, B., Swenson, J., Urbano, L., Stott, C., Taylor, J. & Willett, M. (2003) Pleistocene hydrogeology of the Atlantic continental shelf, New England. *Geol. Soc. Am. Bull.* **115**(11), 1324–1343.

Ray, N. & Adams, J. M. (2001) A GIS-based vegetation map of the world at the last glacial maximum (25,000–15,000 BP). *Internet Archaeology* **11** (http://intarch.ac.uk/journal/issue11/rayadams_toc.html).

Robbin, D. M. (1984) A new Holocene sea level curve for the upper Florida Keys and Florida reef tract. In: *Environments of South Florida: Present and Past* (ed. by P. J. Gleason), 437–458. Miami Geological Society, Memoir 2.

Robinson, C., Gibbes, B. & Li, L. (2006) Driving mechanisms for groundwater flow and salt transport in a subterranean estuary. *Geophys. Res. Lett.* **33**, L03402.

Slomp, C. P. & Van Capellen, P. (2004) Nutrient inputs to the coastal ocean through submarine groundwater discharge: controls and potential impact. *J. Hydrol.* **295**(1-4), 64–86.

Smith, L. & Zawadzki, W. (2003) A hydrogeologic model of submarine groundwater discharge: Florida intercomparison experiment. *Biogeochem.* **66**(1-2), 95–110.

Spalding, M. D., Blasco, F. & Field, C. D. (eds) (1997) *World Mangrove Atlas*. The International Society for Mangrove Ecosystems, Okinawa, Japan.

Sumich, J. L. (1999) *An Introduction to Marine Life*, seventh edn. WCB/McGraw-Hill, New York, USA.

Taniguchi, M., Burnett, W. C., Cable, J. E. & Turner, J. V. (2002) Investigations of submarine groundwater discharge. *Hydrol. Processes* **16**(11), 2115–2129.

Thieler, E. R., Schwab, W. C., Gayes, P. T., Pilkey, O. H., Jr, Cleary, W. J. & Scanlon, K. M. (1999) Paleoshorelines on the US Atlantic and Gulf continental shelf: evidence for sea-level stillstands and rapid rises during deglaciation. In: *Non-steady State of the Inner Shelf and Shoreline: Coastal Change on the Time Scale of Decades to Millennia*, Conference Proceedings (University of Hawaii, Honolulu, November 1999), 207–208.

Uchupi, E. & Mulligan, A. E. (2006) Late Pleistocene stratigraphy of Upper Cape Cod and Nantucket Sound, Massachusetts. *Mar. Geol.* **227**(1-2), 93–118.

Uchupi, E., Driscoll, N., Ballard, R. D. & Bolmer, S. T. (2001) Drainage of late Wisconsin glacial lakes and the morphology and late Quaternary stratigraphy of the New Jersey–southern New England continental shelf and slope. *Mar. Geol.* **172**(1), 117–145.

Willard, D. A. & Holmes, C. W. (1997) Pollen and geochronological data from south Florida: Taylor Creek Site 2. *US Geol. Survey Open-file Report 97-35*.

Willard, D., Bernhardt, C. E., Holmes, C. W., Landacre, B. & Marot, M. (2006) Response of Everglades tree islands to environmental change. *Ecol. Monogr.* **76**(4), 565–583.

2 PHYSICAL APPROACHES

Investigations of the brackish karst springs on the Croatian Adriatic Sea coast

OGNJEN BONACCI & IVANA GABRIĆ
Faculty of Civil Engineering and Architecture, University of Split, Matice hrvatske15, 21000 Split, Croatia
obonacci@gradst.hr

Abstract The Croatian Adriatic Sea coast is a highly and deeply karstified area. Thus its coastal aquifers are open to seawater intrusion. There are many coastal karst springs, and the vast majority of them are brackish year round, or during periods of low summer discharges. Increasing development of these areas threatens to cause a shortage of fresh water. Consequently, there is an increased need for developing water reserves in the coastal aquifers and understanding the mechanism of seawater intrusion. This paper gives an overview of eight selected coastal brackish karst springs along the Croatian Adriatic Sea coast. An explanation of their functioning, an overview of measures taken for the prevention of seawater intrusion in some of them, and the practical success of those measures is given.

Key words coastal karst; brackish karst springs; seawater intrusion; Croatia

INTRODUCTION

Watersheds of the karst springs along the Croatian Adriatic Sea coast are situated in the Dinaric carbonate karst massif. The whole area is significantly open to seawater intrusion deep into the coastal aquifers. The development of karst processes in these areas and their hydrogeological consequences have been significantly influenced by changes of the erosion baseline throughout geological history. About 20 000 years ago the sea level was approx. 100 m below the present level. In such conditions many permanently or temporarily brackish springs have emerged on the Croatian coast.

The major characteristic of these springs is the variable concentration of chlorides during the year with an unfavourable distribution with respect to water supply needs. The increased intrusion of seawater during the warm and droughty summer periods is the result of a concomitant lower groundwater level in the coastal aquifers.

A substantial prerequisite for spring protection is knowledge of the mechanism of seawater intrusion. The dynamic of freshwater and seawater mixing is significantly different for each karst spring. Therefore, extensive investigations of each spring have to be made, resulting in varying decisions regarding appropriate spring protection. Generally, there are three possible ways of protecting brackish karst springs (Bonacci & Roje-Bonacci, 1997; Breznik, 1998): (a) the construction of a grout curtain; (b) the artificial raising of the spring water level; and (c) the interception from a karst aquifer further from the seawater intrusion zone.

The intrusion of seawater into coastal karst aquifers is present along the entire Mediterranean coastline and other coastal karst areas in the world. Many hydrologists, hydrogeologists and other experts have been involved in the solution of this problem and numerous investigations, analyses, mathematical models and attempts regarding their development have been made (Lambrakis *et al.*, 2000; Maramathas *et al.*, 2003; Arfib & de Marsily, 2004; Blavoux *et al.*, 2004; Pulido-Bosch *et al.*, 2005; Guhl *et al.*, 2006).

This paper gives an overview of eight brackish karst springs or spring zones on the Croatian coast, their basic hydrogeological characteristics and information about their seawater intrusion processes, and planned and performed measures for their protection.

OVERVIEW OF EIGHT CROATIAN COASTAL BRACKISH KARST SPRINGS

Hydrogeological characteristics

Almost half of the Croatian territory is made of carbonate rocks, which belong to the Dinaric karst (Herak *et al.*, 1969). The hydrogeological map given in Fig. 1 shows the domination of carbonate rocks in the coastal area with small patches of true and relative barriers. These conditions constitute a favourable prerequisite for brackish spring formation and seawater intrusion deep into the coastal aquifers. The Croatian Adriatic Sea coast is characterised by a large number of permanent or periodical brackish and submarine springs. Brackish springs appear at sea level, below and above sea level, and even 20 m above sea level. The locations of the eight brackish, coastal, and karst springs analysed in this paper are indicated in Fig. 1 .

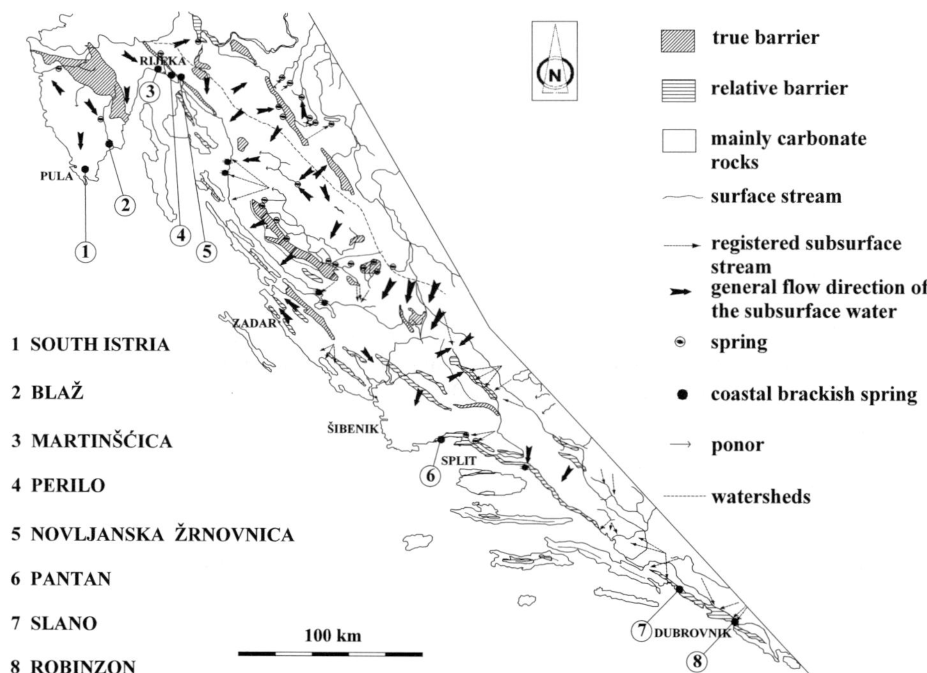

Fig. 1 Hydrogeological map of the Croatian Adriatic Sea coast with indicated positions of brackish karst springs analysed herein (Herak *et al.*, 1969).

South Istria spring zone

The southern Istrian peninsula is formed of a series of carbonate deposits. Limestone is dominant, with the occasional appearance of dolomite layers, lenses, breccias and marls (Fig. 2). The hydrogeology of this karst region can be described as an open coastal zone with a series of small coastal springs with discharges up to 24 L/s. Several wells with a total capacity of 200 L/s have been drilled into the coastal karst aquifers for the water supply of the town of Pula. Many of these wells are situated adjacent to the coast where the groundwater level does not exceed 3 m a.s.l. (Bonacci & Roje-Bonacci, 2000). Intensive exploitation causes the lowering of the groundwater level in dry periods, which results in seawater intrusion into the aquifers, exploitation wells and coastal karst springs. The problem is intensified by many private and uncontrolled wells, which use water for irrigation.

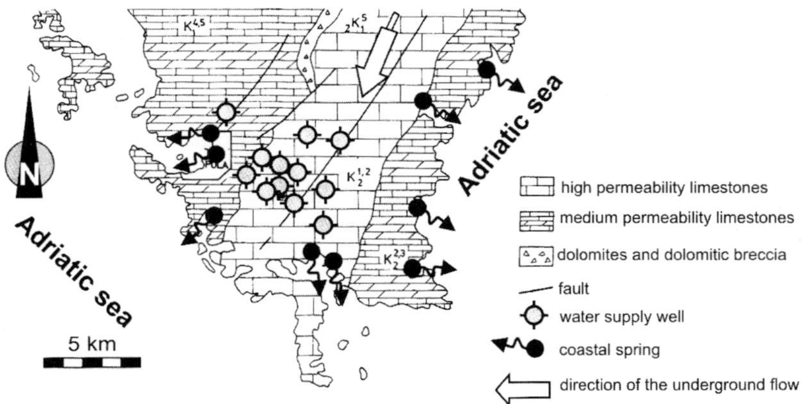

Fig. 2 Hydrogeological map of the Southern Istrian spring zone (COST Action 621, 2004).

Blaž Spring

Blaž Spring is the main spring in a karst spring zone, consisting of about 20 permanent and temporary springs situated along the coast of the Raša Bay in a line 500 m long (Fig. 3). This represents the contact zone between carbonate rocks and impermeable flysch layers. The average spring discharge outflow is about 1.6 m^3/s and the maximum discharge is 2.6 m^3/s (Bonacci & Roje-Bonacci, 1997). Due to the intrusion of seawater and high chloride concentration during the summer period, water from it cannot be used for water supply. Seawater intrusion occurs every year during hot and droughty summer periods. Figure 3 shows the positions of 20 boreholes drilled in the spring vicinity for the purpose of analysing the relationship between freshwater and seawater in the aquifer. Seawater penetrates the aquifer directly through the karst conduit connected to the Adriatic Sea. Seawater intrusion into the Blaž Spring could possibly be prevented by blocking the karst conduit through which the main quantity of seawater intrudes into the spring.

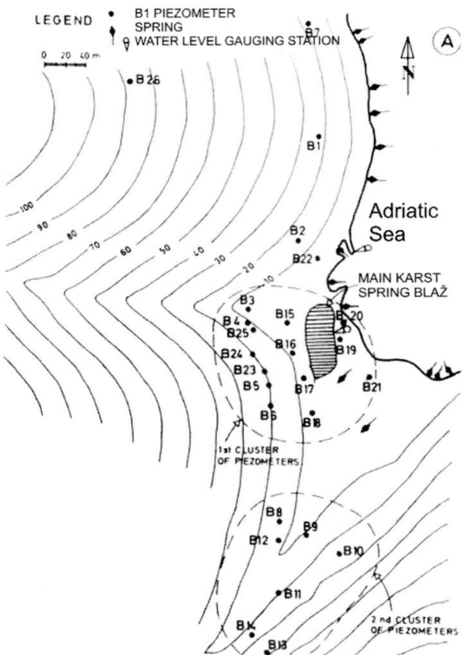

Fig. 3 Blaž Spring location (Bonacci & Roje-Bonacci, 1997).

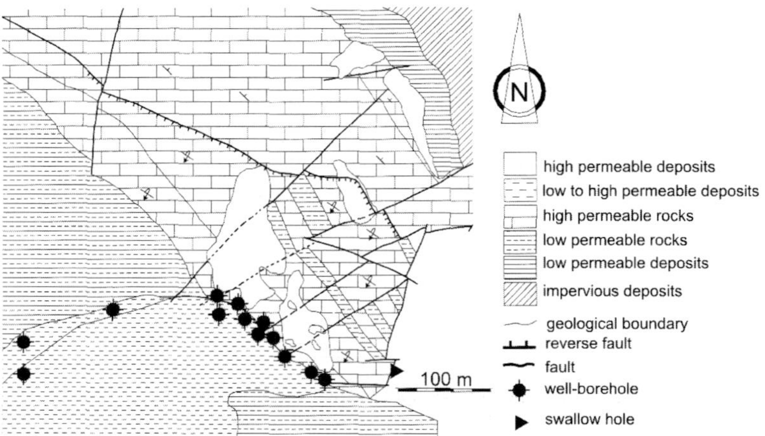

Fig. 4 Hydrogeological map of the Martinšćica Spring zone (COST Action 621, 2004).

Martinšćica Spring

The Martinšćica Spring is one of the springs exploited for the water supply of the city of Rijeka, with a maximum discharge of 400 L/s. The water supply system consists of several wells shown in Fig. 4. However, the wells are not in operation due to increased salinity during the summer dry periods. Piezometers drilled farther in the karst massif showed that the boundary between freshwater and seawater depends on local geological conditions and changes in the groundwater level during the time of operation. As the local dolomite layers are nearly impermeable, brackish water from deeper layers could not have any influence on freshwater flowing through the layer of limestone inside dolomites. This hydrogeological condition gives the possibility of successful spring exploitation, but under the condition of controlling the dynamic reserves of freshwater.

Perilo Spring

Perilo Spring is also one of the coastal karst springs whose water is used for the water supply of the city of Rijeka. A pumping facility has been placed in its karst aquifer farther inland (Fig. 5). Its pumping capacity in the dry summer periods is 240 L/s. In those periods the intrusion of seawater into the spring can occur, being caused by overexploitation and low groundwater levels. The small width of the permeable limestone placed within impermeable flysch layers gives the possibility of grout curtain construction. However, controlled water pumping connected with the continuous monitoring of the groundwater level has resulted in not taking any invasive measures.

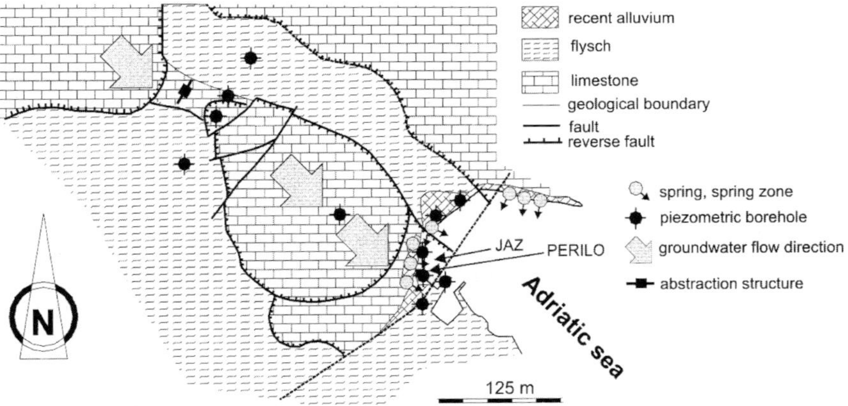

Fig. 5 Hydrogeological map of the Perilo Spring zone (COST Action 621, 2004).

Novljanska Žrnovnica Spring zone

The Novljanska Žrnovnica Spring zone is situated along the northern part of the Croatian Adriatic Sea coast. Its water is used for water supply of the tourist area of the Novljansko-Crkvenička Riviera. Its catchment is mainly underlain by Lower Cretaceous high permeability limestone (Fig. 6). Limestone breccias cross the area and function as a hydrogeological barrier. The dynamic of groundwater circulation and its interaction with seawater is complex, which under natural conditions causes the periodical intrusion of seawater. In order to decrease seawater intrusion, a grout curtain

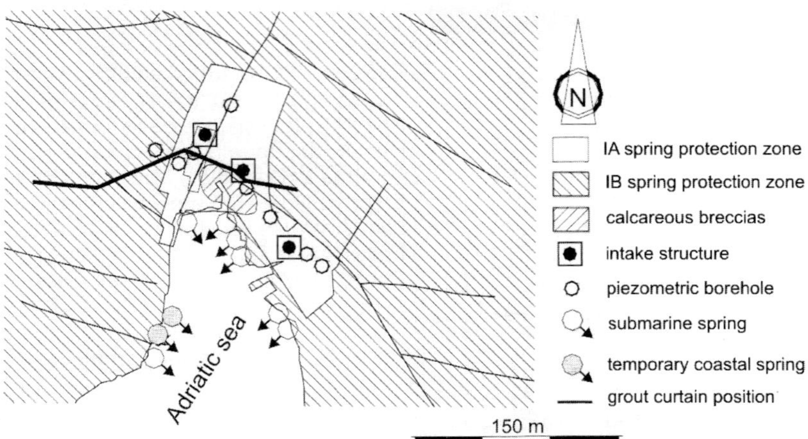

Fig. 6 Hydrogeological map of the Novljanska Žrnovnica spring zone (COST Action 621, 2004).

has been constructed (Biondić *et al.*, 2005). Its purpose is to prevent contact between seawater and freshwater in the shallow zone. Monitoring the results carried out in deep boreholes has revealed that the mixing zone between fresh and seawater is at a depth of 90 to 110 m below sea level. This indicates that the efficiency of the grout curtain is not absolute, and the intrusion of seawater into the aquifer still occurs in extremely dry years.

Pantan Spring

The Brackish karst spring Pantan is situated in Kaštela Bay in the vicinity of the town of Trogir. Its catchment area is underlain by highly permeable limestone rocks. The spring exit is situated at the contact between limestone and flysch layers at an altitude of 2 m a.s.l. The winter discharge is up to 10 m^3/s and in dry periods declines to 1.3–2 m^3/s. Chloride concentration varies between 50 and 10 000 mg/L. During the wet winter months salinity is low, while in the dry summer months it is very high. Close to the Pantan Spring lies the temporary Slanac karst spring, as well as the two submarine springs of Arbanija and Slatina (Bonacci *et al.*, 1995). The Slanac Spring exit is situated at an elevation of 20 m a.s.l. Numerous investigations there have resulted in hypotheses regarding a seawater intrusion mechanism (Mijatović, 1984; Fritz, 1994; Bonacci, 1995). There are many different concepts for the protection of the Pantan Spring, but, because the mechanism of seawater intrusion has yet to be completely explained, spring protection measures have never been undertaken.

Slano Spring

Drinking water for the town of Slano is pumped from two wells which belong to the Slano karst spring aquifer (Fig. 8). The concentration of chlorides in the Slano spring water increases during the dry summer period. Seawater intrusion also appears after

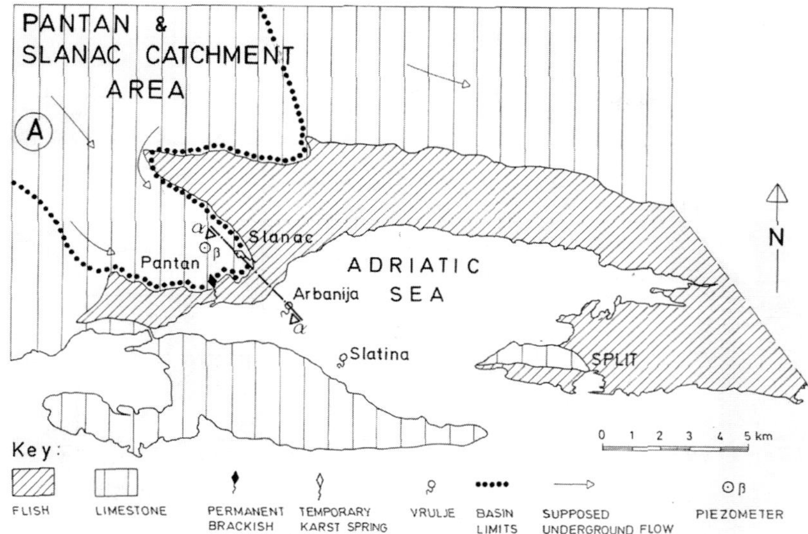

Fig. 7 Hydrogeological map of a wide area around the Pantan spring (Bonacci *et al.*, 1995).

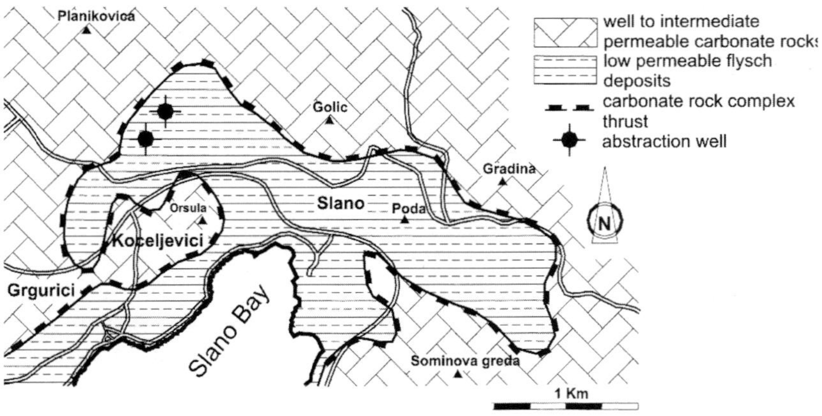

Fig. 8 Hydrogeological map of Slano Spring (COST Action 621, 2004).

the first intensive precipitation event in autumn. This phenomenon can be explained by the influence of turbulence in the mixing zone between seawater and freshwater. After a relatively short period of time, the situation stabilises, and the concentration of chlorides rapidly decreases. The applied solution is controlled water pumping synchronised with the groundwater level conditions.

Robinzon Spring

The Robinzon Spring is situated on the southern part of the Adriatic Sea coast near the town of Dubrovnik. The spring exit is 70 m from the coast. The maximum and

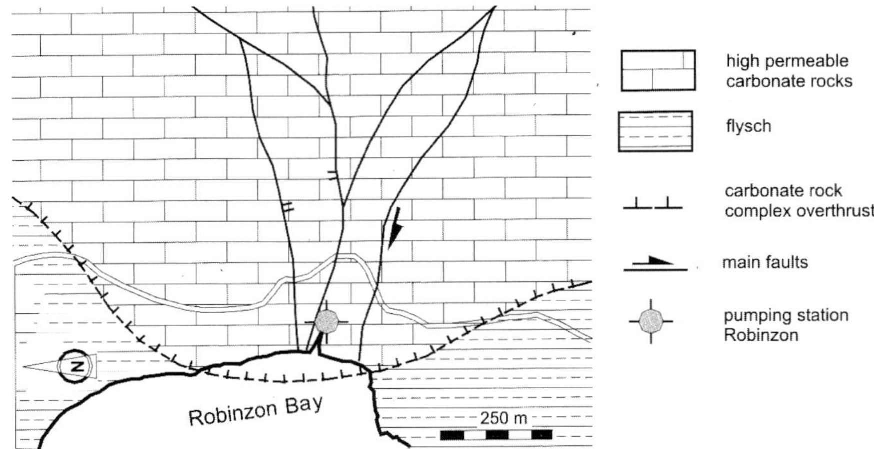

Fig. 9 Hydrogeological map of Robinzon Spring (COST Action 621, 2004).

minimum discharges are about 2 m^3/s and 165 L/s, respectively. The Robinzon Spring emerges on a tectonic contact between very permeable carbonate layers and impermeable flysch layers. Water discharging from the spring is influenced by the seawater, especially during the dry summer periods. The proposed idea was to abstract freshwater from the karst aquifer further from the seawater intrusion zone. For this purpose, the conductivity was measured in boreholes drilled in the aquifer farther inland. As the result of numerous investigations, the construction of a grout curtain was proposed, but so far this solution has yet to be implemented.

DISCUSSION

The overview of the eight Croatian coastal brackish karst springs given in this paper has shown that they often emerge in the contact zone between limestone and flysch layers. Their aquifers are situated in the karst massif, which is mainly composed of limestone with the occasional appearance of dolomite and flysch layers and lenses, dolomite breccias, and marls. Morphological, hydrogeological and geological conditions cause an unstable equilibrium between fresh and seawater during different hydrological situations. This is the main reason that each of these springs has different and very specific conditions of seawater intrusion into their aquifers. For some springs a probable direct connection exists between the sea and the spring exit through unknown karst conduits (Blaž, Pantan). Along the Croatian Adriatic Sea coast, karst spring zones, consisting of many closely distributed and functioned springs, are a normal occurrence. Most of them have a relatively small outflow discharge, whereas only one of them has substantially greater discharge.

The basic characteristics of the eight karst springs described above are shown in Table 1. There are: (a) constantly brackish springs; (b) periodically brackish springs during dry, summer periods; and (c) springs which are brackish only in situations occurring during extremely dry years. Only the last mentioned group of karst springs,

Table 1 Characteristics of coastal brackish karst springs and zones explained in this paper.

Spring	Capacity (L/s)	Abstraction	Seawater intrusion characteristics	Proposed measures	Successfulness of measure	Classification according to the COST Action 621 (2004).			
						Permeability	Structure	Seawater intrusion	Exploitation
South Istria	200	Wells in hinterland	Due to overexploitation in dry periods	Controlled exploitation	Successful	High	Free-unconfined	Moderate-high and variable-seasonal	High
Blaž	1600–2600	Not used for water supply	Every year in dry periods	Grout curtain	Not constructed	High	Free-unconfined	Variable-seasonal	Under-exploited
Martinšćica	400	Wells	Due to overexploitation in dry periods	Controlled exploitation	Successful	High	Free-unconfined	Variable-seasonal	Moderate
Perilo	240	Abstraction structure in hinterland	Due to overexploitation in dry periods	Grout curtain/controlled exploitation	Not constructed/ successful	High	Free-unconfined	Variable-seasonal	Moderate to high
Novljanska Žrnovnica	400	Spring interception with pumping	Seawater intrusion in extremely dry years	Grout curtain/controlled exploitation	Partly successful/ successful	High	Free-unconfined	No intrusion	Moderate
Pantan	1300–10 000	Not used for water supply	Seawater intrusion in dry periods	(1) interception in hinterland; (2) grout curtain (3) spring level rising	No measures applied	High	Free-unconfined	Variable-seasonal	Under-exploited
Slano	6	Two exploitation wells	Overexploitation in dry periods and first high freshwater wave	Controlled exploitation	Partly successful	High	Free-unconfined	Variable-seasonal	Moderate
Robinzon	165–2000	Spring interception with pumping	Seawater intrusion in dry periods	Grout curtain	Not constructed	High	Free-unconfined	Variable-seasonal	Moderate

which are brackish only rarely, can be used for a relatively reliable water supply. However, these springs should be constantly monitored and analysed with the aim of increasing the reliability of their water supply. In the four last columns of Table 1 the classification according to the COST Action 621 (2004) of the eight Croatian karst springs is given.

It should be noted that completely successful protective measures of seawater intrusion into karst spring coastal aquifers and karst springs have not been obtained for any of the presented examples. This means that the mechanism of seawater intrusion has not been completely explained. In some cases there are, to a certain extent, clear hydrogeological conditions which lead to the possibility of an efficient invasive measure, such as the construction of a grout curtain (Blaž, Perilo, Novljanska Žrnovnica, Robinzon). A partly successful grout curtain has been constructed only in the case of the Novljanska Žrnovnica Spring zone, where, in combination with strictly and continuously controlled water pumping, it presents successful spring protection. No other grout curtains have yet been built at any of the other springs.

The most efficient measure for freshwater pumping is the abstraction of water in the karst massif away from the zone of contamination (South Istria, Martinšćica, Perilo and Slano). In this case, water abstraction must be carefully controlled to maintain the equilibrium between freshwater and seawater. One substantial prerequisite is a detailed knowledge of the hydrogeological and hydraulic characteristics which govern the seawater intrusion. In the case of abstracting water from aquifers, the main problem is overexploitation. Agricultural development, which is planned to be intensified, also presents potential dangers. In the case of coastal aquifer overexploitation, the contact zone between sea and freshwater will be moved deeper into the karst massif, resulting in new karst groundwater pollution.

CONCLUSION

Croatia belongs to the group of countries with a large proportion of coastal karst aquifer exploitation. However, seawater intrusion makes large quantities of this water unsuitable for water supply, especially during the hot summer season. For the further development of these coastal areas, it is very important to protect their springs and aquifers from seawater intrusion.

In recent practice many of the karst springs which are potentially useful for water supply were investigated. By monitoring and studying hydrological, hydrogeological, geological and morphological relationships using different parameters (discharge, salinity in the spring and in piezometers, groundwater and spring water levels, dimensions of conduits, geological layers and their characteristics, etc.), detailed hypotheses of the seawater intrusion mechanisms were developed.

Generally, the problem of karst spring protection from seawater intrusion is very complex, invasive technical measures are very expensive, and at the same time their success is uncertain. A very common solution is water abstraction from the karst aquifer with controlled interception, i.e. controlling the unstable equilibrium of fresh groundwater and seawater. It should be noted that the construction of grout curtains in Croatian coastal karst areas is usually avoided. It is a very expensive technical measure in the karst environment, and at the same time its efficiency is problematic. An additional complication is the fact that, in many cases, where the seawater intrusion occurs in very deep layers, the grout curtain must be deeper than 100 m.

A universal solution for the coastal karst spring protection of seawater intrusion does not exist. The hydrological and hydrogeological functioning of each spring is unique. The efficient exploitation and protection of coastal karst aquifers open to seawater intrusion is a complex task, which needs a high level of knowledge about aquifer functioning, as well as of the unstable equilibrium of freshwater and seawater. This overview of eight karst springs along the Croatian Adriatic Sea coast shows that the most applied measure is controlled interception farther inland from the spring. This does not mean it is always the best solution. Better knowledge of the springs' functioning would result in better solutions for their protection. Improved coastal karst aquifer management-based on more complex knowledge of freshwater and salt water dynamics in vulnerable parts of these aquifers, should result in increases in the available quantities of drinking water.

REFERENCES

Arfib, B. & de Marsily, G. (2004) Modelling the salinity of an inland coastal brackish karstic spring with a conduit-matrix model. *Water Resour. Res.* **40**, W11506, doi:10.1029/2004WR003147.

Blavoux, B., Gilli, E. & Rousset, C. (2004) Watershed and origin of the salinity of the karstic submarine spring Port-Miou. *C. R. Geoscience* **336**, 523–533.

Biondić, B., Biondić, R. & Kapelj, S. (2005) The sea water influence on karstic aquifers in Croatia. In: *Ground Water Management of Coastal Karstic Aquifers.* Final report, COST Action 621 (ed. by L. Tulipano), 303–311. Office for Official Publications of the European Communities, Luxemburg.

Bonacci, O. (1995) Brackish karst spring Pantan. *Acta Carsologica* **XXIV**, 97–107.

Bonacci, O. & Roje-Bonacci, T. (1997) Sea water intrusion in coastal karst springs: example of the Blaž Spring. *Hydrol. Sci. J.* **42**(1), 89–100.

Bonacci, O. & Roje-Bonacci, T. (2000) Heterogeneity of hydrological and hydrogeological parameters in karst: examples from Dinaric karst. *Hydrol. Processes* **14**(14), 2423–2438.

Bonacci, O., Fritz F. & Denić, V. (1995) Hydrogeology of Slanac spring, Croatia. *Hydrogeol. J.* **3**(3), 31–40.

Breznik, M. (1998) *Storage Reservoirs and Deep Wells in Karst Regions.* A. A. Balkema, Rotterdam, The Netherlands.

COST Action 621 (2004) *The Main Coastal Karstic Aquifers of Southern Europe: A Contribution by Members of the COST-621 Action "Groundwater Management of Coastal Karstic Aquifers"* (ed. by J. M. Calaforra) (EUR 20911), Office for Official Publications of the European Communities, Luxemburg.

Fritz, F. (1994) On the appearance of a brackish spring 30 m above sea level near Trogir (southern Croatia). *Geologia Croatica* **47**(2), 215–220.

Guhl, F., Pulido-Bosch, A., Pulido-Leboeuf, P., Gisbert, J., Sanchez-Martos, F. & Vallejos, A. (2006) Geometry and dynamics of the freshwater–seawater interface in a coastal aquifer in southeastern Spain. *Hydrol. Sci. J.* **51**(3), 543–555.

Herak, M., Bahun, S. & Magdalenić, A. (1969) Pozitivni i negativni utjecaji na razvoj krša u Hrvatskoj. *Krš Jugoslavije* **6**, 45–71 (in Croatian).

Lambrakis, N., Andreou, A. S., Polydoropoulos, P. & Georgopoulos, E. (2000) Nonlinear analysis and forecasting of a brackish karstic spring. *Water Resour. Res.* **36**(4), 875–884.

Maramathas, A., Maroulis, Z. & Marinos-Kouris, D. (2003) Brackish karstic springs model: application to Almyros spring in Crete. *Ground Water* **41**(5), 608–619.

Mijatović, B. F. (1984) Problems of sea water intrusion into aquifers of the coastal Dinaric Karst. In: *Hydrogeology of the Dinaric Karst* (ed. by G. Castany, E. Groba Castany, E. Groba & E. Romijn), 115–142. International Contributions to Hydrogeology, AIH 4, Hannover, Germany.

Pulido-Bosch, A., Vallejos, A., Pulido-Leboeuf, P., Molina, L. & Calaforra, J. M. (2005) Some considerations about karst coastal aquifers, on the example of two Andalusian cases. In: *Ground Water Management of Coastal Karstic Aquifers.* Final report. COST Action 621 (ed. by L. Tulipano), 271–281. Office for Official Publications of the European Communities, Luxemburg.

Geoelectric and geochemical studies for hydrological characterization of Sagar Island, South 24 Parganas, West Bengal, India

R. K. MAJUMDAR & D. DAS
Department of Geological Sciences, Jadavpur University, Kolkata-700032, West Bengal, India
ranjit_mazumdar2000@yahoo.co.in

Abstract Integrated geoelectric and geochemical investigation were carried out in the Sagar Island region to assess the prevailing groundwater conditions and chemical quality of groundwater. Geologically, the area is constituted of alluvial and marine sediments of Quaternary age. Vertical electrical soundings (VES) in the area of investigation mostly show five layers consisting of topsoil, saline water, brackish water, a clay layer and freshwater-bearing zones. The VES findings show the potential freshwater-bearing zone to be of appreciable thickness at depths of 175.0 to 220.0 m under confined conditions. The surface true resistivity contour map shows the intrusion of saline water in the southern part of Sagar Island at shallower depths. The results of VES studies significantly correspond with the borehole data. Chemically, the fresh groundwater is Na-HCO_3 type with TDS ranging from 465 to 645 mg/L. The water is safe for drinking and domestic purposes but unsuitable for irrigation purposes. The concentrations of arsenic, iron, lead and mercury in the samples are below the recommended limit for drinking water of the World Health Organization (WHO).

Key words Sagar Island; Vertical Electric Sounding (VES); litho-resistivity relation; seawater contamination (SWC); freshwater aquifer

INTRODUCTION

Sagar Island, the largest island in the Ganga Delta (21°37′N to 21°57′N, 88°2′35″E to 88°11′E), is elongated in the N–S direction (~30 km) and has varying width in the E–W direction The southern portion of the Island widens to ~12 km (Fig. 1, insets). It is bordered to the north, west, east and south by the Hooghly, Gabtala and Muriganga rivers and the Bay of Bengal, respectively (Fig. 1). Sagar Island has a flat topography with no significant variation in elevation from mean sea level (~3 m). The island covers 235 km^2 and is home to 46 villages, a population of 0.20 million (Majumdar *et al.*, 2002), and the "Kapil Muni" Hindu temple (near T1, in Fig. 1(a)). The village of Gangasagar has been selected as a tourist centre by the Government. For drinking water, the island is solely dependent upon the deeper aquifers at between 180 and 330 m below ground level (b.g.l.) (Das, 1991). Fresh groundwater in the deeper aquifers occurs under confined conditions and is tapped by means of small diameter tubewells fitted with hand pumps. Shallow dug wells in this area produce saline water. A geoelectric (resistivity) survey has been conducted to identify various lithological units, evaluate the groundwater condition, find potential aquifer zones, and determine the possibility of seawater contamination in and around the Gangasagar area.

(a)

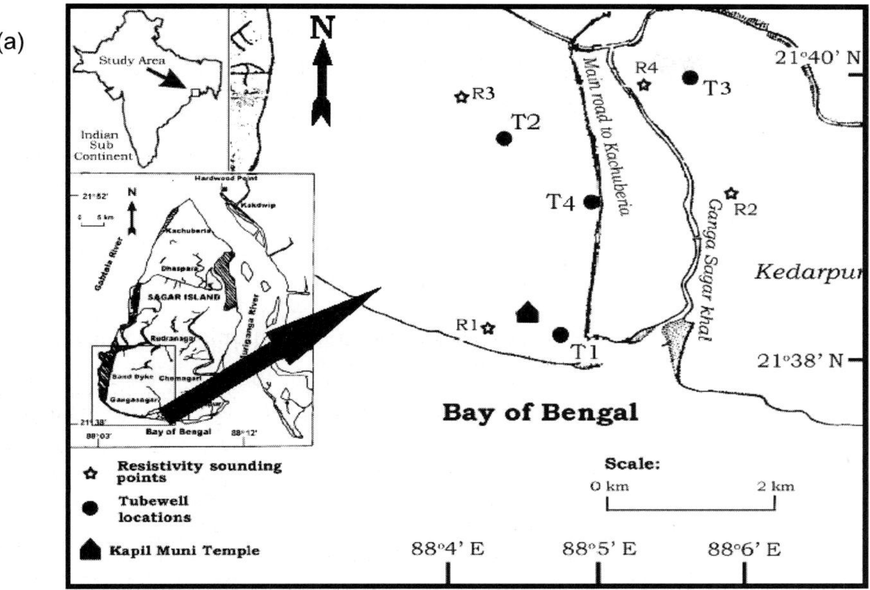

(b)

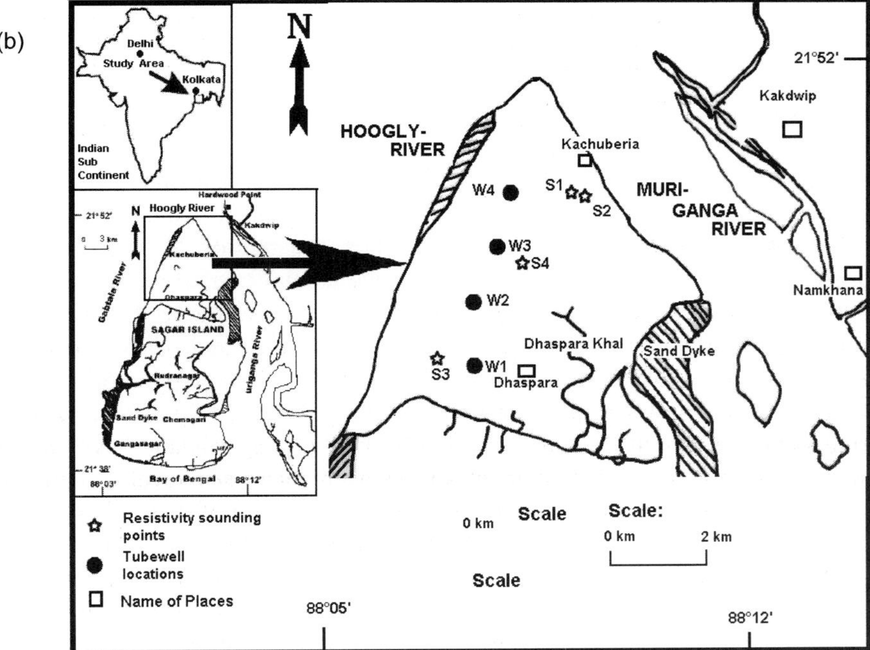

Fig. 1 (a) Southern Part of Sagar Island. Tubewell locations are as follows: T1—beside Hanuman Temple; T2—northern part of Paschim Nutungheri village; T3—Lalpur village; T4—beside Sridham Gangasagar High School. (b) (Northern Part of Sagar Island. Tubewell locations are as follows: W1—Eastern-Mandirtala; W2—inside Bamankhali Market; W3—Eastern part of Patharpratima village; W4—Pakhirala village beside main road.

GEOLOGY OF THE STUDY AREA

Sagar Island lies at the southernmost part of the Indo-Gangetic Plain, which is the largest alluvial tract in the world, and the Quaternary alluvial fill of this plain is carried and deposited by the River Ganga and its tributaries/distributaries. The Quaternary sediments of the Bengal Plain are composed of flood plain and deltaic deposits that are subdivided into two major groups (Roychoudhuri, 1974): younger Holocene and older Pleistocene alluvium. The southern extremity of the Bengal Plain is characterized by the presence of an extensive coastal belt, which Chakrabarti (1995) divided into two environmental zones. Sagar Island, being a part of this coastal belt, falls under the "macrotidal Hooghly estuary" zone. It has been established that Hooghly estuary is characterized by a broad expanse of syn-depositional fluviotidal and marine coastal sediments (viz. sand, silt and clay) deposited during the Flandrian Transgression around 6000 years BP (the on-lapping sequence) and subsequent delta progradation (the off-lapping sequence) (Chakrabarti, 1995). From the early part of this century, it underwent a destructive phase (Chakrabarti, 1992). The present day configuration of Sagar Island reveals that a number of small isolated islands, earlier separated by tidal creeks, are now welded almost into a single landmass due to gradual reduction of the width of the tidal creeks.

Coastal marshes, mangrove swamps, tidal flats, mudflats, sand dunes or ridges, marine terraces, and tidal inlets are all coastal features of this island (Paul & Bandyopadhyay, 1987).

Hydrogeology

Sagar Island is criss-crossed by numerous tidal creeks and man-made canals. These creeks, and tanks, are the main sources of surface water, which has a high hardness and salinity (Chakrabarti, 1995). In rainy seasons, the salinity of the water of the tanks decreases and becomes brackish. The average annual rainfall of Sagar Island is about 200 cm, with a mean temperature of 22°C. The area is characterized by the presence of fluvio-tidal and marine coastal facies deposits (Chakrabarti, 1991), where freshwater aquifers occur between the depths of 180 and 330 m. The upper aquifer zone contains saline water in its upper part and brackish water in its lower part. The lower aquifer zones are confined and separated from the overlying brackish aquifers by an impermeable clay layer of ~20 m thickness. The piezometric head in the freshwater zones varies from 1 to 4 m b.g.l. with the hydraulic gradient being generally towards the sea.

Geoelectric resistivity investigation

The geoelectric resistivity method has been extensively used for structural, hydrological and geothermal investigations (Stewart *et al.*, 1983; Yadav & Abolfazli, 1998; Majumdar *et al.*, 2000; Pal & Majumdar, 2001; Majumdar & Pal, 2005). Here, a Schlumberger vertical electrical sounding (VES) study was carried out in the southern and northern parts of the Sagar Island region to ascertain the vertical distribution of the water bearing zones within the aquifer system.

DATA ACQUISITION AND INTERPRETATION

The VES investigation was conducted at seven locations with a maximum electrode spacing of 1200 m (Fig. 1) using resistivity equipment DDR-4 manufactured by Integrated Geo Instruments and Services Pvt. Ltd, Hyderabad. The resistivity sounding curve is interpreted by a one dimensional (1-D) inversion technique using the "RESIST" software. Preliminary values of the model parameters are obtained by matching the VES field curves with the theoretical master curves and auxiliary point charts. These model parameters are subsequently used as input (starting model) in RESIST for further refinement of the results of the 1-D inversion algorithm. The resistivities of different layers and corresponding thicknesses are reproduced by a number of iterations until the model parameters of all VES curves are totally resolved with minimum RMS error. The 1-D inversion model parameters can serve as the starting model for 2-D and 3-D approaches to improve the approximation of the sub-surface geology. In such cases, 1-D interpretation is usually found to be fairly consistent with those observed in 2-D and 3-D inversions (Monteiro Santos *et al.*, 1997; Olayinka & Weller, 1997). The results of VES are interpreted in terms of the subsurface geology and aquifer characteristics under prevailing hydrodynamic conditions.

DISCUSSION OF RESULTS

Inversion results for all eight VES points were interpreted and subsequently correlated to resolve the lithological conditions. The nature and distribution of different lithologies along with the variation in resistivity reveals the presence of four to five layers in the region. The first layer is interpreted as alluvial clayey soil. The second and third layers represent saline and brackish water zones. Both these layers are composed of clay with silt and sand lenses. The fourth layer is impermeable clay. The most important layer is the lowermost, which is interpreted to be a sandy, freshwater zone. The freshwater zone is at a depth of 175–206 m at six sites and at 218–220 m at two sites. The freshwater zone has high resistivity values (41.0–57.3 Ωm) while the overlying saline and brackish layers have low values (0.4–1.0 Ωm, and 6.0–9.7 Ωm, respectively). Field samples have confirmed the presence of saline water within a few metres b.g.l.

A generalized borehole lithology prescribed by the Public Health Engineering Department (PHED, 1994) and Central Ground Water Board (CGWB, 1987) near Kapil Muni Temple (near VES R1) shows the presence of a clay layer with silt and sand lenses from the surface down to 204 m. Beyond 204–271 m, the formation is made up of medium to fine sand with clay intercalation, which is a potential freshwater zone. The layer parameters interpreted from the VES studies correlate with the borehole data (Fig. 2). The true surface resistivity contour maps (Fig. 3(a)–(d)) show the intrusion of saline water in the southern part of Sagar Island at a shallow depth (20 m). Using the results of VES and borehole lithologs, a litho-resistivity relationship has been established (Table 1).

Hydrogeochemistry

A geochemical investigation was carried out to assess the suitability of the water for irrigation and drinking purposes and document any seawater/groundwater interaction.

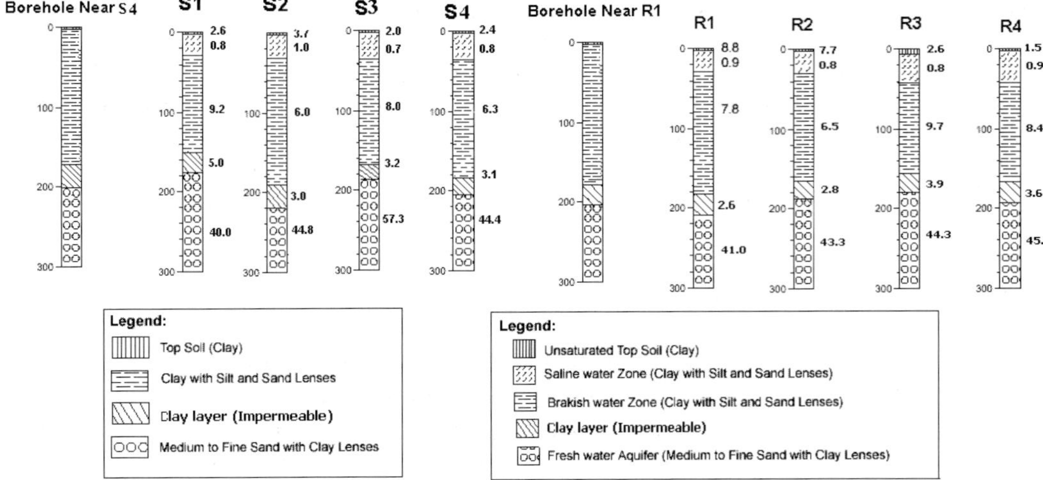

Fig. 2 Borehole lithology described by CGWB in Patharpratima Village (near to VES location S4 – northern part of Sagar Island) and combined borehole lithology by (CGWB & PHE) near Kapil Muni Ashram (near to VES location R1 – southern part of Sagar Island) and layer parameters depicting VES interpretation for different locations. Numbers on the left- and right-hand side of the logs show depth from ground level (m) and true resistivity values (Ωm), respectively.

Table 1 Litho-resistivity relationship.

Probable lithology	Resistivity range
Unsaturated top soil (Clay)	1.5 to 7.7 Ωm
Saline water zone (Clay with silt and sand lenses)	0.7 to 1.0 Ωm
Brackish water zone (Clay with silt and sand lenses)	6.0 to 9.7 Ωm
Clay, grey, sticky layer (impermeable)	2.6 to 5.0 Ωm
Freshwater aquifer (Medium to fine sand with clay lenses)	41.0 to 57.3 Ωm

Eight groundwater samples were collected from tubewells from ~180 to ~330 m between 16 and 20 March 2005 (T1–T4) and 14 and 18 March 2006 (W1–W4). The water samples were collected in pre-cleaned transparent plastic bottles. Before collection of water samples, the bottles were thoroughly rinsed with the same. Concentrations of major cations, anions and some hazardous elements (present in Table 2) were measured in the chemical laboratory of the Center for Study of Man and Environment, Salt Lake, India.

DISCUSSION OF RESULTS

Total dissolved solids (TDS) contents (465–645 mg/L) clearly shows that all the samples are freshwater. TDS *vs* electrical conductivity (EC) plots show a linear trend (Fig. 4) with strong correlation ($R^2 = 0.992$). The ratio of TDS and EC is 0.603, which is close to the ratio (0.627) for water from sands of the Gangetic alluvium and Tarai-Bhabar (Chaterji & Karanth, 1963).

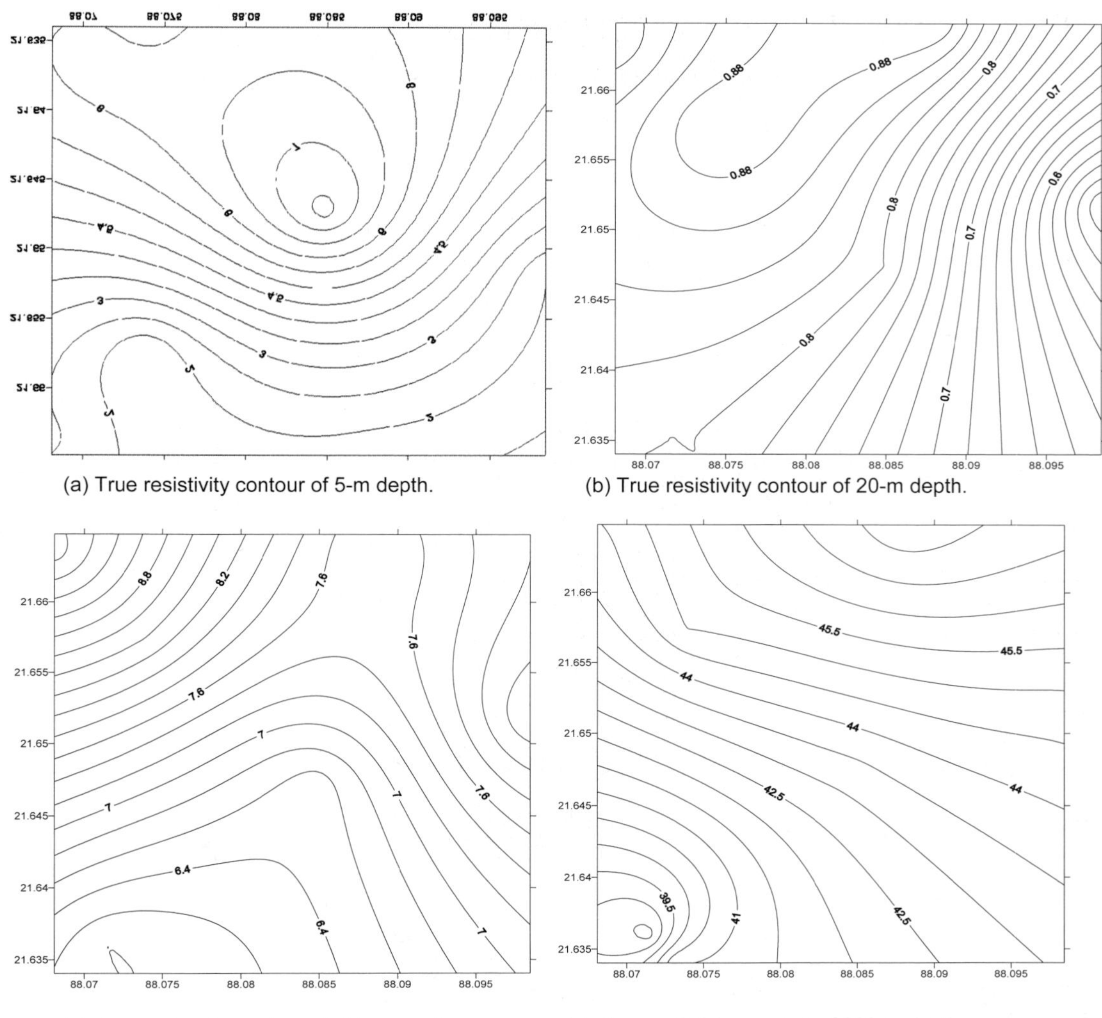

(a) True resistivity contour of 5-m depth. (b) True resistivity contour of 20-m depth.

(c) True resistivity contour of 120-m depth. (d) True resistivity contour of 200-m depth.

Fig. 3 True surface resistivity contour maps, Sagar Island.

Total hardness (2.497 Ca + 4.115 Mg) values indicate the waters are moderately soft (Sawyer & McCarty, 1967) and, according to WHO guidelines (2004), suitable for domestic use. The Sodium Adsorption Ratio (SAR) for all samples was calculated ($Na^+/\{(Ca^{++}+Mg^{++})/2\}^{1/2}$, all values are in meq/L) and plotted against EC (Richards, 1954) (Fig. 6) to determine the suitability of the water for irrigation purposes. The plots suggest the groundwater in the southern parts of Sagar Island are unsuitable for irrigation purpose in terms of the sodium (SAR) and salinity hazard (EC), but the groundwater from deep tube wells in the northern part of Sagar Island are suitable in terms of the SAR and Soluble Sodium Percentage (SSP).

The soluble sodium percentage [$\{(Na^+)/(Na^++K^++Ca^{++}+Mg^{++})\} \times 100$, all values are in meq/L] values (63.07–85.42) were also high, indicating the dominance of Na^+

Table 2 Chemical composition of groundwater in the Sagar Island.

Sample no.	Electrical conductivity at 25°C (μs/cm)	TDS (mg/L)	Carbonate (as $CaCO_3$, mg/L)	Bi-carbon-ate (as $CaCO_3$, mg/L)	Chloride (Cl, mg/L)	Sulphate (SO_4, mg/L)	Sodium (Na, mg/L)	Potassium (K, mg/L)	Calcium (Ca, mg/L)
T1	940.0	574.0	40	360.0	96.0	8.1	219.0	25.0	8.01
T2	1040.0	645.0	80	420.0	84.0	7.1	225.0	20.0	9.6
T3	810.0	498.0	30	385.0	44.0	0.4	190.0	16.0	11.2
T4	750.0	470.0	60	345.0	50.0	0.4	191.0	17.0	9.6
W1	820.0	502.0	0.0	353.5	76.0	0.20	155.0	20.0	22.5
W2	840.0	510.0	0.0	365.5	60.0	0.20	150.0	20.0	20.9
W3	920.0	560.0	0.0	295.8	120.0	12.1	145.0	22.0	25.7
W4	750.0	465.0	0.0	295.8	74.0	0.30	115.0	13.0	28.9

Sample no.	Magnes-ium (Mg, mg/L)	Arsenic (As, mg/L)	Iron (Fe, mg/L)	Lead (Pb, mg/L)	Mercury (Hg, mg/L)	Total hardness (mg/L, CaCO3)	SAR	Soluble sodium percent age	Residual sodium carbonate	Seawater contamina tion
T1	7.8	<0.01	0.72	<0.01	<0.001	150.0	13.2	84.29	2.75	0.72
T2	8.8	<0.01	0.38	<0.01	<0.001	60.0	12.62	85.42	3.93	0.46
T3	7.8	<0.01	0.39	<0.01	<0.001	60.0	10.67	84.44	2.59	0.33
T4	6.8	<0.01	0.35	<0.01	<0.001	52.0	11.53	85.11	3.05	0.34
W1	10.70	<0.01	0.40	<0.01	<0.001	100.0	6.74	78.44	3.79	0.36
W2	13.62	<0.01	0.35	<0.01	<0.001	108.0	6.27	76.49	3.83	0.28
W3	15.57	<0.01	0.42	<0.01	<0.001	128.0	5.56	72.85	2.29	0.69
W4	20.46	<0.01	0.33	<0.01	<0.001	156.0	4.00	63.07	1.73	0.43

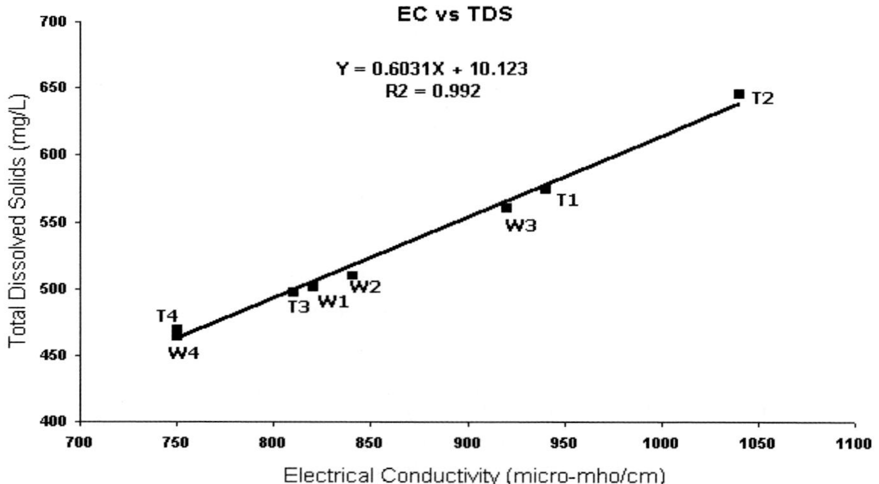

Fig. 4 Variations of TDS with electrical conductivity (EC) for groundwater samples. (at 25°C).

in the major cations. Residual sodium carbonate [$(HCO_3^- + CO_3^{--}) - (Ca^{++} + Mg^{++})$, all values are in epm] values (1.73 to 3.93) suggest that only the groundwater in the northern part of Sagar Island is suitable for irrigation.

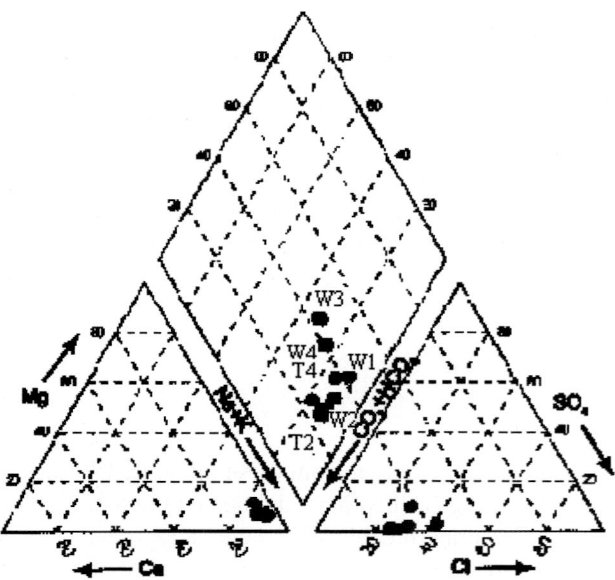

Fig. 5 Piper's trilinear diagram for groundwater samples.

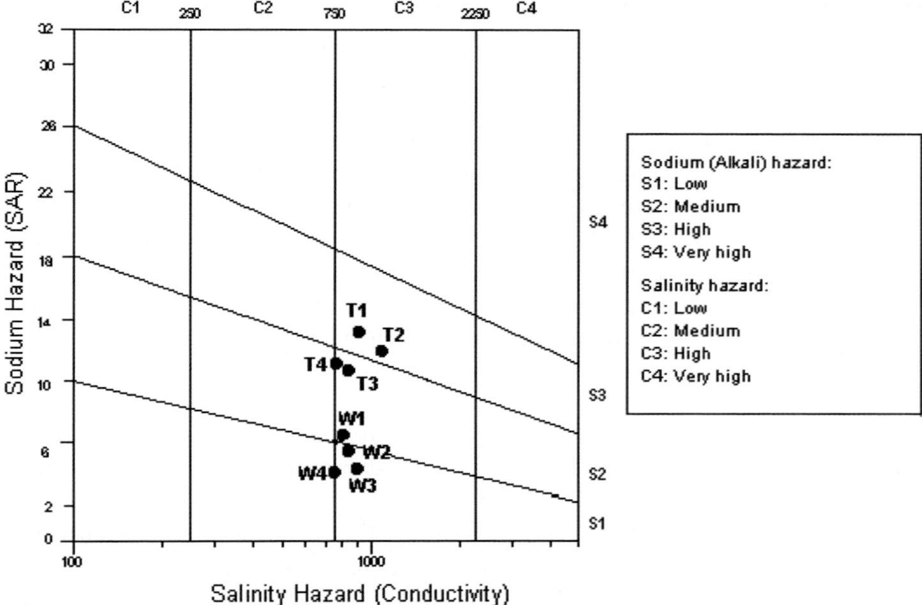

Fig. 6 Richard's diagram for classification of irrigation waters (EC *vs* SAR).

The possibility of seawater contamination is examined using the ratio $Cl^-/(CO_3^{--} + HCO_3^-)$ (all values are in epm), as suggested by Revelle (1941). The ratio is ~243 for seawater from Bay of Bengal and 0.5 for freshwater (Majumdar *et al.*, 2002).

The same ratio in the groundwater samples (0.28–0.72, with an average of 0.45). indicates that nearly all the samples are free of seawater contamination. Only the sample (T1, nearest to the sea) with a ratio of 0.72 represents slightly contaminated water. Concentrations of arsenic, iron, lead and mercury in the samples were found to be safe for drinking water purposes.

It is important to note that the Na content is higher than that generally found (20 mg/L) in groundwater (WHO, 2004). It ranges from 115–225 mg/L, with a mean of 173.7 mg/L. In two samples near the sea (T1 and T2), the Na content is higher than 200 mg/L, the acceptable limit for the drinking water recommended by the WHO (2004).

CONCLUSIONS

The following conclusions can be drawn from these integrated studies:

1. Vertical electrical soundings (VES) have delineated the topsoil, the saline/brackish groundwater zones, an impermeable clay layer and freshwater aquifers in sub-surface geological formations. The freshwater zones are at a depth of 175–206 m at six sites and 218–220 m at two sites.
2. VES findings suggest potable groundwater zones of appreciable thickness, which can be tapped for drinking water purposes. The potential groundwater-bearing zone is under confined conditions. The groundwater conditions of the region as interpreted from the VES study correlate well with the borehole data.
3. A litho-resistivity relationship was established that can be used for estimating lithologies in other unexplored areas under similar hydrodynamic conditions.
4. There is no evidence of seawater mixing with groundwater. Only one sample (T1), nearest to the sea, is slightly contaminated with seawater.
5. Groundwater of this area is of the Na-HCO_3 type.
6. The chemical quality of groundwater is safe for drinking, domestic purposes, and partly suitable for irrigation purposes.
7. Concentrations of arsenic, iron, lead and mercury in the samples are below the recommended limit for drinking water of WHO (2004).

Acknowledgements The authors are grateful to the Department of Science and Technology, Government of India, for financial support and also to the Department of Geological Sciences, Jadavpur University, for infrastructural facilities. Thanks are also due to Sri A. Mondal, Sri B. Das and Sri S. Das, students of the Department of Geological Sciences, Jadavpur University, for helping in survey work.

REFERENCES

Central Ground Water Board (CGWB) (1987) Unpublished borehole data.

Chakrabarti, P. (1991) Morphostratigraphy of coastal Quaternaries of West Bengal and Subarnarekha delta, Orissa. *Indian J. Earth Sci.* **18**(3-4), 219–225.

Chakrabarti, P. (1992) Geomorphology and Quaternary geology of Hooghly estuary, West Bengal, India. PhD Thesis, University of Calcutta, India.

Chakrabarti, P. (1995) Evolutionary history of the coastal Quaternaries of the Bengal plain, India. *Proc. Indian. Nat. Sci. Acad. A* **61**(5), 343–354.

Chaterji, G. C. & Karanth, K. R. (1963) A note on the relationship between resistivity and quality of ground water in sands of the Gangetic alluvium and Tarai-Bhabar formations in India. *Rec. Geol. Surv. India* **92**(2), 279–292.

Das, S. (1991) Hydrological features of Delta's of India. *Mem. Geol. Soc. India.* **22**, 183–225.

Majumdar, R. K. & Pal, S. K. (2005) Geoelectric and borehole lithology studies for groundwater investigation in alluvial aquifers of Munger District, Bihar. *J. Geol. Soc. India* **4**, 463–474.

Majumdar, R. K., Majumdar, N. & Mukharjee, A. L. (2000) Geoelectric investigations in Bakreswar geothermal area, West Bengal, India. *J. Appl. Geophys.* **45**, 187–202.

Majumdar, R. K., Mukharjee, A. L., Roy, N. G., Sarkar, K. & Das, S. (2002) Groundwater studies on south Sagar Island region, South 24-Parganas, West Bengal. In: *Analysis and Practice in Water Resources Engineering for Disaster Mitigation* (Proc. of International Conference on Water Related Disasters, ICWRD-2002, held at Kolkata, December 2002), vol. 1, 175–183. New Age International (P) Ltd, India.

Monteiro Santos, F. A., Andrade Afonso, A. R. & Mendes Victor, L. A. (1997) Study of the Chaves geothermal field using 3D resistivity modeling. *J. Appl. Geophys.* **37**, 85–102.

Olayinka, A. I. & Weller, A. (1997) The inversion of geoelectrical data for hydrological application in crystalline basement areas of Nigeria. *J. Appl. Geophys.* **37**, 103–115.

Paul, A. K. & Bhandyopadhyay, M. K. (1987) Morphology of Sagar Island, a part of Ganga delta. *J. Geol. Soc. India* **29**, 412–423.

Pal, S. K. & Majumdar, R. K. (2001) Determination of ground water potential zones using iso-resistivity maps in alluvial areas of Munger district, Bihar. *Indian J. Earth Sci.* **1-4**, 16–26.

Public Health and Engineering Department (PHED) (1994) Unpublished bore hole data.

Roychoudhuri. M. K. (1974) Geology and mineral resources of the States of India., Part I, West Bengal. *Geol Surv. India. Miscellaneous Publication* **30**, 1–30.

Revelle, R. (1941) Criteria for recognizing sea water in groundwater. *Am. Geophys. Union Trans.* **22**, 593–597.

Richards, L. A. (ed.) (1954) *Diagnosis and Improvement of Saline and Alkali Soils.* Hand Book 60, US Dept of Agriculture.

Sawyer, C. N. & McCarty, P. L. (1967) *Chemistry for Sanitary Engineers* (second edn). McGraw-Hill, New York, USA.

Stewart, M., Layton, M. & Lizanec, T. (1983) Application of resistivity surveys to regional hydrologic reconnaissance. *Ground Water* **21**(1), 42–48.

World Heath Organization (WHO) (2004) *Guidelines for Drinking-Water Quality, 1. Recommendations* (third edn). WHO, Geneva, Switzerland.

Yadav, G. S. & Abolfazli, H. (1998) Geoelectrical sounding and their relationship to hydraulic parameters in semiarid regions of Jalore, northwestern India. *J. Appl. Geophys.* **39**, 35–51.

Investigation of submarine groundwater discharge using several methods in the inter-tidal zone

TOMOTOSHI ISHITOBI[1], MAKOTO TANIGUCHI[1], YU UMEZAWA[1], SHIGERU KASAHARA[2], SHIN-ICHI ONODERA[3], MASAKI HAYASHI[4], KUNIHIDE MIYAOKA[5] & MITSURU HAYASHI[6]

1 *Research Institute for Humanity and Nature, 457-4 Motoyama, Kamigamo, Kita-ku, Kyoto 603-8047, Japan*
tomotoshi@chikyu.ac.jp

2 *Sohgoh kagaku Inc., 1-4-8 Minami-Shin-Machi, Chuo-ku, Osaka 540-0024, Japan*

3 *Faculty of Integrated Arts and Sciences, Hiroshima University, 1-7-1 Kagamiyama, Higashi-Hiroshima 739-8521, Japan*

4 *Graduate School of Biosphere Science, Hiroshima University, 1-7-1 Kagamiyama, Higashi-Hiroshima 739-8521, Japan*

5 *Faculty of Education, Mie University, 1577 Kurimamachiya-cho, Tsu 514-8507, Japan*

6 *Research Center for Inland Seas, Kobe University, 5-1-1 Fukae-Minami-Machi, Higashi-nada-ku, Kobe 658-0022, Japan*

Abstract To accurately estimate the flux of terrestrial groundwater discharge into the sea, a study using several methods was done in the coastal zone of Osaka Bay, Japan. The seepage-meter method and the measurement of temperature near the seabed were applied based on the hypothesis that seawater temperature in summer would decrease, reflecting the extent of active mixing with colder fresh groundwater. As a result, it was confirmed from the seepage-meter method that submarine groundwater discharge rates decreased with the distance from the coast. Evaluations of groundwater discharge rates from seabed temperature showed similar values to the results using the seepage meter, which means that the values were reasonable. Finally, the total groundwater discharge flux from this beach was estimated at 36.7% of the river discharge rate.

Key words submarine groundwater discharge; inter-tidal zone; seepage meter; seabed temperature

INTRODUCTION

Recognition of the importance of submarine groundwater discharge (SGD) is increasing in the studies of water and dissolved-material transport from the land to the sea. According to recent studies from hydrology and coastal oceanography, SGD consists of both terrestrial groundwater and recirculated seawater (Taniguchi *et al.*, 2002; Burnett *et al.*, 2003). However, there are many uncertainties about SGD, because it is invisible and difficult to evaluate quantitatively. Therefore, further study is needed to better estimate submarine fresh–groundwater discharge (SFGD) in coastal areas.

Our study area is located in the coastal zone of Omaehama, Osaka Bay, Nishinomiya city, Japan (Fig. 1). Osaka Bay is located next to Osaka city, the second capital of Japan. The annual precipitation and air temperature are about 1300 mm/year and 16.5°, respectively. Several rivers flow into the Osaka Bay and the largest of them

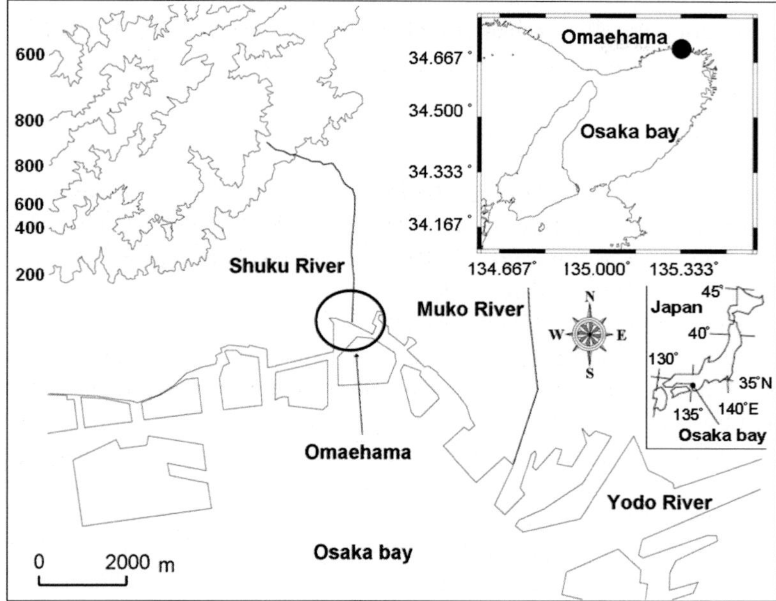

Fig. 1 The study area for measuring SGD.

is the Yodo River with an average discharge rate of about 250 m^3/s. The smaller, the Shuku River, also directly flows into the Osaka Bay, and the river mouth is located near the centre of this beach. The area near Omaehama is well-known because groundwater is abundant and there are many Japanese sake (alcoholic drink) breweries, which need good quality water for making the Japanese sake. Therefore, it was worth monitoring how much fresh groundwater actually contributes to the terrestrial water flux into the ocean in this area. However, some problems such as deposition of organic matter and degradation of the seabed are occurring with the recent developments of the coastal area, although Omaehama is a natural coastal area near a mega-city. Therefore, some environmental assessments have been undertaken in this area (Kasahara, 2005). However, quantitative research on groundwater discharge to the sea has not been done here, despite the possibility that the groundwater transports a certain amount of contaminating materials. The purposes of this study are: (1) to quantitatively evaluate groundwater discharge to the sea by using several methods including a seepage meter and a fibre-optic cable, and (2) to compare groundwater and river discharge.

STUDY METHODS

According to some previous studies, it is thought that SGD rates decrease with the distance from the coast (Bokuniewicz, 1992; Taniguchi *et al.*, 2002). Therefore, seepage meters were deployed along a transect set perpendicularly to the coast line to evaluate the SGD rates from the seabed (Fig. 2). Lee-type seepage meters (Lee, 1977) were used in the inter-tidal zone (A, B and C) and continuous heat flow type seepage meters (Taniguchi & Iwakawa, 2001) were used in the subtidal zone (D and E). Electric conductivity measurements of SGD were done with CT (conductivity-temperature, "Compact-CT" Alec Electronics, Co., Ltd) sensors in the chamber of the seepage meter to evaluate the amount of terrestrial groundwater in SGD. Measurements of tidal change were made concurrently using a Diver sensor (Daiki-rikakogyo, Co., Ltd).

In general, the temperature of terrestrial groundwater is lower than seawater in the summer season. Therefore, it is assumed that the temperature near the seabed will be lower than its surroundings if groundwater discharge from the seabed is actively occurring. In fact, there are some studies which mention a good correlation between SGD and temperature (Banks *et al.*, 1996). In this study, therefore, temperature measurements on the seabed and coastal land surface were taken using a fibre-optic cable (Hitachi, Co., Ltd) to evaluate the change of temperature of the pore water caused by SGD. However, it was also considered that the seabed temperature could be easily influenced by various factors such as the sunlight and marine currents. Therefore, temperature measurements under the seabed (5 cm below the seabed surface) were also collected using a temperature data logger ("tidbit" HOBO, Co., Ltd). The locations of these instruments at the site are shown in Fig. 2.

The measurement period was from 21 to 24 August 2006. Three measurements using the Lee-type seepage meters (A, B, and C) were done at low and high tide. Measurements by continuous a heat-type seepage meter (D and E) were taken every 1 minute during this period. Temperature measurements of the seabed by a fibre-optic cable, and under the seabed by the temperature data logger, were taken every 5 minutes from 6:30 to 18:30 and every 1 minute from 13:00 to 18:30 on 23 August. In addition to these measurements, river discharge rates were also estimated by multiplying the velocity and the cross-section of the river near the mouth on 22 and 23 August to compare the discharge of groundwater from this beach.

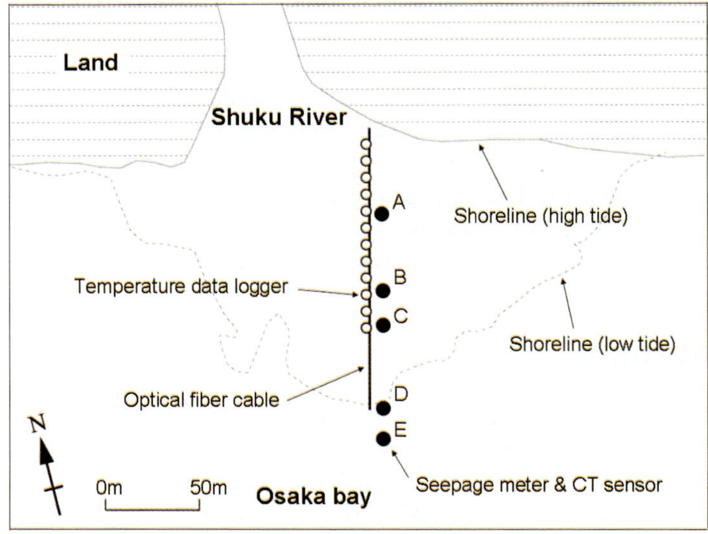

Fig. 2 Location of the instruments measuring SGD.

RESULTS AND DISCUSSION

Results

Table 1 shows the measurement results of discharge rates and electric conductivity of the SGD compensated at 25°. The SGD rates at location C was higher than those at the other locations, and electric conductivity was about 40 mS/cm at all the locations.

Figure 3 shows results of temperature measurements on the seabed and under the seabed by the fibre-optic cable and temperature data logger at high tide (18:00 on 23 August). Temperature on the seabed was higher near to the coast, and decreases with distance from the coast. Temperature under the seabed here was about 31° and lower than those on the seabed by about 2–3°.

River discharge rates were estimated to be 3456 m^3/d and 7776 m^3/d on 22 and 23 August, with an average rate of 5616 m^3/d.

Table 1 Measurement results of discharge rates and electric conductivity (EC) of SGD.

Point	Distance (m)	Average depth (m)	SGD (cm/d)	EC (mS/cm)
A	45	0.5	11.04 ± 3.95	39.24 ± 1.65
B	95	0.6	11.41 ± 2.12	40.74 ± 0.93
C	110	0.7	37.64 ± 7.48	40.63 ± 1.00
D	150	1.0	4.76 ± 6.63	40.59 ± 0.37
E	165	3.1	2.72 ± 9.84	39.47 ± 0.65

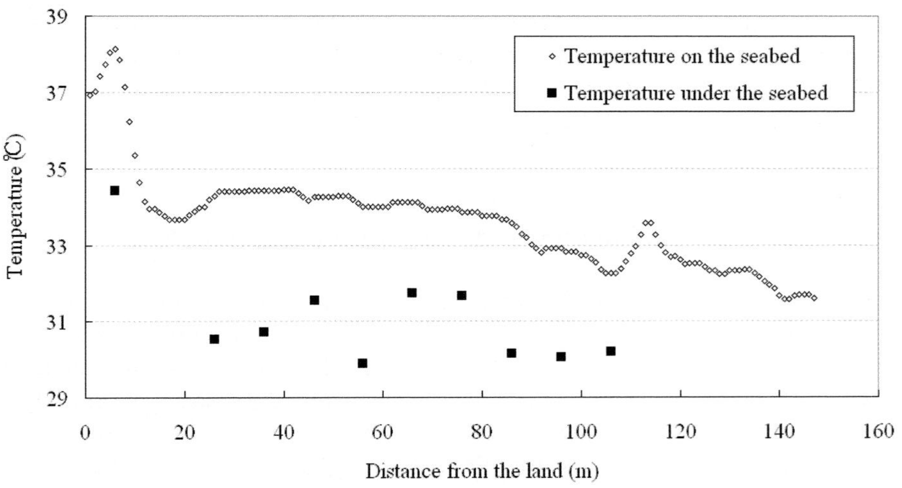

Fig. 3 Results of temperature measurements on the seabed and under the seabed.

Evaluation of terrestrial groundwater discharge rates at each point

SGD can consist of terrestrial groundwater (SFGD = Submarine Fresh Groundwater Discharge) and recirculated seawater (Taniguchi *et al.*, 2002). Separation of the two components was accomplished by using SGD rates and the electric conductivity of SGD (Table 1), the known conductivities of terrestrial groundwater (0.38 mS/cm) and seawater (46.01 mS/cm) as expected end-members to estimate the amount of SFGD from the land to the sea. Figure 4 shows the result of the separation and it is seen that SFGD rates decrease with the distance from the coast, and the highest SFGD rate was still found at location C. Negative correlation between SFGD rates and distance from the land has frequently been seen in previous studies (Bokuniewicz, 1992). If the result at location C is excluded, a regression line that is calculated from the results of each point gives a high correlation coefficient. However, Taniguchi *et al.* (2006) indicate that higher SFGD rates are found near the seaward edge of the inter-tidal zone, and it is assumed that the location of the fresh–salt water interface under the seabed correlates with the position of high SFGD rates. Therefore, the location of the fresh–salt water interface is likely the cause of the high SFGD rates at location C.

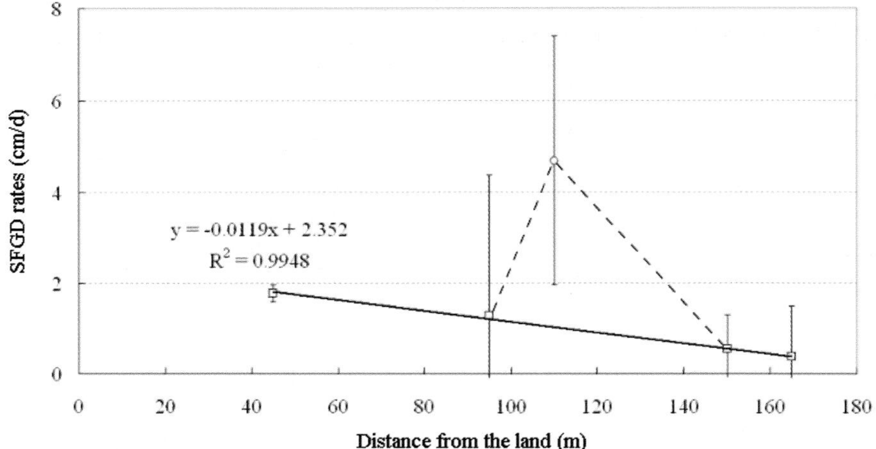

Fig. 4 Average of SFGD rates by seepage meters.

Evaluation of groundwater discharge flux per unit shoreline length

A scale-up of the observed SFGD rates (Fig. 4) is needed to evaluate the total groundwater discharge rate from the land to the coastal zone at Omaehama. Therefore, a scale-up of the observed SFGD rates was attempted in order to estimate the total "area" SFGD rates using the "point" SFGD rates evaluated by the seepage meters (Fig. 4). The intercept of the regression line on the Y-axis can be used to estimate how far terrestrial groundwater discharge occurs from the coastline. Therefore, the estimated SFGD rate of 2.35 cm/d along the shoreline (0 m) gradually decreased with the distance from the shore, and reached zero flux at 197 m offshore from the coast. It is estimated by integration of the SFGD rates estimated from the regression line that SFGD flux per unit shoreline length is 23 100 cm^3/cm/d. To consider the observed higher SFGD rate at location C, however, estimation using the modified regression line (broken line in Fig. 4) was also attempted, because it was also likely that the SFGD flux was large at a point a little away from the coast (Taniguchi *et al.*, 2006). From this result, the SFGD per unit shoreline length was estimated as 33 200 cm^3/cm/d, which is about 1.5 times higher than the first estimate of the groundwater discharge rates.

Scale-up of groundwater discharge rates from results of temperature measurement

For the evaluation of accurate terrestrial groundwater discharge flux from Omaehama, scaling-up of SFGD rates using the results of the temperature measurements was also applied to compare with the results for SFGD flux from seepage meters. Here we applied two techniques for this evaluation, one using a regression line between temperature on the seabed and SFGD rates in the same location, and the other using a regression line between SFGD rates and the difference in temperature between the seabed and under the seabed. The latter is based on the assumption that the difference in temperature decreases if the groundwater discharge rate is higher.

Figure 5(a) shows the relationship between temperature on the seabed and SFGD rates, and Fig. 5(b) shows the relationship between SFGD rates and difference in temperature between the seabed and 5 cm under the seabed. Although the numbers of data points are limited, it can be seen that temperature and SFGD rates have negative correlations and Fig. 5(b) showed better correlation than Fig. 5(a).

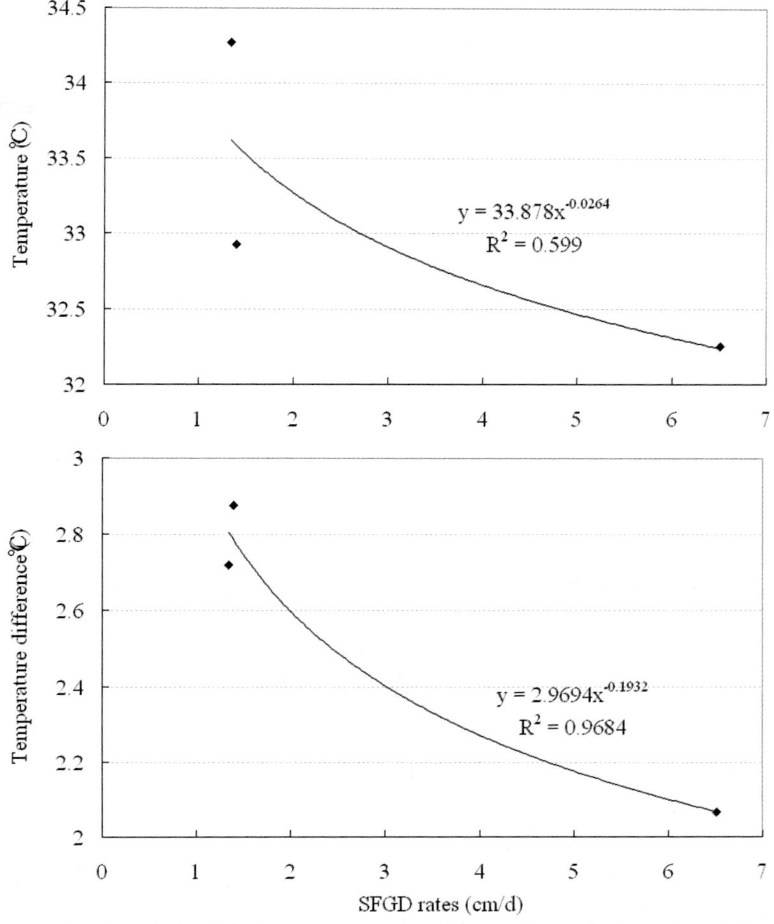

Fig. 5 (a) Relationship between temperature on the seabed and SFGD rates. (b) Relationship between SFGD rates and difference in temperature between the seabed and under the seabed.

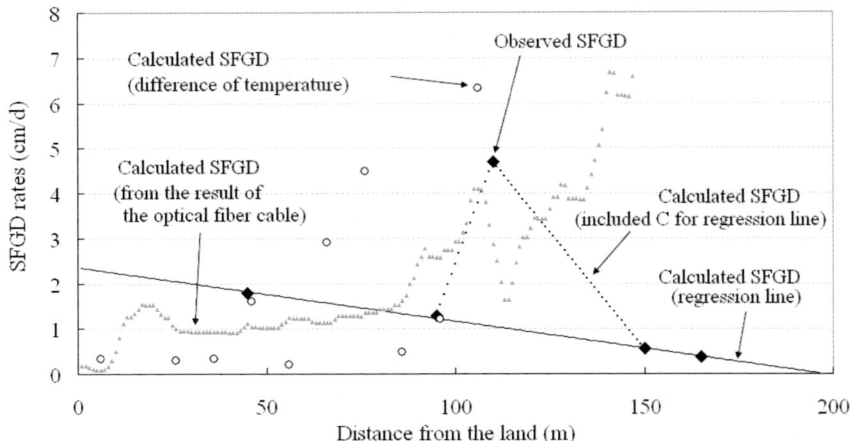

Fig. 6 SFGD rates estimated from several different methods.

Figure 6 shows SFGD rates calculated using the three regression lines from Figs 4, and 5(a) and (b). In this result, differences in the spatial variation of SFGD rates are related to the difference in evaluation techniques. However, all the results indicate that the location of the higher SFGD rates is at the point 110 m from the coast.

Comparison of SFGD rates by each technique was done in the region out to 106 m from the coast, because the data sets for all SFGD rates coexist in this region. From this comparison, the SFGD flux per unit shoreline length are 18 100 cm^3/cm/d by the regression line of the observed SFGD from seepage meters (except for C), 19 700 cm^3/cm/d by the regression line of the observed SFGD (including C), 14 700 cm^3/cm/d by the regression line of the SFGD rates and temperature on the seabed, and 15 300 cm^3/cm/d by the regression line of SFGD rates and differences in temperature.

From these results, the evaluation of the SFGD rates per unit shoreline length appears valid, because no great difference was found between calculated SFGD rates up to 106 m offshore using several different methods.

Comparison of groundwater discharge and river discharge

The processes of water movement from land to sea can be divided into river discharge and groundwater discharge. In order to compare groundwater and river discharge, the total groundwater discharge rates to the sea from Omaehama was roughly estimated by multiplying the SFGD rates per unit coast line out to 197 m from the land, and the length of the coast at Omaehama (620 m). As a result, the total groundwater discharge rate to the sea is 2062 m^3/d, equivalent to 36.7% of the river discharge (see result). According to previous studies, the groundwater discharge rate is assumed to be only a few to 10% of the river discharge on a global scale (Taniguchi *et al.*, 2002). Therefore, this study has shown that the ratio of groundwater to total water discharge to the sea is relatively high in this area. As well as the river discharge, terrestrial groundwater discharge can strongly influence the coastal zone of Omaehama with regard to both water and material (contaminant) transport. However, the estimation of groundwater discharge rates in this study is rough, and more accurate estimation including spatial variations of SGD rates is expected from future work.

Acknowledgements Special thanks are due to T. Uemura, The Graduate University for Advanced Studies, for helping with this field research.

REFERENCES

Banks, W. S., Paylor, R. L. & Hughes, W. B. (1996) Using thermal-infrared imagery to delineate ground-water discharge. *Groundwater* **34**(3), 434–443.

Bokuniewicz, H. J. (1992) Analytical descriptions of subaqueous groundwater seepage. *Estuaries* **15**, 458–464.

Burnett, W. C., Bokuniewicz, H., Huettle, M., Moore, W. S. & Taniguchi, M. (2003) Groundwater and pore water inputs to the coastal zone. *Biogeochemistry* **66**, 3–33.

Kasahara, S. (2005) Habitat environment of benthic species and groundwater flow at the tidal flat in urban area –Kohro-en beach as a case study. In: *Beneficial Use and Various Problems of Groundwater* (Proc. Symposium on the Groundwater Geo-Environment 2005), 33–38 (in Japanese).

Lee, D. R. (1977) A device for measuring seepage flux in lakes and estuaries. *Limnol. Oceanogr.* **22**, 140–147.

Taniguchi, M. & Iwakawa, H. (2001) Measurements of submarine groundwater discharge rates by a continuous heat – type automated seepage meter in Osaka Bay, Japan. *J. Groundwater Hydrol.* **43**(4), 271–277.

Taniguchi, M., Burnett, W. C., Cable, J. E. & Turner, J. V. (2002) Investigation of submarine groundwater discharge. *Hydrol. Processes* **16**. 2115–2129.

Taniguchi M. Ishitobi, T. & Shimada, J. (2006) Dynamics of submarine groundwater discharge and freshwater–seawater interface. *J. Geophys. Res.* **111**. C01008, doi:10.1029/2005JC002924.

A New Focus on Groundwater–Seawater Interactions
(Proceedings of Symposium HS1001 at IUGG2007, Perugia, July 2007). IAHS Publ. 312, 2007.

Seasonal changes in the groundwater–seawater interaction and its relation to submarine groundwater discharge, Ise Bay, Japan

KUNIHIDE MIYAOKA
Department of Geography, Mie University, 1577 Kurima-machiya, Tsu, Mie 514-8507, Japan
miyaoka@edu.mie-u.ac.jp

Abstract The purpose of this study is to elucidate the conditions of seasonal changes in the fresh and salt water distribution, and the control factors of the relationship between the groundwater flow system and submarine groundwater discharge (SGD) at Ise Bay, Japan. The results indicate that the groundwater levels and qualities have different seasonal change patterns in each depth at the measurement sites. Deep freshwater discharges as SGD in the irrigation season. The water quality of the SGD changes is affected by groundwater–seawater interaction. Seasonal changes in the groundwater–seawater interaction are controlled by geology, recharge water, and tidal conditions.

Key words geological conditions; groundwater flow system; resistivity; seasonal changes; submarine groundwater discharge (SGD); tidal conditions

INTRODUCTION

Anthropogenic influences in inland areas can reach coastal areas where groundwater discharge areas are located (Tokunaga *et al.*, 2003). Rice fields are one of the most common types of land use in monsoon Asia. Rice fields are widely distributed over the coastal zone at Ise Bay, so the recharge rate from surface water to groundwater in the irrigation season is larger than during the non-irrigation season. This suggests that the nutrient flux and groundwater discharge rate have seasonal changes in the coastal zone located at the groundwater discharge area. Taniguchi *et al.* (2002) show that submarine groundwater discharge (SGD) is important to the seawater quality and ecosystem in the coastal area. Taniguchi *et al.* (2006) state that the SGD volume is large near to the coast, and is gradually reduced with distance from the coast. These previous studies are important for elucidating the actual conditions of three dimensional groundwater–seawater interaction. However, these studies do not consider the effects of seasonal factors such as rainfall and irrigation on the seasonal changes of groundwater–seawater interactions. The purposes of this study are twofold: to elucidate the actual conditions of seasonal changes in groundwater–seawater interaction, and to evaluate the factors that are related to groundwater–seawater interaction and SGD through the seasons.

STUDY AREA

Figure 1 show the location of the study area and the distribution of land use, geomorphology and geology. The study area is the coastal plain located at the mouth

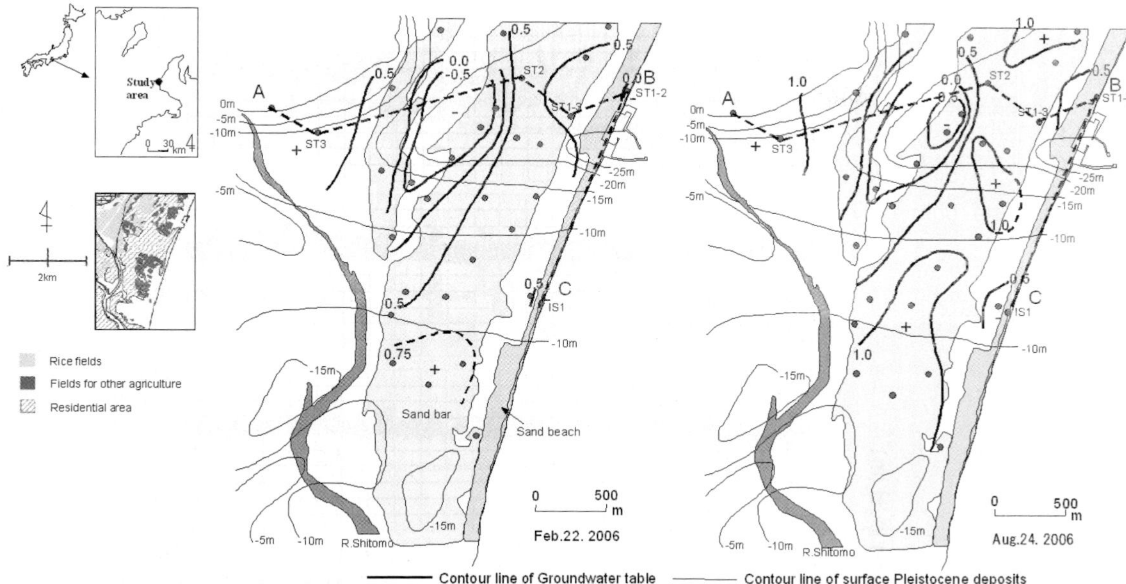

Fig.1 The location of the study area and cross-section measurement line, and the distribution of groundwater table contour lines, surface contour lines of Pleistocene deposits.

of the Shitomo River. Two sandbars are present, and there are residential areas and agricultural fields. Two lagoons are present, and there are rice fields and wetlands. An alluvial plain is present on the western side of the inland sandbar. A Pleistocene valley is present in the alluvial and coastal plain (Geological Survey of Japan, 1987, 1995). This valley is thought to be a palaeo river channel (Miyaoka & Yoshizaki, 2006). Rice fields are widely distributed across the alluvial plain and lagoon. The irrigation season is from March to August, and non-irrigation season is from October to February in the study area.

METHODS

Monitoring wells at depths of 5, 10, 20, and 30 m were installed at two sites in the tidal zone and four sites inland along the Pleistocene valley. Each well site in the tidal zone has a different geology; site ST1-2 is located in thick Holocene deposits where a palaeo river channel sand and gravel layer exists in the deep layer; site IS1 is located in thin Holocene deposits. A conductivity–temperature–depth (CTD) sensor was installed in each well, and continuous measurements of electrical conductivity, water temperature, and water level were made at 15-min intervals from 1 December 2005. Hourly measurements of resistivity from onshore to offshore were made using the Schlumberger method from low tide to the next low tide during the monthly spring tide beginning February 2006.

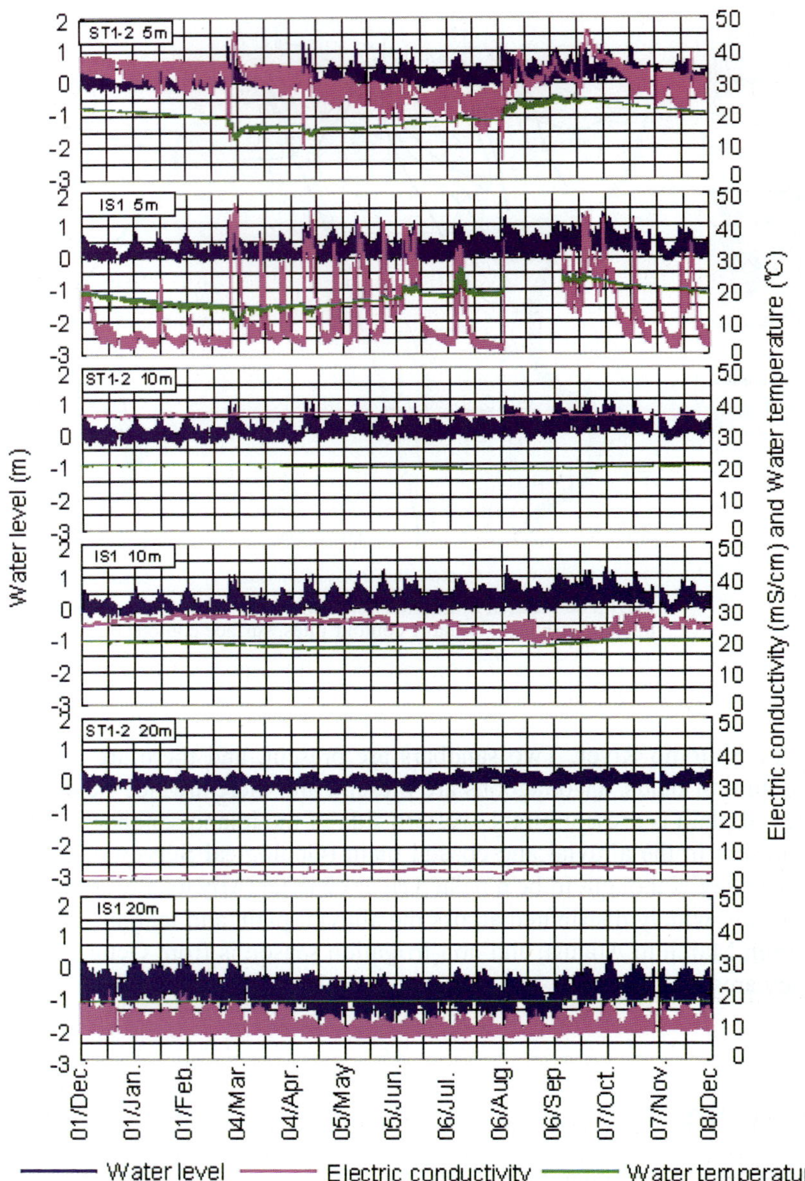

Fig. 2 Seasonal change of groundwater level, electrical conductivity and water temperature.

RESULTS AND DISCUSSION

Horizontal groundwater distribution measurements

Figure 1 shows the seasonal spatial variation of the water table in the non-irrigation and irrigation seasons. Water levels in the irrigation season are higher than in the non-irrigation season, so groundwater is being affected by irrigation and/or rainfall. The highest water level is in the sandbar area, and a trough and lowest point in the water table is in the lagoonal area. The location of the trough in the water table corresponds to the location of the geological valley of the Pleistocene deposits (Fig. 1). These suggest that the water table is controlled by geomorphological, geological and/or land use conditions.

Seasonal changes in the water level and electrical conductivity have been measured at two sites located in different geological conditions (Fig. 2). The patterns of the water levels and electrical conductivity are quite different, even at the same depth, at these sites. There are some important points to note here. First, freshwater exists in the deep layer at site ST1-2, but not at site IS1. Second, brackish water exists in the shallow layer at site IS1 in the non-irrigation season, and during neap tide in the irrigation season. Third, the range of water levels in the deep layer is larger than in the shallow, and is similar to the tidal change from high to low tide at both sites. These many different points suggest that the groundwater flow system, groundwater potential distributions, seawater intrusion, and other factors are controlled by the geological, geomorphological, and land use conditions.

Vertical groundwater distribution in the irrigation and non-irrigation seasons in the hydrogeological water path

The longitudinal-profile is located along the water table trough and geological Pleistocene deposits in the valley as shown in Fig. 1. The longitudinal-profile along the A–B line of the groundwater potential is shown in Fig. 3, and the longitudinal-profile of electrical conductivity in the non-irrigation and irrigation season is shown in Fig. 4.

In the deep layer along the coast, the groundwater potential shows seasonal changes between the irrigation and non-irrigation seasons, and between the high and low tides. High groundwater levels are present during the high tide through the year, but this location is different in that a high potential is present at a depth of 20 m during the non-irrigation season, but that changes to a depth of 30 m during the irrigation season.

At the upstream area, an electrical conductivity of 0.5 to 1.0 mS/cm is present at a depth of 10 to 20 m, i.e. in the Pleistocene deposits. The same range of electrical conductivity occurs at 20 to 30 m depth along the coast, and this electrical conductivity distribution is similar to the groundwater potential distribution (Fig. 3). High electrical conductivity occurs at the depth of 10 m, and is stable through the seasons, so it is suggested that salt water is constantly intruding to the inland area at this depth. At a depth of 5 m the electrical conductivity is high, the same as at the 10 m depth value in the non-irrigation season, but it gradually becomes lower when the irrigation season begins. The lowest electrical conductivity is present at the end of the rainy season (the end of July). According to the landform conditions in Figs 1 and 3, a sandbar is a recharge area of the local groundwater flow system. Therefore the groundwater at 5 m depth is also affected by some land surface conditions.

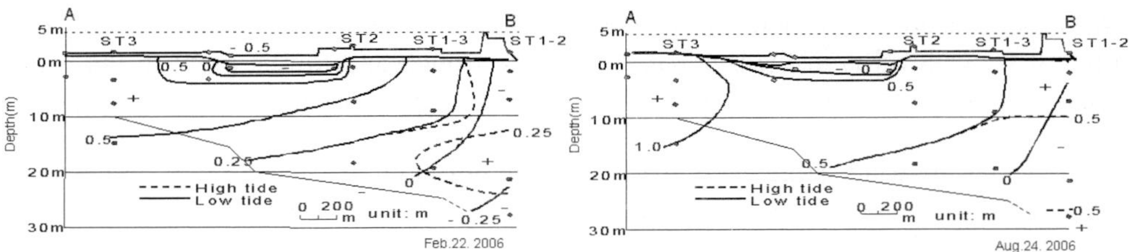

Fig. 3 Longitudinal-profiles of groundwater potential in the non-irrigation and irrigation season.

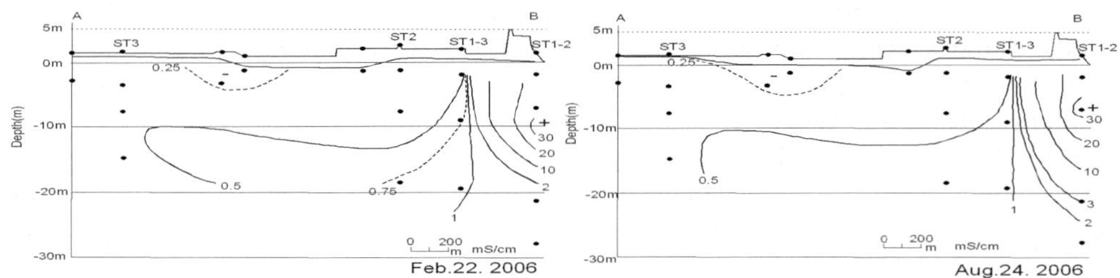

Fig. 4 Longitudinal-profiles of electrical conductivity in the non-irrigation and irrigation season.

Vertical groundwater distribution at the coastal zone during the irrigation and non irrigation season

The cross-section B–C of the groundwater potential is shown in Fig. 5. Upward flow occurs from deeper than 20 m in the thick Holocene deposits. Otherwise, downward flow is shown in all layers in the thin Holocene deposits. According to the seasonal change of electrical conductivity in site ST1-2 and site IS1 (Fig. 2), the freshwater distribution is different from the tidal conditions and irrigation conditions at each site. This shows that interactions between surface water, groundwater, and seawater have some patterns; i.e. via the recharge area, groundwater discharge, seawater intrusion, and geological conditions. This suggests a regional groundwater flow system exists in

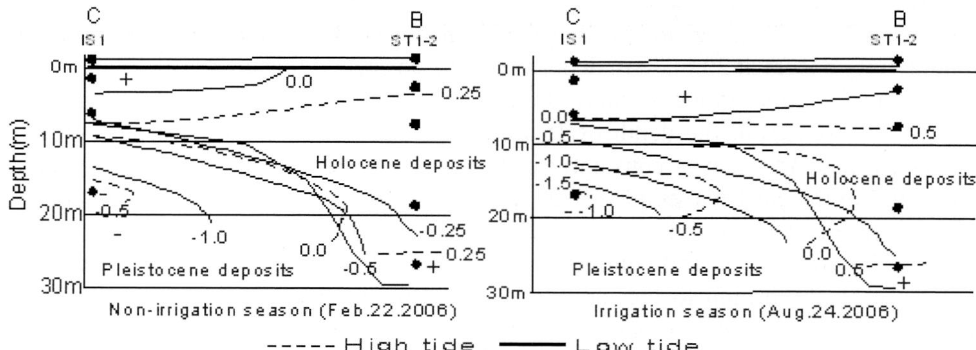

Fig. 5 Cross-section of groundwater potential in the non-irrigation and irrigation season.

Holocene deposits where the palaeo river channel is located. Finally, seawater may intrude easily into the shallow inland layer there because of the trough in the water table.

Estimation by resistivity measurements of the location of SGD and its seasonal variation

The resistivity and electrical conductivity distribution from high tide through low tide during both the non-irrigation (15 March 2006) and irrigation seasons (13 July 2006) were evaluated in order to estimate the seasonal change of the interaction between groundwater and seawater, and the SGD locations (Fig. 6). Both measurements were carried out under spring tide conditions. The seepage face is shown at the same location during low tide in both seasons, but the electrical conductivity in July is lower (29.5 mS/cm) than that in March (45.5 mS/cm). The seepage face is buried under seawater during high tide, but at the deep layer, freshwater rises up to the shallow layer during high tide. Furthermore, this freshwater zone reaches the seepage face at low tide

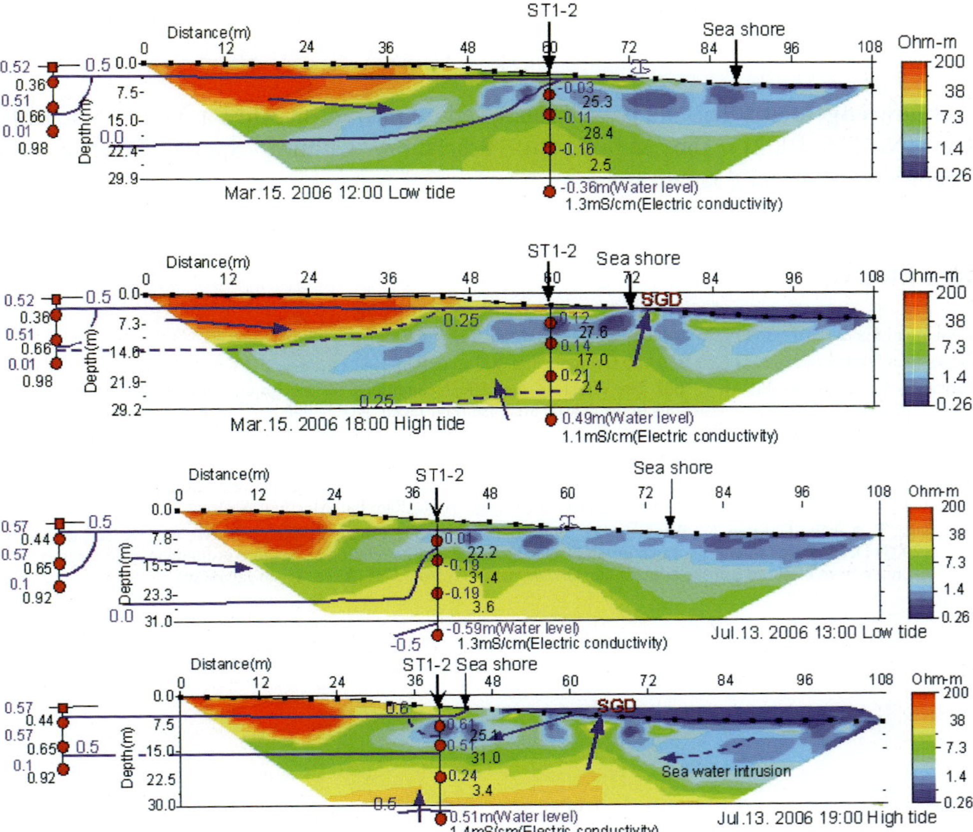

Fig. 6 Cross-section of resistivity distributions at the coastal zone in the high tide and low tide, in the non-irrigation season and irrigation season.

during the irrigation season. But during the non-irrigation season, deep freshwater does not reach the shallow depth and the seepage face. These seasonal variations of the freshwater flow condition show that the groundwater discharge rate is greater during the irrigation season. During high tide, freshwater flow is prevented by the seawater pressure, so strong upward flow occurs and discharges at the seepage face as SGD. At low tide, this seepage face acts as the shallow groundwater discharge point.

CONCLUSIONS

This study elucidated the seasonal changes of the horizontal and vertical distribution of fresh and salt water, and the relation between their seasonal changes and that of the SGD. The interaction between groundwater and seawater changes is affected by the seasonal condition. That is, the distributions of fresh and salt water in the aquifers depend on the geological conditions, especially the location of the Pleistocene valley in the coastal area. These seasonal variations in the freshwater flow condition shows that the groundwater discharge rate is large during the irrigation season. During the high tide, the seawater pressure interrupts freshwater flow, so strong upward flow occurs and discharges at the seepage face as SGD. The water quality of the SGD changes between high and low tide, and between the non-irrigation and irrigation seasons.

Acknowledgements This study was supported by a Grant-in-Aid for Young Scientists (A) (No.17681002), The Ministry of Education, Culture, Sports, Science and Technology (MEXT), Japan.

REFERENCES

Geological Survey of Japan (1987) 1:50 000 Geological Map, Tsu-Tobu.

Geological Survey of Japan (1995) 1:50 000 Geological map, Tsu-Seibu.

Miyaoka, K. & Yoshizaki, M. (2006) The effect of geological conditions on the groundwater flow system and the groundwater–seawater interactions in the coastal area, Ise Bay, Japan. AGU 2006 Fall Meeting, Abstract.

Taniguchi, M., Burnett, W. C., Cable, J. E. & Turner, J. V. (2002) Investigation of submarine groundwater discharge. *Hydrol. Processes* **16**, 2115–2129.

Taniguchi, M., Burnett, W. C. & Aureli, A (2006) Global assessments of submarine groundwater discharge and groundwater resources under the pressures of humanity and climate change. AGU 2006 Fall Meeting, Abstract.

Tokunaga, T., Nakata, T., Mogi, K., Watanabe, M., Shimada, J., Zhang, J., Gamo, T., Taniguchi, M., Asai, K. & Saegusa, H. (2003) Detection of submarine fresh groundwater discharge and its relation to onshore groundwater flow system: an example from offshore Kurobe alluvial fan. *Groundwater Hydrol. J.* **45**, 133–144 (in Japanese).

Basin-wide groundwater flow study in a volcanic low permeability bedrock aquifer with coastal submarine groundwater discharge

JUN SHIMADA[1], DAISUKE INOUE[2], SOU SATOH[3], NAOHIKO TAKAMOTO[1], TOMOYA SUEDA[1], YOSHITAKA HASE[1], SHOW IWAGAMI[4], MAKI TSUJIMURA[4], TOMOTOSHI ISHITOBI[5] & MAKOTO TANIGUCHI[5]

1 *Grad. School of Science & Technology, Kumamoto Univ., Kumamoto 860-8555, Japan*
jshimada@sci.kumamoto-u.ac.jp

2 *NTT Docomo Co., Fukuoka 810-0004, Japan*

3 *Geosphere Environmental Technology Co. Ltd, Tokyo 103-0027, Japan*

4 *Grad. School of Life and Env. Sci., Inst. of Geosci., Univ. of Tsukuba, Tsukuba, 305-8572, Japan*

5 *Res. Inst. for Humanity and Nature, Kyoto 603-8047, Japan*

Abstract The purpose of this study is to reveal the basin-wide groundwater flow system in a Quaternary pyroclastic bedrock aquifer by using several hydrological methods: catchment hydrometric observations, environmental isotope study of the spring water and observation borehole levels including in-land, on-shore and offshore boreholes, basin-wide groundwater potential monitoring, geophysical methods to understand the aquifer distribution, water balance for the representative paired catchments in the study basin including micro-meteorological evapotranspiration measurements, direct submarine groundwater discharge (SGD) measurements by automatic seepage meters, and three-dimensional groundwater flow simulations based on observed hydrological data. The results clearly show that topographically driven groundwater flow systems with different flow dynamics and residence times exist in the study catchments and strongly support the hydrological characteristics of local springs and a river discharge system including coastal SGD. Also, a stagnant fresh groundwater system exists under the present sea bed which is completely separate from the land-based groundwater flow systems, and is thought to be a kind of remnant palaeo groundwater recharged during the previous regression era. This has no direct relation to the present SGD in the area.

Key words groundwater flow system; environmental isotopes; groundwater potential; submarine groundwater discharge; observation borehole; remnant palaeo groundwater

INTRODUCTION

Quaternary pyroclastic flow deposits are widely distributed in Japan, but their groundwater flow systems have not been studied much. The relatively steep morphology and humid temperate hydrological conditions of Japan should create active groundwater flow including coastal groundwater seepage-out. A 5.2 km^2 mountainous pyroclastic catchment has been selected in Uto peninsula, Kumamoto, Japan (Fig. 1) for this purpose, and the following four major themes have been taken to reveal the regional groundwater flow system:

1. Precise hydrometric study in the headwater catchments relating the rainfall–runoff process.

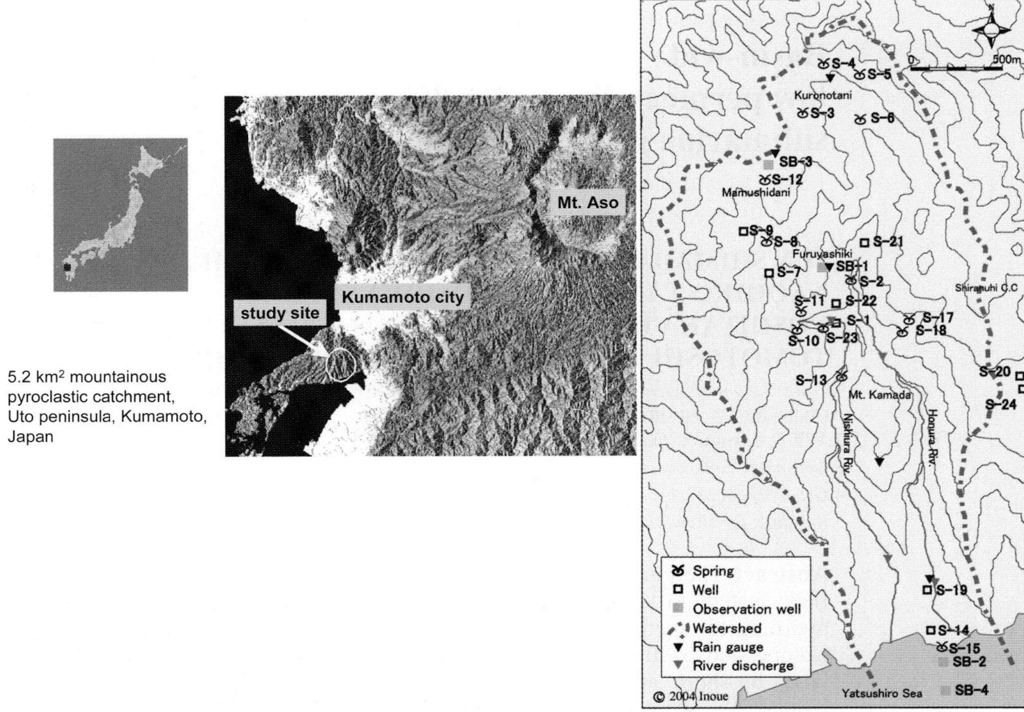

Fig. 1 Location of study site and major hydrological feature in the study catchments.

2. Isotope hydrological study with the help of groundwater potential measurements using deep observation boreholes of different depth and the self potential survey in the study catchments.
3. Self potential survey, resistivity survey, seepage meter measurements and groundwater potential measurements in the observation boreholes have been conducted in the on-shore and off-shore area during the tidal fluctuation period.
4. Three dimensional groundwater flow simulation in the study catchments, including river water discharge and evapotranspiration.

The study started in 2002 and a major hydrological observation system including four river discharge gauging stations and nine boreholes of different depth were installed in the study basin for monitoring purposes (Fig. 1). The groundwater flow regime characteristics observed during the four year period since, allow study of the relation between coastal groundwater seepage and the inland groundwater flow system and a multi-scale groundwater flow system has been confirmed.

STUDY AREA

The study basin is about 4 × 2 km in area with an elevation range of 0–400 m, which we call the Shiranui study site. It mainly consists of Tertiary/Quaternary lava and volcanic tuff-breccias. The Palaeozoic sedimentary rock, a hydrogeological basement boundary to the above volcanic rocks, exists about 200 m below the ground surface.

The major land uses of this basin are citrus farming, paddy field and bamboo/cedar forest. There are many natural springs in the mid-basin area, and an artesian spring in the tidal flat area of the coastal zone is also a feature. We established four observation borehole sites, including the tidal flat and the offshore site below the present sea bed, to monitor the groundwater potential and to collect water samples. Two sites in the mountain area have several boreholes with different depths down to 200 m. Another two sites in the coastal and offshore area have 50 m deep boreholes divided into several zones by multi-packer systems to obtain groundwater potential data and water samples at the different depths (Fig. 1). Two major river systems are gauged to monitor the discharge rate and nearly 20 natural springs and domestic wells are also seasonally sampled to understand the groundwater flow system of the study area.

GROUNDWATER FLOW SYSTEM IN THE STUDY BASIN

Seasonal change of groundwater potential and stable isotope content in the groundwater of the observed boreholes

Figure 2 shows the seasonal fluctuation of the groundwater potential for the observation boreholes. In the case of SB3 boreholes, which are located at the highest elevation in the basin, the shallow well (25 m; SB3-1) has groundwater only in the wet season and clearly shows perched groundwater characteristics. However, the 120 m deep well (SB3-3) did not show any seasonal trend. However, the elevation of the water level of this deep borehole (SB3-3) is clearly higher than that of the mid basin observation well, and it is considered that these headwater areas function as the groundwater recharge area of the basin-wide regional groundwater system. In the case of the SB1 boreholes, the shallowest (SB1-3, 25 m deep) shows a perched aquifer

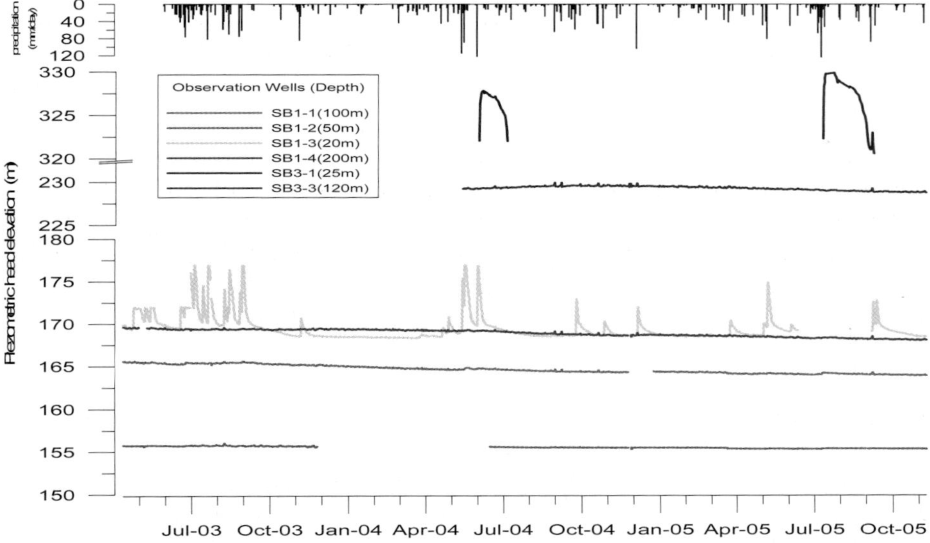

Fig. 2 Seasonal fluctuation of groundwater potential in the observation boreholes.

fluctuation. While the other three boreholes with much deeper depths have no seasonal trend, a clear upward groundwater flux depends on the difference of their groundwater potentials. This upward flux can explain the relatively large discharge rate of the local spring in the mid basin area (Fig. 5). The observation well at the coastal area (SB2) with a multi-packer system shows daily fluctuations associated with tidal movement, but only a little seasonal trend. This coastal well also shows an upward groundwater flux.

Figure 3 shows the stable oxygen isotope content in groundwater/spring water sampled at different altitudes in the study catchments. Seasonally fluctuating spring water, which has a relatively short residence time (within one year), and represents relatively small catchments with different altitudes, clearly shows an isotopic altitude effect in the study area. Using the isotopic altitude trend from Fig. 3, the isotopic content of spring water and borehole groundwater with no seasonal isotope trend, was used to estimate recharge elevations, which were found higher than the sampling altitudes.

Using the observed groundwater data, two dimensional groundwater potential distributions have been constructed in Fig. 4. The groundwater system of the studied area is clearly affected by both the local topography and geology. The flow characteristics obtained are also supported by previously estimated recharge altitudes, which depend on the stable isotope content of groundwater.

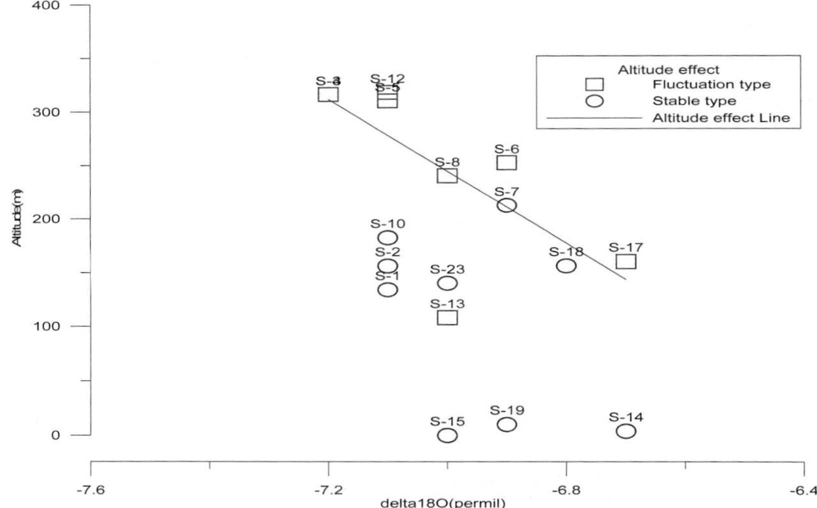

Fig. 3 Stable Oxygen isotope content in groundwater sampled at different altitudes.

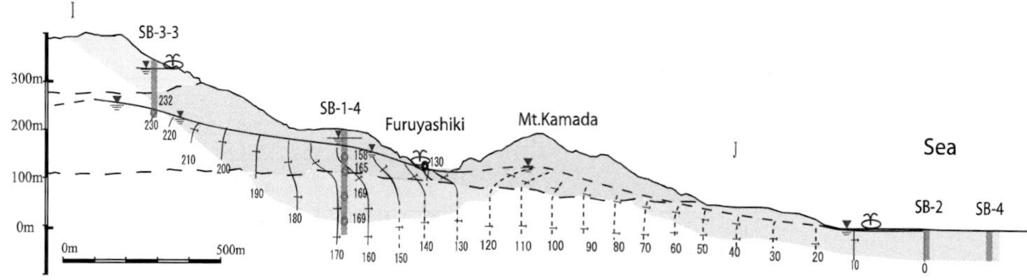

Fig. 4 Cross sectional groundwater potential distribution of the study basin.

Rainfall–runoff characteristics of the head water catchments

Three representative small spring sites were selected to monitor rainfall–runoff processes in the headwater catchments in the study area: Mamushi (S-12), Yozaemon (S-2), and Suzure (S-18), Fig. 1. Specific discharge rates for each catchment are shown in Fig. 5 for the two year period. The Mamushi site discharges only during the wet season, and responds quickly to rainfall events, while the other two sites have continuous discharge throughout the year. The Yozaemon site, situated in the mid-basin, shows discharge fluctuations with the highest base runoff of the three monitored catchments. Annual runoff percentages for these catchments are 9% for Mamushi, 156% for Yozaemon, and 72% for Suzure. The annual runoff percentage differences explain the differences in groundwater conditions of the catchments. Mamushi site is a perched groundwater catchment, which also works as a regional-scale groundwater recharge area, while Yozaemon site works as a regional groundwater discharge area in the mid-basin.

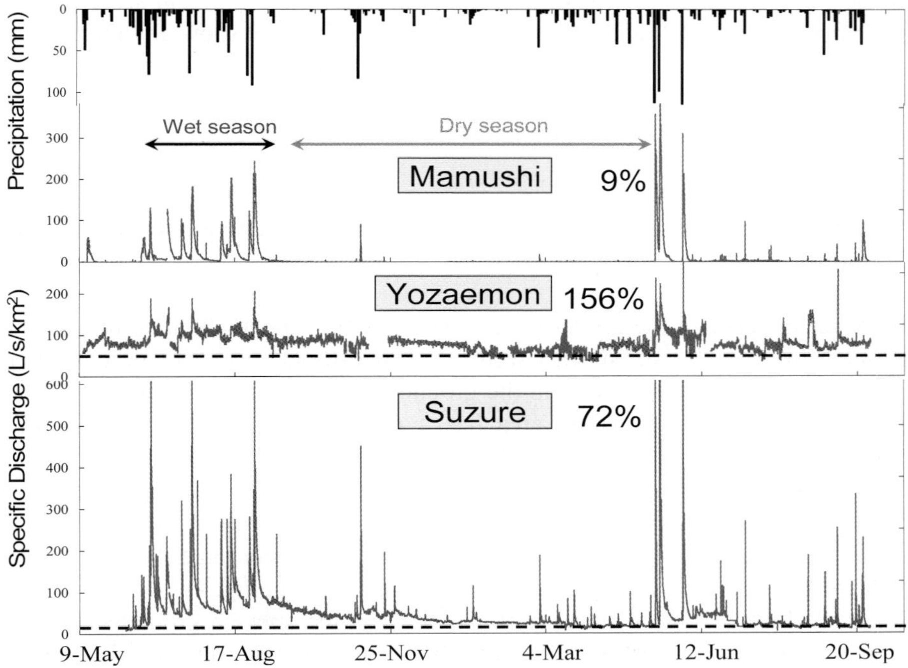

Fig. 5 Comparison of hydrograph in three representative head water catchments.

Basin-wide self potential (SP) survey

Surface self potential surveys have been used for better understanding of subsurface fluid movement in geothermal areas, as well as for the detection of groundwater flow (Fournier, 1989; Fagerlund & Heinson, 2003; Revil *et al.*, 2004), showing characteristics of groundwater recharge to discharge areas. Figure 6 shows the regional scale SP distribution along the valley line in the study area. The SP at the top basin area in both

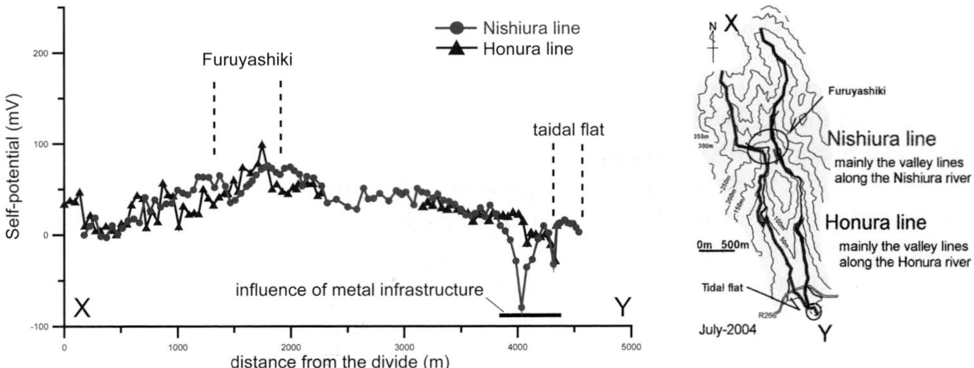

Fig. 6 Self potential distribution along the valley line of Nishi-ura and Hon-ura rivers.

valleys show a relatively low voltage, which gradually increases toward the mid-basin of the Furuyashiki area, where it is characterised by the regional-scale groundwater discharge (Fig. 5). Below the Furuyashiki area, SP does not show any increasing trend, but shows a weak decreasing tendency toward the coastal discharge areas. This regional-scale SP trend could be explained by the three dimensional groundwater flow system in the study area (Sato *et al.*, 2005).

Apparent groundwater age determination using radioactive tritium and carbon isotopes

To better understand the groundwater flow system in the study area, the groundwater samples from wells and springs were used to estimate apparent groundwater ages. Figure 7 shows tritium content in the groundwater along a N–S transect of the basin. The higher recharge elevation area has a relatively higher tritium content, which represents a relatively young age, while the lower discharge elevation or deep groundwater, has a low tritium content which represents an old age. Because the tritium method has a detection age limit of only 100 years, C^{14} analysis has also been carried out in this area, as shown in Fig. 8. The estimated apparent age distributions support well the groundwater flow system presented previously in Fig. 4.

If we look at Fig. 8 more carefully, we see that sub-sea groundwater with over 2000 year ages is not directly related to the present-day groundwater flow regime, but it should be considered to be a "remnant" palaeo-fresh groundwater.

REGIONAL WATER BALANCE AND SUBMARINE GROUNDWATER DISCHARGE

Water balance of the paired catchments in the study basin

For understanding the regional groundwater flow system, it is necessary to know the precise water balance. Two parallel river basins were gauged (Fig. 1) to calculate the

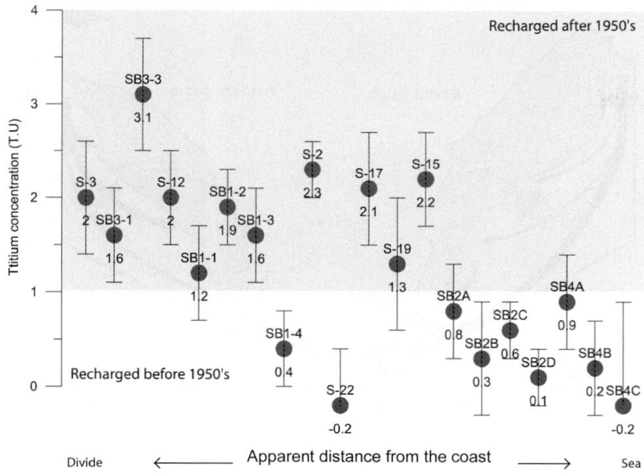

Fig. 7 Tritium content in the groundwater along the N-S line of the study basin.

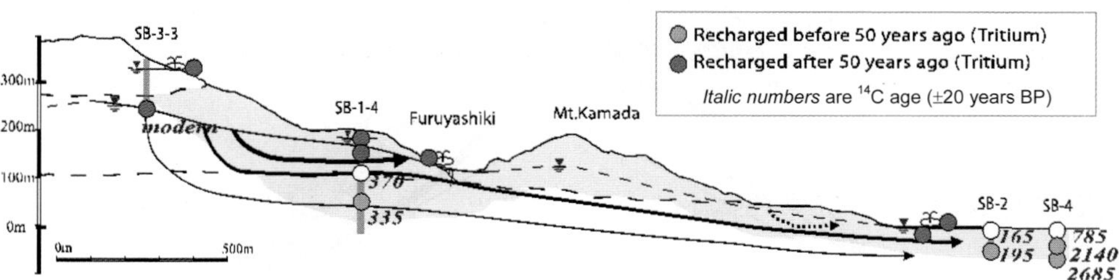

Fig. 8 Apparent groundwater ages from tritium and radioactive carbon isotopes.

regional water balance which measured the actual evapotranspiration at five meteorological stations in the study catchments during the three year period July 2003 to September 2006. Here, the water year is defined as from October to the next September. Figure 9 shows a schematic water balance of the study catchments for each of the three year periods (Sueda *et al.*, 2006). In spite of similar amounts of precipitation and evapotranspiration, the eastern Hon-ura River basin always shows a higher groundwater discharge rate than the western Nishi-ura River basin. This is due to the regional geological setting affecting the groundwater flow system in the study area.

Evaluation of the spatial distribution of submarine groundwater discharge

Continuous heat-type automated seepage meters (Taniguchi *et al.*, 2003) were installed at about 50, 100, and 150 m distance offshore from the coastal line at high tide. All the seepage meters were located between the high tide coast and the low tide coast, which is the representative area for the fresh submarine groundwater discharge. A previous seepage study was carried out at this study area along the transect line from the landward coast to the open sea (Taniguchi *et al.*, 2005). The spatial distribution of

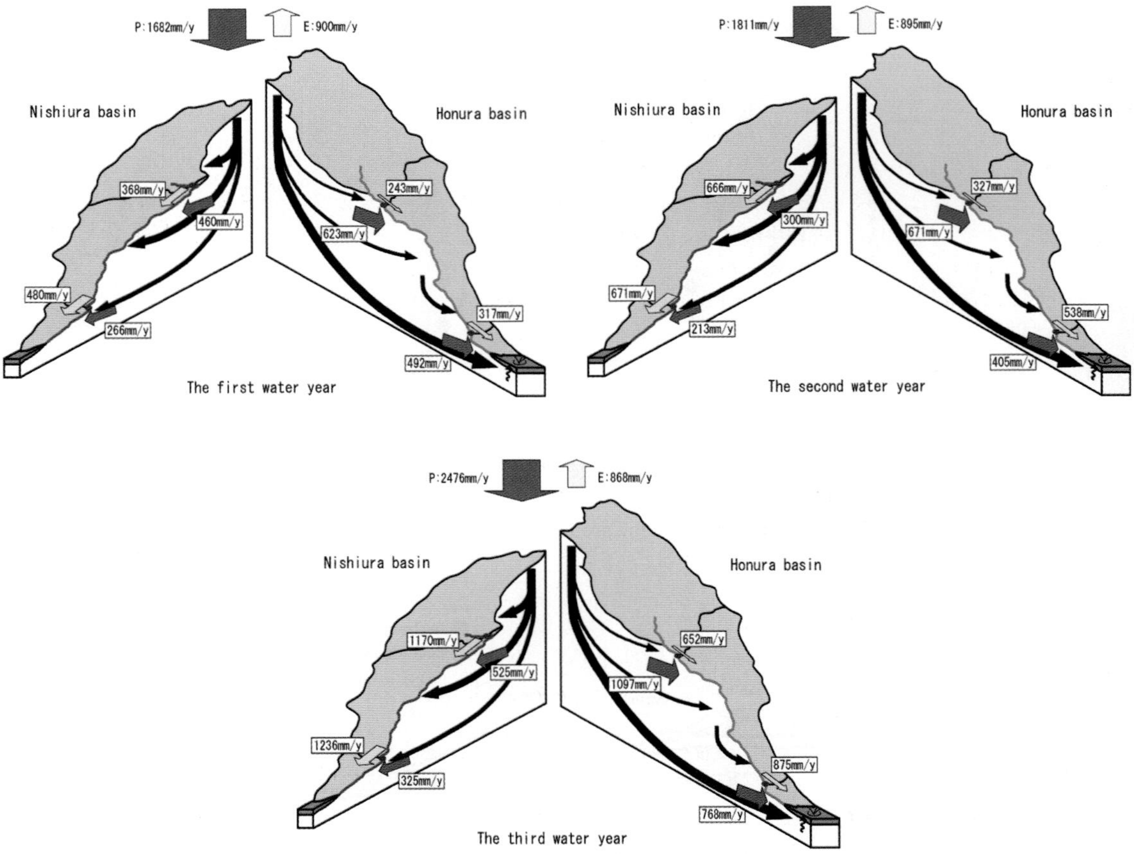

Fig. 9 Schematic water balance for the paired river basins in the study area.

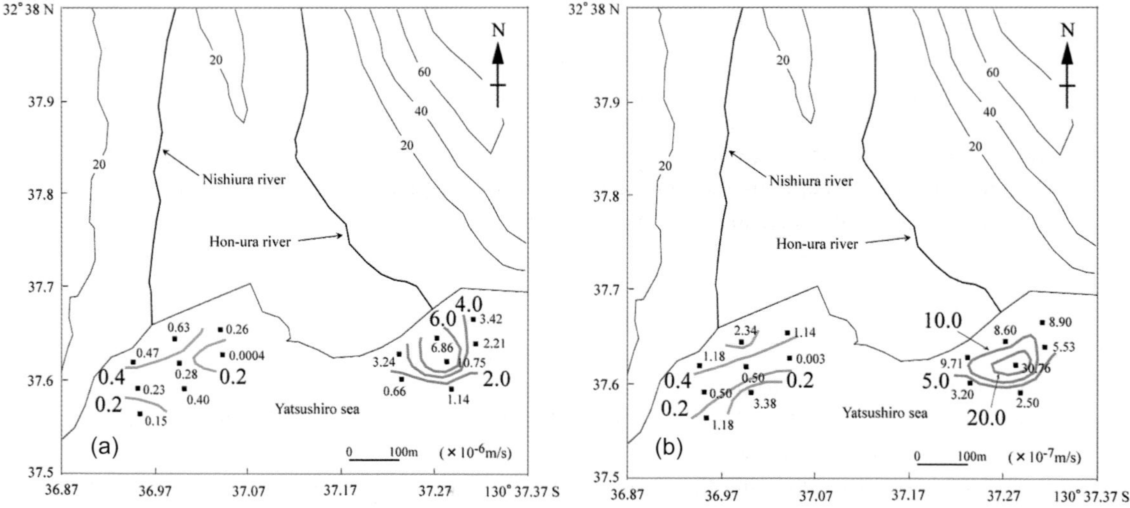

Fig. 10 Spatial distribution of: (a) SGD, and (b) SFGD in the coastal area of the study basin (modified after Taniguchi *et al.*, 2006).

daily-averaged SGD (Fig. 10(a)) shows that the SGD rate offshore of Hon-ura basin is about ten times larger than that in Nishi-ura basin. The contour lines of SGD are parallel to the coast, indicating that the dominant process of SGD depends on the distance from the coast. Based on the observed SGD conductivities during the SGD monitoring, the spatial distribution of the terrestrial fresh groundwater discharge (SFGD) is shown as Fig. 10(b). As can be seen from this figure, SFGD is also larger at offshore Hon-ura basin than that in Nishi-ura basin. The spatial integration of SFGD as measured by seepage meters agrees relatively well with the groundwater discharge estimated by water balance calculations in the paired sub-catchments (Taniguchi, *et al.* 2006).

3-D GROUNDWATER FLOW SIMULATIONS

In order to understand the three dimensional groundwater flow regime, a steady state three dimensional groundwater flow simulation was conducted in the study area. The model code used is the Visual Modflow Pro(Version 3.1) by Waterloo Hydrogeologic, and the boundary conditions and model validations are as follows:

The Palaeozoic sedimentary rock is selected as a hydrogeological basement boundary which exists about 200 m below the ground surface.

1. Impermeable side boundary is selected at the same location with topographical river basin boundary.
2. Groundwater recharge rate is 900 mm/year, depending on the water balance calculation of the study basin.
3. The model used does not cover the perched groundwater system and the seawater–freshwater boundary condition in the coastal area.
4. Permeability distribution used in the model is adjusted as to match the estimated groundwater residence time, to fit the estimated apparent groundwater ages.
5. Model results are validated by the observed groundwater potential at the observation boreholes, the location of the highest permanent spring connecting to the two major river systems in the study catchments, and the groundwater flow line and its residential time.

Figure 11 shows an example of the model results along the N–S vertical cross-section for the groundwater potential distribution and the flow line with estimated residence times. Except for the perched groundwater system in the headwater region and the stagnant palaeo fresh groundwater under the present sea bed, the regional-scale groundwater flow system can be explained by these results.

DISCUSSION AND POSSIBLE FUTURE WORK

The conceptual diagram of the subsurface environment of the studied area is shown in Fig. 12. In the headwater region, the perched groundwater system with accompanying local seasonal spring discharges has developed even at high elevations. This local flow system, with a residence time of a few years, is mainly caused by the local geology. However, the area itself mostly works as a major recharge area either for the

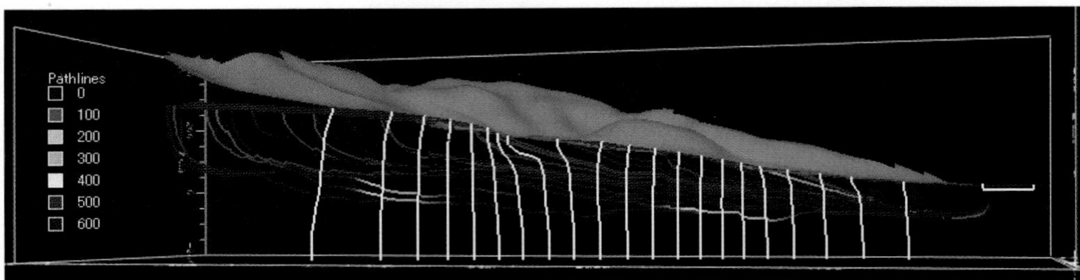

Fig. 11 Groundwater potential distribution and flow line along the cross section of the study basin by the 3D groundwater flow simulation.

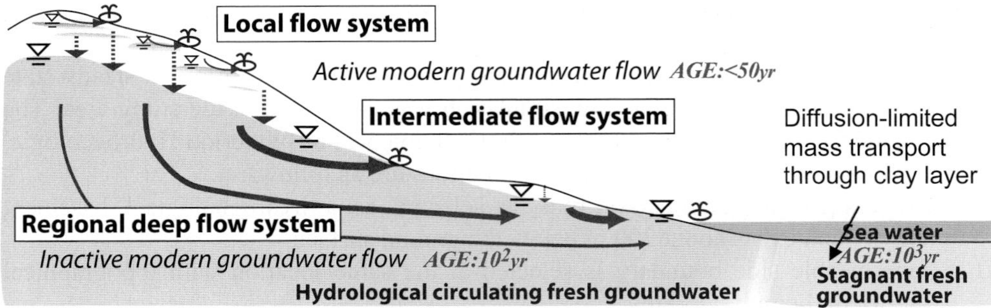

Fig. 12 Conceptual diagram of the subsurface environment of the study area.

intermediate groundwater flow system or for the regional deep groundwater flow system. These concepts are confirmed by the groundwater potential distribution, the recharge elevation analysis using stable isotopes, specific discharge rates of local springs, river water discharge-related regional water budget study, regional scale SP survey, and the radioactive isotope age analysis of groundwater in the study area.

The intermediate groundwater, aged less than 50 years, mostly flows out of the mid basin springs, and supports the base flow of two local rivers, while the regional groundwater, aged 100+ years, mostly flows out of the coastal area, including natural springs and submarine groundwater discharges at the tidal flat zone. The evidence to support those regional-scale groundwater flow systems is the results of the resistivity survey, SP survey, groundwater potential distribution of the observation boreholes, water chemistry and environmental isotope contents in groundwater, and on-site seepage measurements.

This study also shows evidence of stagnant fresh groundwater under the present sea bottom. This seems to be remnant palaeo fresh groundwater formed during a different geological era (Tokunaga *et al.*, 2006). An inland sea, such as the Ariake Sea and the Yatushiro Sea, is expected to have similar subsurface conditions as in the present study area because of the similarity of the local geological settings. In this case, the effective submarine groundwater discharge might be limited only to the area close to the coast. To confirm this, we need to develop a technique to detect the submarine groundwater discharge not at a single point, but as a much wider range of its spatial distribution.

CONCLUSIONS

The basin-wide groundwater flow system in the Quaternary pyroclastic bedrock aquifer was studied using many hydrological methods. The major results of this study are:

1. The shallow seasonally fluctuating perched groundwater flow system dominates in the headwater catchments and also contributes to the regional-scale groundwater recharge area.
2. The intermediate groundwater flow system has been confirmed by the groundwater potential distribution, the isotope content, the annual runoff ratio in the headwater catchments and the regional-scale SP survey.
3. The coastal area shows distinctive submarine groundwater seepage which has been confirmed by resistivity survey and direct seepage measurements. This submarine groundwater discharge, which is mainly supplied by the large-scale regional groundwater flow system was evaluated quantitatively by a basin-scale water balance.
4. Under the present sea floor, remnant palaeo fresh groundwater formed during a different geological era is present.
5. The schematic groundwater flow system of the studied pyroclastic aquifer has been confirmed and is mainly controlled by local topography and geology.

Acknowledgements This study was financially supported by JSPS 14208064.

REFERENCES

Fagerlund, F. & Heinson, G. (2003) Detecting subsurface groundwater flow in fractured rock using self-potential (SP) methods. *Environ. Geol.* **43,** 782–794.

Fournier, C. (1989) Spontaneous potentials and resistivity surveys applied to hydrogeology in a volcanic area: case history of the Chaine des Puys (Puy-de-Dome, France). *Geophys. Prospecting* **37**, 647–668.

Inoue, D., Shimada, J., Hase, Y. & Miyaoka, K. (2005) Groundwater flow system study in volcanic low permeability bedrock basin. Abstract H23E-1478, AGU Fall Meeting 2005.

Revil, A., Naudet, V. & Meunier, J. D. (2004) The hydroelectric problem of porous rocks: inversion of the position of the water table from self-potential data. *Geophys. J. Int.* **159**, 435–444.

Sato, S., Shimada, J. & Gotoh, T. (2005) Use of self potential (SP) method to understand the regional groundwater flow system. Abstract H23E-1482, AGU Fall Meeting 2005.

Shimada, J., Watanabe, K., Taniguchi, M., Miyaoka, K. & Onodera, S. (2003) Tidal fluctuation of the coastal groundwater seepage revealed by intensive electric resistivity survey. Abstract A137, IUGG/IAHS, Sapporo 2003.

Shimada, J., Inoue, D., Satoh, S., Gotoh, T., Hase, Y., Tsujimura, M., Taniguchi, M. & Miyaoka, K. (2004) Groundwater flow system study in the Quaternary pyroclastic flow aquifer including coastal groundwater seepage-out. Joint AOGS 1st annual meeting, July 2004, Singapore, Abstract vol. 1, 761.

Sueda, T., Shimada, J., Ohba, K., Maruyama, A., Takamoto, N., Nakamura, K. & Takano, I. (2006) The evaluation of regional water balance based on the actual evapotranspiration in mountainous igneous rock aquifer in Uto Peninsula, Japan. Proc. Int. Symp. Interrelations Between Seawater and Groundwater in the Coastal Zone and their Effect on the Environmental Nutrient Load Toward the Sea (Dec. 2006, Kumamoto Univ.), 96–98.

Taniguchi, M., Burnett, W. C., Cable, J. E. & Turner, J. V. (2003) Assessment methodologies of submarine groundwater discharge. In: *Land and Marine Hydrogeology* (ed. by M. Taniguchi *et al.*), 1–23, Elsevier, New York, USA.

Taniguchi, M., Ishitobi, T., Shimada, J. & Takamoto, N. (2006) Evaluations of spatial distributions of submarine groundwater discharge. *Geophys. Res. Lett.* **33**, L06605, doi:10,1029/2005GL025288.

Tokunaga, T., Kimura, Y., Shimada, J. Inoue, D. & Hase, Y.(2006) Evidence and the flow regime of the submarine fresh groundwater in the Yatsushiro bay, Japan. *Proc. Int. Symp. Interrelations Between Seawater and Groundwater in the Coastal Zone and their Effect on the Environmental Nutrient Load Toward the Sea* (Dec. 2006, Kumamoto Univ.), 30–37.

Tokunaga, T., Kimura, Y., Shimada, J., Sano, A. & Hishiya, T.(2006) Diffusion-limited chloride migration revealed by stable chlorine isotope profile and dating of groundwater age at Yatsushiro bay, Kumamoto, Japan. Abstract 041B-0386 AGU Fall Meeting 2006.

Comprehensive evaluation of the groundwater–seawater interface and submarine groundwater discharge

MAKOTO TANIGUCHI[1], TOMOTOSHI ISHITOBI[1], WILLIAM C. BURNETT[2] & JUN SHIMADA[3]

1 *Research Institute for Humanity and Nature, Kyoto 603-8047, Japan*
makoto@chikyu.ac.jp

2 *Department of Oceanography, Florida State University, Tallahassee, Florida 32306, USA*

3 *Kumamoto University, Kumamoto 860-8555, Japan*

Abstract Comprehensive studies of the groundwater–seawater interface and submarine groundwater discharge (SGD) have been made at Yatsushiro, Kumamoto, Japan, and other areas, by use of automated seepage meters with conductivity sensors to evaluate SGD and fresh/saline components of SGD continuously, and resistivity measurements to evaluate the relationship between temporal changes in the location of the saltwater–freshwater interface and SGD composition. The processes of SGD differ between the landward and offshore sides of the saltwater–freshwater interface. SGD in the nearshore can be mainly explained by connections of terrestrial groundwater, while offshore SGD is controlled mostly by oceanic process such as recirculated saline groundwater discharge. Global evaluations of SGD based solely on observational data (>25 000 automated flux measurements) showed that fresh groundwater discharge is estimated to be 2600 km^3/year (from the coast to 200 m offshore) and is equivalent to 7% of the global river flux.

Key words submarine groundwater discharge; terrestrial groundwater; recirculated saline water; saltwater–freshwater interface; global assessment of groundwater discharge

INTRODUCTION

Submarine Groundwater Discharge (SGD) is increasingly recognized as an important pathway for water and dissolved materials moving from the land to the ocean (Moore, 1996; Burnett *et al.*, 2001; Taniguchi *et al.*, 2006). Recent field work has revealed that SGD contains Submarine Fresh Groundwater Discharge (SFGD) and Recirculated Saline Groundwater Discharge (RSGD) (Taniguchi *et al.*, 2002), and that there are tidal effects on SGD (Taniguchi, 2002; Kim & Hwang, 2002).

Saltwater–freshwater interfaces have been intensively studied in hydrological sciences for many years, because saltwater intrusion due to excessive groundwater pumping is a serious problem for water resources in coastal areas. SGD from aquifers on the land into the ocean, and seawater intrusion from the ocean into aquifers on land, are complementary processes, functioning in opposite flow directions as a result of the hydraulic gradient across the coastal freshwater–saltwater interface being directed away from shore, or towards shore, respectively. The quality of SGD at locations along a shore-perpendicular transect depends on the location of the freshwater–saltwater interface (Taniguchi *et al.*, 2006a). However, there are many uncertainties on the relationship between SGD and the location of the interface.

Some global-scale estimates are based on hydrological or hydrogeological assumptions (Nace, 1970; COSODII, 1987). Others are based on water balance methods, which are the most widely used for global-scale estimation of the role of SGD. Global estimates of SGD vary widely, but most estimates range from 6 to 10% of surface water inputs (Taniguchi *et al.*, 2002). However, thus far there has been no attempt to estimate global SGD based solely on experimental data.

The purposes of this study are to evaluate: (1) the processes of SFGD and RSGD in the coastal zone, (2) the relationship between SGD and the freshwater–saltwater interface, and (3) a global assessment of SGD made solely by use of seepage meters

METHODS

Comprehensive studies on SGD and saltwater–freshwater interfaces have been made in Yatsushiro Sea in Kyushu Island, Japan (Fig. 1). Aquifers in this study area consist of permeable Quaternary volcanic rocks (andesite lava and tuff breccia) and pyroclastic flow deposits. The basin area is 4.5 km^2, and the length of the basin from the top (elevation is 400 m above sea level) to the coast is 4 km. The annual precipitation is about 1800 mm year^{-1}, and average annual air temperature is about 17°C. The Yatsushiro Sea is an inland sea, and the average tidal change is from 3 to 5 m.

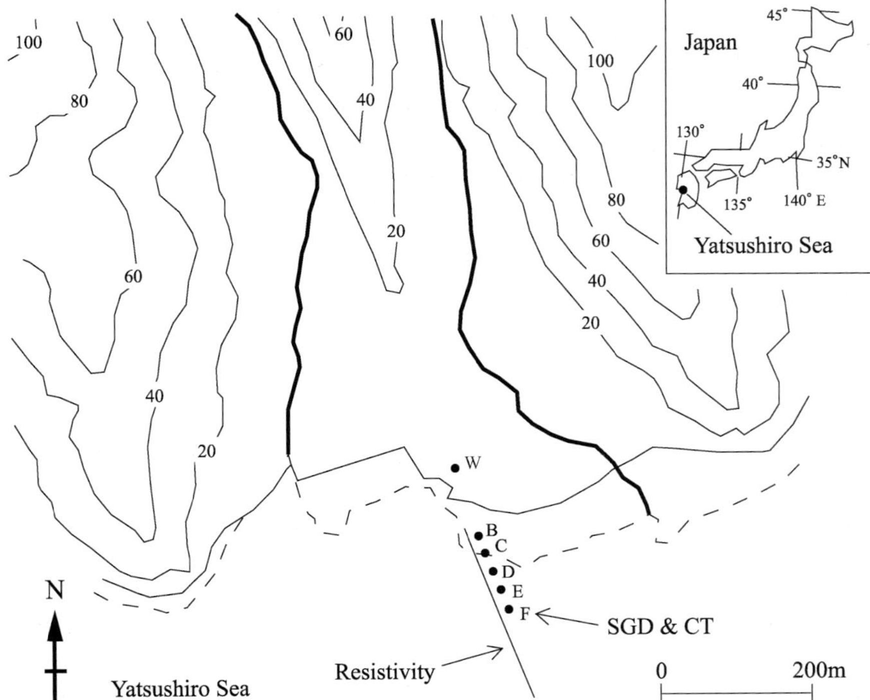

Fig. 1 Locations of seepage meters and resistivity cable at Shiranui, Kumamoto, Japan. (The broken line shows the location of the coast at low tide).

Five continuous heat-type automated seepage meters (Taniguchi & Iwakawa, 2001) were located at 30, 60, 90, 120 and 150 m distance offshore from the coastal line at high tide in Yatsushiro, Japan (Fig. 1). Seepage meters B and C are located between the high tide coast and low tide coast. The automated seepage meter is based on the effect of heat convection due to water flow by measuring the temperature gradient of the water flowing between the downstream and upstream positions in a horizontal flow tube with a diameter of 1.3 cm, which is connected to the chamber. The principle of the automated seepage meter is described in detail by Taniguchi & Iwakawa (2001) and Taniguchi *et al.* (2003). Measurements of SGD using the continuous heat type automated seepage meter have been made every 10 minutes from 2 to 7 August 2003. Tidal (sea) levels were recorded every 10 minutes at F using a pressure transducer which was attached to the outside of the seepage meter chamber. Conductivities and temperatures of water within the chambers were also measured continuously by Conductivity-Temperature-Depth (CTD) sensors (DIK 603A CTD, Daiki Rika Kogyo, Co., Ltd) which were installed inside the five seepage meters.

Resistivity under the seabed and land surface at the transect line, which is perpendicular to the coast (Fig. 1) were measured by Sting R1 IP/Swift (AGI). There are 28 probes along the 270 m transect (interval length between probes was 10 m). The Wenner method and RES2DINV ver. 3.50 (Geotomo Software) were used for the resistivity analyses.

We also provide here a global assessment of the magnitude of SGD by using data directly obtained in the field via automated seepage meters (Taniguchi *et al.*, 2003). We analysed measurement data from 10 countries, 17 locations, 97 observation points, and more than 25 000 individual data points. In each case, we evaluated the average total SGD. In most cases (13 out of 17 locations) we also had continuous conductivity measurements inside the measurement chambers and were able to separate the fresh (Submarine Fresh Groundwater Discharge, SFGD), and saline (Recirculated Saline Groundwater Discharge, RSGD) flow components. These were examined as functions of distance from shore.

RESULTS AND DISCUSSION

Location of saltwater–freshwater interface and SGD changes

The cross-section results of resistivity measurements along the transect line (Fig. 1) at the lowest and highest tides on 18 September 2003, are shown in Fig. 2(a) and (b), respectively. The darker colour shows fresher water (higher resistivity), and lighter colour shows saltier water (lower resistivity). As can be seen from Fig. 2(a), seepage meters B, C and D are located at a relatively fresher seepage area (darker colour in figure); on the other hand, E and F are located in a relatively saltier area (lighter colour). The freshwater–saltwater interface is not usually sharp, but has a transient zone. Although we cannot define the location of the sharp interface, we can see the fresher water seepage face at B, C and D. However, the locations E and F may be seaward of the freshwater–saltwater interface.

Changes in tide, SGD rate, and temperature of SGD at D (near shore) and F (offshore) are shown in Fig. 3(a) and (b), respectively. As can be seen from Fig. 3(a),

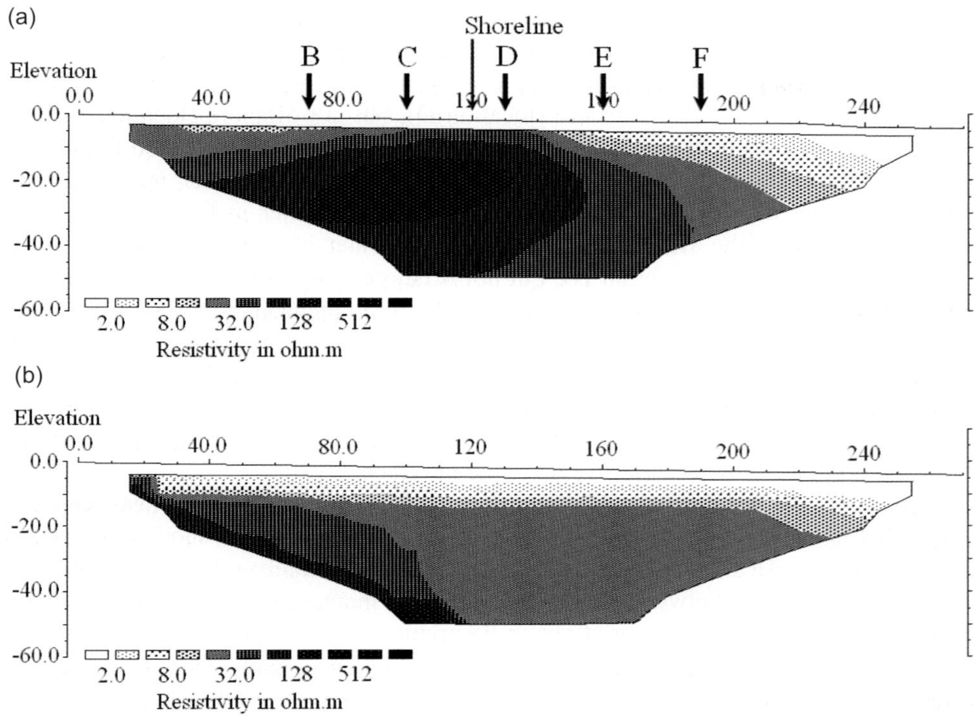

Fig. 2 Resistivity at: (a) high tide, and (b) low tide.

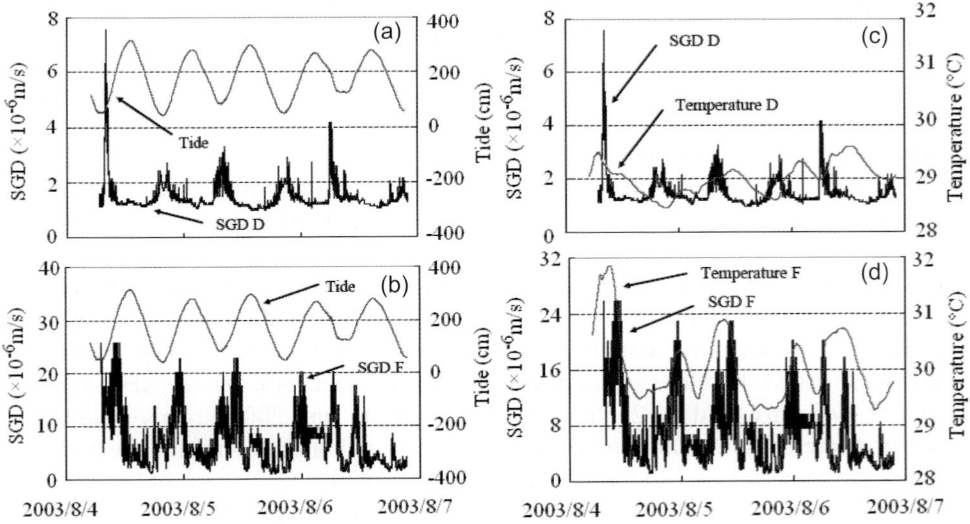

Fig. 3 Changes in: (a) tide and SGD rate at D, (b) SGD rate at F, (c) temperature of SGD at D, and (d) temperature of SGD at F.

negative correlations were found between tide and SGD at location D, however there is a time lag between tidal change and SGD changes at Location F (Fig. 3(b)). These results show that the changes in the hydraulic gradient between land and sea directly cause the changes of SGD nearshore (location D), but not offshore (location F). The negative correlations between temperature of SGD and SGD rate were also found near shore (location D), but not offshore (location F). The groundwater temperature is relatively cooler than that of the seawater during the summer in the field site. This result also supports the view that terrestrial groundwater is directly connected to the coastal water nearshore (location D), but not offshore (location F). These differences were found not only between D and F, but also between the nearshore group (B, C and D) and offshore group (E and F). As can be seen from Fig. 2, the nearshore group (B, C and D) is located landward from the seawater–freshwater interface, on the other hand, offshore group (E and F) is located seaward from the interface. Therefore, the SGD process is different on the landward side of the seawater–freshwater interface compared to that from the seaward side of the interface.

Recently collected detailed and continuous seepage flux and salinity measurements, together with tidal change and precipitation data at Shiranui, Japan, show that changes of Submarine Fresh Groundwater Discharge (SFGD) are likely related to changes in the hydraulic gradient between land and sea after an increase of groundwater recharge resulting from precipitation. Variations in Recirculated Saline Groundwater Discharge (RSGD), however, are related to tidal change and thus presumably caused by seawater exchange due to tide/wave pumping. Therefore, RSGD causes a phasing out of total SGD with the tide. Our automated measurements of groundwater flux suggest that precipitation and wave pumping appear to be important controls of terrestrial (fresh) and marine-induced (recirculated seawater) subterranean flows, respectively.

Global assessment of SGD

Observed SGD rates by seepage meters show that the total SGD distribution generally decreases, depending upon the distance from shore. The higher SGD flow closer to shore is likely a result of both SFGD, which is terrestrial fresh groundwater discharge, and RSGD due to wave set-up being elevated. SFGD observations clearly show that the ratio of the fresh groundwater discharge to the total discharge decreases systematically with distance from shore. This is attributed to a general decreasing hydraulic connection between terrestrial groundwater and seawater. The average flux of SFGD from the coast to 200 m offshore was 0.059 m/d. Therefore, fresh groundwater discharge per unit shoreline length from the coast to 200 m offshore is 12 m^3 m^{-1} d^{-1}. Using an estimated shoreline length of 600 000 km, we calculate that the SFGD flux to the world's ocean would be 2600 km^3/year within the first 200 m. This estimate of SFGD from the coast to 200 m offshore agrees well with a previous global estimation for fresh groundwater of 2400 km^3/year (or 11 m^3 m^{-1} d^{-1}) by hydrograph separation and water balance methods (Zektser, 2000). Although comparisons of SFGD between water balance methods and direct measurements have been done before on a local scale (Taniguchi *et al.*, 2006b), this is the first attempt on a global scale.

Figure 4 shows the magnitude of SFGD (solid bars) and RSGD (open bars) per unit shoreline length and location map of all known SGD observations cited in the

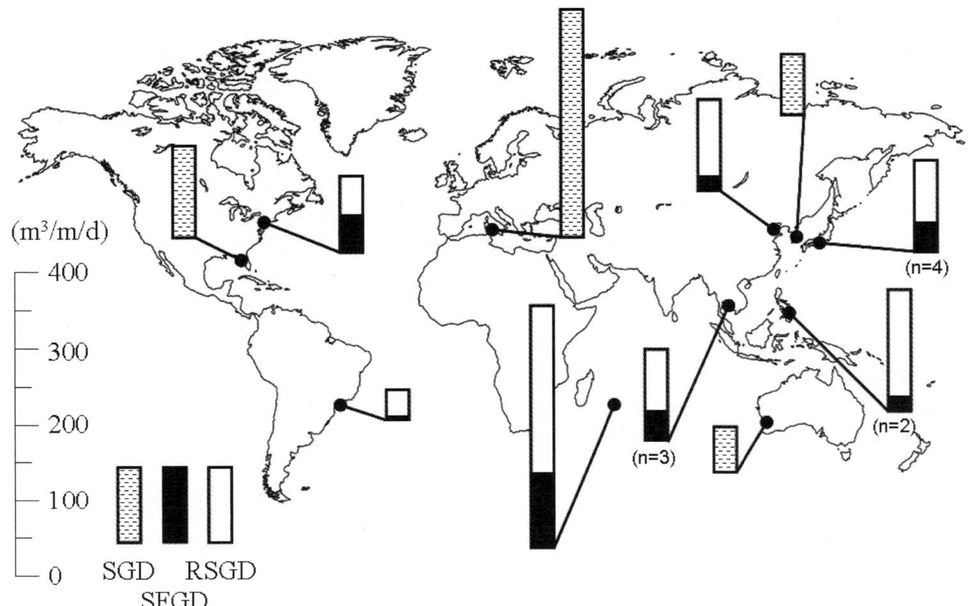

Fig.4 Global assessment of SGD, SFGD, and RSGD.

literature (closed circles). Total SGD (crossed-hatched symbol) given when separation of fresh and saline components was not possible. The numbers next to each circle refer to locations given in Taniguchi *et al.* (2002).

The geographical distribution of SFGD and RSGD estimates per unit shoreline length shows considerable spatial heterogeneity. Higher SFGD was found in east and south-east Asia, Long Island (New York), and Mauritius, areas where precipitation is also high and/or the sediments are very permeable (Emery, 1968). On the other hand, areas with low SFGD are found in deltaic areas such as near the mouths of the Yellow River (China) and Chao Phraya River (Thailand). The two areas we have measured that have the very highest SGD (= SFGD + RSGD) per unit shoreline length (>300 m^3 m^{-1} d^{-1}) are located on the coasts of Sicily and Mauritius, both characterized by high rainfall, steep topography, and an absence of well-developed rivers

CONCLUSION

The main conclusions of this study are:

1. SGD variations within the saltwater–freshwater interface had negative correlations with tidal variations, because of the connections of terrestrial groundwater on the land and the ocean. On the other hand, there is a time lag between tidal change and SGD changes offshore from the interface.
2. The negative correlations between temperature of SGD and SGD rate were also found landward from the freshwater–saltwater interface, but not offshore. This is attributed to the direct connection of nearshore coastal water to the terrestrial groundwater.

3. The processes of SGD differ between seaward and landward of the saltwater–freshwater interface. SGD landward of the interface can be explained mainly by connections of terrestrial groundwater; however, SGD seaward of the interface is controlled mostly by oceanic process such as recirculated saline groundwater discharge.
4. Global evaluations of SGD based solely on observational data showed that fresh groundwater discharges is estimated to be 2600 km^3/year (from the coast to 200 m offshore) and is equivalent to 7% of the global river flux. There is heterogeneity in the distribution of SGD depending on hydrogeology and precipitation.

Acknowledgements This work was financially supported by the following organizations: Research Institute for Humanity and Nature (RIHN), SCOR/LOICZ, UNESCO-IHP/IOC, IAEA, JSPS, and NSF.

REFERENCES

Burnett, W. C., Taniguchi M. & Oberdorfer, J. A. (2001) Assessment of submarine groundwater discharge into the coastal zone. *J. Sea Res.* **46**, 109–116.

COSOD II (1987) *Fluid Circulation in the Crust and the Global Geochemical Budget.* Report of the Second Conference on Scientific Ocean Drilling (Strasbourg, France, July, 1987).

Emery, K. O. (1968) Relict sediments on continental shelves of the world. *Am. Assoc. Petroleum Geol.* **52**, 445–464.

Kim, G. & Hwang, D. W. (2002) Tidal pumping of groundwater into the coastal ocean revealed from submarine Rn-222 and CH_4 monitoring. *Geophys. Res. Lett.* **29**, 1–4.

Moore, W. S. (1996) Large groundwater inputs to coastal waters revealed by ^{226}Ra enrichments. *Nature* **380**, 612–614.

Nace, R. L. (1970) World hydrology: status and prospects. In: *World Water Balance*, vol. I.(IAHS-UNESCO symposium, Reading, August 1970), 1–10. IAHS Publ. 92. IAHS Press, Wallingford, UK.

Taniguchi, M. (2002) Tidal effects on submarine groundwater discharge into the ocean. *Geophys. Res. Lett.* **29**, 1–3, 10.1029/2002GL014987.

Taniguchi, M. & Iwakawa, T. (2001) Measurements of submarine groundwater discharge rates by a continuous heat – type automated seepage meter in Osaka Bay, Japan. *J. Groundwater Hydrol.* **42**, 271–277.

Taniguchi, M., Burnett, W. C., Cable, J. E. & Turner, J. V. (2002) Investigations of submarine groundwater discharge. *Hydrol. Processes* **16**, 2115–2129.

Taniguchi, M., Burnett, W. C., Cable, J. E. & Turner, J. V. (2003) Assessment methodologies of submarine groundwater discharge. In: *Land and Marine Hydrogeology* (ed. by M. Taniguchi, K. Wang, & T. Gamo), 1–23. Elsevier, Amsterdam, The Netherlands.

Taniguchi, M., Ishitobi, T. & Shimada, J. (2006a) Dynamics of submarine groundwater discharge and freshwater–seawater interface: *J. Geophys. Res.* **111**, C01008, doi:10.1029/2005JC002924.

Taniguchi, M., Ishitobi, T. & Shimada, J. (2006b) Evaluation of spatial distribution of submarine groundwater discharge, *Geophys. Res. Lett.* **33**, L06605, doi:10.1029/2005GL025288.

Zektser, I. S. (2000) *Groundwater and the Environment.* Lewis Publishers, Boca Raton, Florida, USA.

Possibilities of geophysical survey for groundwater contamination and subsurface pollution determination and monitoring in the coastal zone

YURIY R. OZOROVICH[1] & EVGENY A. KONTAR[2]

1 *Space Research Institute, Russian Academy of Sciences, Moscow, Russia*
yozorovi@iki.rssi.ru

2 *P.P. Shirshov Institute of Oceanology, Russian Academy of Sciences, Moscow, Russia*

Abstract One of the important challenges facing coastal zone managers today is how to identify, measure and monitor submarine groundwater discharge and seawater intrusion, and how to evaluate its influence on cumulative impacts of coastal land-use decisions over distance and time. A new geophysical technique can help to solve the problem and provide direct monitoring of groundwater–seawater interactions in coastal aquifers. The Transient Electromagnetic Method, TEM, allows subsurface sounding to 300 m depth.

Key words coastal geophysics; hydrogeology; oceanography; groundwater–seawater interaction

INTRODUCTION

The focus of this report is to describe the current understanding of groundwater in freshwater–saltwater environments using a new geophysical method. One of the important challenges facing coastal zone managers today is how to identify, measure and monitor submarine groundwater discharge (SGD) and seawater intrusion (SWI) and how to evaluate its influence on cumulative impacts of coastal land-use decisions over distance and time (Burnett *et al.*, 2002, 2003; Kontar, 2002; Kontar *et al.*, 2002; Kontar & Ozorovich, 2006; Taniguchi *et al.*, 2006). Numerous field studies have yielded a wealth of information on the occurrence and intrusion of saltwater into coastal–zone freshwater aquifers. Several geochemical and geophysical techniques are used to directly or indirectly monitor saltwater in coastal aquifers. Because of the very high concentration of chloride in seawater (typically about 19 000 mg/L) the chloride concentration of groundwater samples has been the most commonly used indicator of saltwater occurrence and intrusion in coastal aquifers. However, other indicators of groundwater salinity, such as the total dissolved-solids concentration or specific conductance of groundwater samples, are also used. Understanding of the multi-variable dimensions of groundwater management in coastal zones can be improved through the application of innovative information technologies (the application of neural networks to data analysis, optimization, pattern recognition, image identification, etc.), and by using new generations of non-invasive techniques for groundwater exploration, which can be successfully combined with nuclear and isotopic techniques (Burnett *et al.*, 2002) and will help to understand the results obtained using these techniques (Kontar & Ozorovich, 2006).

This report is a result of research conducted by a team of Russian scientists (E. A. Kontar, L. I. Lobkvsky, I. A. Garagash, I. Ya. Rakitin, Yu. R. Ozorovich and

F. A. Babkin). It is the opinion of the team that further development of electromagnetic sounding as a geophysical method (Kontar & Ozorovich, 2006) in conjunction with nuclear and isotopic techniques for investigation of SGD (Burnett *et al.*, 2002) is useful for this type of research. This method can provide a clear understanding of the fluctuation of the interface between freshwater and groundwater in the coastal zone over time and its spatial dimensions.

The new operational monitoring system, MARSES TEM, has capabilities and advantages that can be used for water search tasks (groundwater table), the definition of waste levels, monitoring changes in subsurface horizons, etc. The system is based on the transient electromagnetic method (TEM) sounding technology, which enables one to conduct subsurface soundings to a depth of 300–500 m. This system can also be applied to solve long-term tasks for monitoring natural subsurface ecosystems, subsurface horizons, soil salinity levels, salinity gradients, and groundwater levels. All of these parameters can be used to track seasonal and climatic changes in selected coastal areas. The distinctive difference in the proposed coastal monitoring system is the measurement of soil humidity along the coast at all depths down to the groundwater level. It is not possible to detect these subsurface horizon humidity parameters by any other monitoring means, such as observation wells or digging pits, etc.

The goal of this contribution is to obtain a fundamental understanding of the physical and chemical processes taking place at the dynamic subsurface freshwater–seawater interface by carrying out detailed studies on a small-scale, at one or two sites in a coastal area. This should provide a conceptual framework for understanding the effects of these processes on a larger scale and over longer periods of time. Insight into the controlling processes and the development and testing of a coupled variable-density flow, multi-species transport, and reaction code will increase the possibilities for sustainable management of coastal aquifers.

At present, a stand-alone portable non-invasive subsurface research instrument based on TEM measurement is available. It is capable of measuring the resistivity of subsurface slices of up to 100–150 metres and works with a stock IBM compatible portable computer using a serial interface. Brief technical details of the instrument are as follows:

- weight: 1.5 kg
- size (mm): 103 × 27 × 310
- working temperature: –20°C to +65°C
- power consumption: more that 50 measurements at maximum depth.

During IAEA SGD CRP (2002–2006) two successful field tests of the MARSES TEM system were conducted: in Sicily, Italy and on Long Island, USA. Only the results obtained during the experiment in Sicily are reproduced here.

RESEARCH BACKGROUND

In the course of Russian space research missions to Mars, soil conditions were found to be similar to the Earth's arid and semi-arid lands. The MARSES TEM was developed for subsurface sounding and mapping applications. More specifically, these instruments relate to methods for the mapping, tracking, and monitoring of: groundwater, groundwater channels, groundwater structures, subsurface pollution plumes, mapping interconnected fracture or porous zones, leaks in earthen dams, leaks in drain fields, monitoring changes in subsurface water flow, changes in ion concentration in the groundwater, monitoring *in situ* leaching of solutions, changes in subsurface redox or reaction fronts, underground chemical reactions, subterranean bio-reactions, or other subsurface water and related geological structures. Recent research has been devoted to the development of new technologies for monitoring the subsurface processes in coastal zones and has revealed significant comparative advantages.

These include:

1. Flexible software development gives an opportunity to adapt this monitoring system to solve different tasks: monitoring of groundwater discharge, identification of saltwater intrusion at the groundwater level, etc.
2. At present the geophysical equipment and instrumentation market does not have a similar system for non-invasive exploration of subsurface ecosystems. The system has been developed for space missions and has met stringent requirements for both hardware and software.
3. A flexible architecture gives an opportunity to build a system by using unified instrumentation and specialized software for resolving particular monitoring tasks.
4. These advantages in system monitoring make possible the development of integrated monitoring systems for different urban, dry land, arid and semi-arid coastal-zone lands.
5. Simplicity of use and maintenance, with the possibility of exchanging working modules during service procedures provides an opportunity to apply this system in any region for up to five years, with no need to develop a service and maintenance station network.
6. The property of TEM at the late stages of transience determines the maximum depth of sounding, which together with its good resolution and high degree of accuracy is the main advantage of the method, as well as the physical and geological parameters acquired during measurements of geo-electric sections, makes the MARSES TEM monitoring system one of the most competitive and innovative for further development and use for monitoring and studying SGD and SWI in coastal zones.

MODERNIZATION AND TESTING OF THE MARSES TEM SYSTEM

Theory

The physical and mathematical bases of TEM are described fully and sufficiently in the literature (e.g. F. Kamenetsky (in Russian) "Electromagnetic geophysical researches" Moscow, GEOS, 1997). Here we shall state only the basic aspects of the theory having direct relation to the technology of soundings with TEM-FAST.

One of the few models of media, for which the formulae are simple and accessible for the analysis of received TEM signals, is the model of the homogeneous half-space. To exemplify the opportunities with TEM, we shall consider asymptotic estimation of signals for late and early stages. The late stages of transient $t = t/\mu(R^2/\rho) >> 1$ for one-turn, round antennas R and r, lying above the homogeneous half-space with resistivity ρ and magnetic permeability of vacuum μ, are described by the formula:

$$E(t)/I = 0.05 \times (\pi^{1/2}\mu^{5/2})/\rho^{3/2}(r^2R^2)\, t^{-5/2} \quad (1)$$

The registered signal is proportional to conductivity $\sigma^{3/2} = 1/\rho^{3/2}$ and to the product R^2r^2. Thus, in TEM the amplitude of signals at late stages is sensitive to changes of conductivity of the section in comparison, for example, with methods of direct current. The signals $E(t)/I$ at $t/\mu(R^2/\rho) >> 1$ do not depend on a site of receiving loop $r < R$. The formula (1) is also true at a height h above the surface of the half-space of the antennae. In late stages of transience, the signal registered in the receiving antenna, is caused by currents induced in the current's ring inside section with effective radius R_{eff} and "attitude" depth $H_{eff} \sim R_{eff}$. $R_{eff} = (t\rho/\mu)^{1/2}$, some times exceeding the radius of the transmitting loop $R_{eff} >> R$.

The vertical magnetic field created by this contour is practically homogeneous within the limits of its area at $h < R_{eff}$, therefore the registered signals are proportional to the derivative of the magnetic field over time, and do not depend on the site of reception. The property of TEM at late stages of transience determines the maximum depth of sounding, which in a combination with good resolution of the method ($E \sim \sigma^{3/2}$) and high degree of accuracy is the main advantage of the method. Early stages of transient $t_0 << 1$ for the coincident antennas $R = r$ do not depend on the resistivity of media:

$$E(t)/I = \mu R/(2t) \quad (2)$$

At early stages ($t_0 << 1$) for small receiving antennas $r/R << 1$ the signals are proportional to the specific resistivity of media ρ and do not depend on time, t:

$$E(t)/I = 3\pi\rho\,(r^2/R^3) \quad (3)$$

Monitoring

MARSES TEM sounding instruments are based on time domain electromagnetic sounding. MARSES TEM is a portable, reliable geo-electrical sounding instruments made to satisfy small space requirements and simple, intuitive usage.

TEM allows subsurface sounding at depths of 300–500 metres. Applications include: groundwater, prospecting deposits, hydrogeological research, geological surveys required before the construction of buildings, ecological research, archeological and subterranean objects search, monitoring of high risk industrial and engineering objects, research and testing of rock samples.

A number of MARSES instruments should be placed in coastal zones with ecological hazards to provide measurements in the areas of interest. Using a radio system (Fig. 1) it will be possible to transmit data to headquarters, where the data will be processed and stored for future use and reference. Using modelling and visual presentation software it will be possible to create graphs, charts, time variation and other easy to understand representations of the monitored area.

The significant advantage of this system in comparison to known ecological hazard monitoring systems is its cost effectiveness. It does not require expensive satellite monitoring of the coastal zone or employment of satellites in data transfer. Another advantage is its compact and ready to use nature that allows rapid measurements at any required area. Moreover, it is significantly cheaper than competitors.

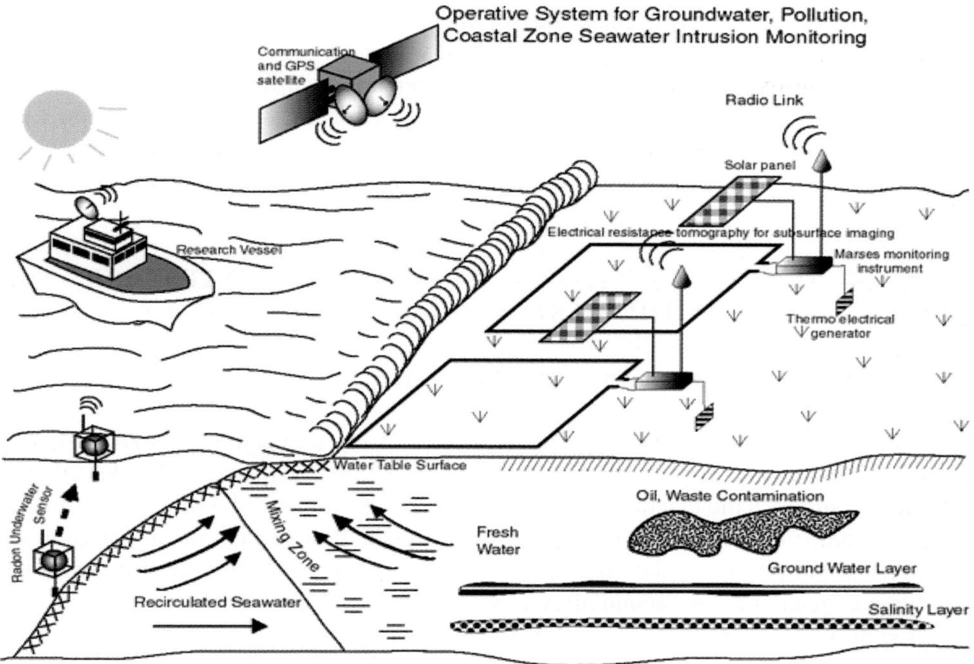

Fig. 1 The coastal zone subsurface monitoring system organization structure.

In contrast with many of the competitors' instruments MARSES TEM has exceptional advantages among small depth sounding devices. Because the MARSES TEM hardware was developed to be employed in space research missions it possesses a unique portability and is dramatically lighter. It fits inside a suitcase, with the antenna, supporting notebook, and batteries. Use of a notebook PC for data processing makes it easier to view the results, process them, prepare reports and send them to headquarters. MARSES TEM supplies reliable data in areas with high EM distortions (about 1 volt and higher), such as in industrial zones and on urban streets. Sounding parameters can be easily adjusted from a PC.

At present there is no mobile operational monitoring system based on non-invasive technology and methodology. Thus development of these innovative monitoring systems has great potential and widespread application.

As this system was developed within the framework of space research missions, it possesses unique methodological and technological features and advantages in comparison to other monitoring systems based on different principles and methods.

RESULTS OF TESTS

We conducted two field tests of the MARSES TEM system: in Sicily, Italy and on Long Island, USA. A distinctive feature of the experiments with sounding of the subsurface horizons was the realization of simultaneous measurements of the geo-electrical sections at two points 70 metres apart for the various cycles of a tidal wave over a period of several days. These measurements have, for the first time, revealed the spatial and time variability of the saltwater–freshwater interface, and show the effect of complex transformation of the salt and fresh water interface in the coastal zone. The analysis of the time variation of the geo-electrical section shows high correlation with the daily tide and nonlinear transformation of the boundary of the saltwater–freshwater interface in the process of its spreading offshore.

GEOHYSICAL SURVEY OF THE SALT–FRESH WATER INTERFACE AND SEA SPRING ZONE IN THE MARINA DI RAGUSA AREA, SICILY

Hydrogeology of the Marina Di Ragusa beach study area

The Marina di Ragusa beach study area is located in the south of Sicily. The study of the geological characteristics of the territory has allowed us to exactly identify the aquifer that supplies the discharge into the sea along this stretch of Sicilian coastline. It is a carbonate aquifer in which karstic phenomena have taken on a determining function in conditioning the underground water circulation that locally assumes artesian characteristics. A second alluvial aquifer has also been identified that is partly supplied through lateral contact and through artesian flow, by a deeper aquifer.

The links between the tectonic morphology and underground circulation are then evident. The supply of the main deep aquifer occurs on the Hyblaean relief. Calculations were performed to assess the effective infiltration in the region's different areas according to the nature of the terrain and the climatic conditions and the river waterbed interchanges, which were revealed to be a determining factor in quantifying the supply volume of the aquifers.

Measurements of the spatial distribution of geoelectrical sections of the saltwater–freshwater interface

Although overuse and contamination of groundwater are not uncommon throughout Sicily the proximity of coastal aquifers to saltwater creates unique issues with respect to groundwater sustainability in the coastal region of Donnalucata. These issues are primarily those of possible saltwater intrusion into freshwater aquifers and changes in the amount and quality of fresh groundwater discharging to coastal saltwater eco-systems.

Further development of electromagnetic sounding as a geophysical method in combination with nuclear and isotopic techniques for investigation of SGD is useful for this type of research. Potential directions for field geophysical methods include the development of improved interpretation techniques, particularly for three-dimensional interpretation and integration with other data sets, and the development of survey techniques for high-resolution measurements of geoelectrical slices of small study areas using simultaneous measurements with four sounding loops for spatial and temporal surveys in 3-D geoelectrical section.

Additional methodical measurements are necessary for the reception of complex geoelectrical data in study areas with physical and chemical parameters of subsurface horizon in the field sites, which include:

1. Chloride content based on water sample measurements at the field sites.
2. Measurements of resistivity in the upper level of the sounding area and the conductivity directly on induction logs measured in the well.
3. Laboratory measurements of resistivity of soil samples from the sounding area.

After implementation of all the requirements stated above it is possible to construct closed measures in the saltwater interface, which will allow connection between data from *in situ* isotope and geophysical measurements.

This method will allow for the expansion of information on spatial and temporal variations in the saltwater interface, and its structural and geological properties, which cannot be obtained by alternative methods.

At the present time there is no method for adequate and direct *in situ* measurement of porosity, salinity and resistivity of soil in subsurface horizons. For this reason methodical and experimental investigations for the exploration of the saltwater interface using different field instrumentation represent an important direction for future research for comparative studies in SGD and saltwater interaction, and vadose zone flow processes, pollution determination and monitoring of the coastal zone.

Acknowledgements The authors would like to thank Pavel P. Povinec, Jean Comanducci and B. Oregioni (IAEA Marine Environment Laboratory, Monaco) for operational support during the expedition in Sicily. The studies were financially supported in part by UNESCO and IAEA.

REFERENCES

Burnett, W. C., Chanton, J., Christoff, J., Kontar, E. A., Krupa, S., Lambert, M., Moore, W., O'Rourke, D., Paulsen, R., Smith, C., Smith, L. & Taniguchi, M. (2002) Assessing methodologies for measuring groundwater discharge to the ocean. *EOS* **83**(117), 122–123.

Burnett, W. C., Chanton, J. P. & Kontar, E. A. (eds) (2003) Submarine groundwater discharge. *Biogeochemistry* **66**.

Kontar, E. A. (2002) Submarine monitors and tracer methods for investigations of groundwater discharge into the coastal zone. In: 34th International Liege Colloquium on Ocean Dynamics (Liege), p. 30.

Kontar, E. A. & Ozorovich, Y. R. (2006) Geo-electromagnetic survey of fresh/salt water interface in the coastal southeastern Sicily. *Continental Shelf Research* **26**, 843–853.

Kontar, E. A., Burnett, W. C. & Povinec, P. P. (2002) Submarine groundwater discharge and its influence on hydrological trends in the Mediterranean sea. In: *Tracking Long-term Hydrological Change in the Mediterranean Sea* (Proc. CIESM Workshop: Monaco), 109–114.

Taniguchi, M., Burnett, W. C., Duevalova, H., Kontar, E. A., Povinec, P. P. & Moore, W. (2006) Submarine groundwater discharge measured by seepage meters in Sicilian coastal waters. *Continental Shelf Research* **26**, 835–842.

Multi-channel resistivity investigations of the freshwater–saltwater interface: a new tool to study an old problem

PETER W. SWARZENSKI[1], SARAH KRUSE[2], CHRIS REICH[1] & WOLFGANG V. SWARZENSKI[3]

1 *US Geological Survey, St Petersburg, Florida 33701, USA*
pswarzen@usgs.gov

2 *Dept of Geology, University of South Florida, Tampa, Florida, USA*

3 *US Geological Survey, retired, 1932 Los Angeles Ave., Berkeley, California 94707, USA*

Abstract It has been well established that fresh or brackish groundwater can exist both near and far from shore in many coastal and marine environments. The often permeable nature of marine sediments and the underlying bedrock provides abundant pathways for submarine groundwater discharge. While submarine groundwater discharge as a coastal hydrogeological phenomenon has been widely recognized, only recent advances in both geochemical tracers and geophysical tools have enabled a realistic, systematic quantification of the scales and rates of this coastal groundwater discharge. Here we present multi-channel electrical resistivity results using both a time series, stationary cable that has 56 electrodes spaced 2 m apart, as well as a 120 m streaming resistivity cable that has two current-producing electrodes and eight potential electrodes spaced 10 m apart. As the cable position remains fixed in stationary mode, we can examine in high resolution tidal forcing on the freshwater–saltwater interface. Using a boat to conduct streaming resistivity surveys, relatively large spatial transects can be rapidly (travel speed ~2–3 knots) acquired in shallow (~1–20 m) waters. Sediment formation factors, used to convert resistivity values to salinity, were calculated from porewater and sediment samples collected during the installation of an offshore well in Tampa Bay, Florida, USA. Here we examine the seabed resistivity from sites within Tampa Bay using both stationary and streaming configurations and discuss their overall effectiveness as a new tool to examine the dynamic nature of the freshwater–saltwater interface.

Key words electrical resistivity; coastal hydrogeology; formation factor; submarine groundwater discharge

INTRODUCTION

Submarine groundwater discharge (SGD) may provide an important additional pathway for material and constituent transport to the sea (Moore, 1996; Burnett *et al.*, 2003, 2006). This SGD may consist of both recycled seawater as well as land-derived fresh groundwater, yet the former component most often dominates a SGD signal (Michael *et al.*, 2005; Swarzenski, 2007). This recycled component may also recharge a coastal aquifer with ambient seawater and water column-derived constituents, including Cl^- and select nutrients. Near-shore dynamic mixing processes, driven in part by tidal forcing and density differences, create a zone of enhanced biogeochemical reactivity (Fig. 1) that has been called a “subterranean estuary” (Moore, 1999; Charette

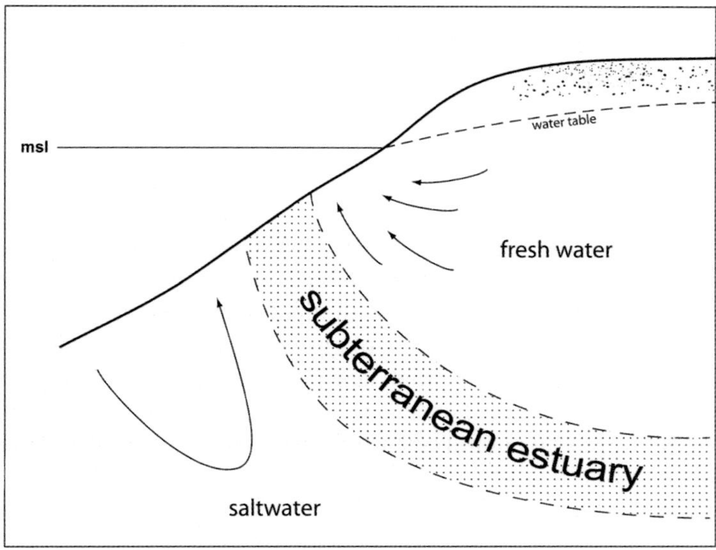

Fig. 1 Idealized hydrogeological cross-section at a land/sea interface, depicting the subterranean estuarine zone.

& Sholkovitz, 2006). While much is known about biogeochemical transport phenomena in surface water estuaries (Boyle *et al.*, 1977; Sholkovitz, 1976, 1977; Swarzenski *et al.*, 2006a), still relatively little is known of such processes and reactions in subterranean estuaries. A useful tool for imaging the spatial scales and mixing dynamics of the freshwater–saltwater interface of a coastal aquifer is surface electrical resistivity.

The direct current (DC) electrical resistivity approach measures potential differences that are produced by directing current flow into sediment. The resistivity (conductivity = $(\text{resistivity})^{-1}$) of a formation is a function of both the geological material and its textural character, including porosity, as well as the resistivity of porewaters contained in the geological matrix. The resistivity contrast between freshwater and saline water is strong compared to the resistivity differences between many common nearshore geological materials (sands, limestone). As a result the freshwater–saltwater interface is typically a dominant feature in the resistivity structure of nearshore sediments, and resistivity methods have been used for decades to map this interface. Resistivity techniques have advanced considerably with the development of high-resolution streaming and stationary marine cable configurations and multi-channel systems (Manheim *et al.*, 2004). These newer systems permit much more rapid multi-pole data acquisition and more user-friendly and accurate data reduction software packages. The advantage of the streaming multi-channel systems (towed behind a boat) is that relatively long transects across a shore face can be studied rapidly. While several groups are currently using such systems in streaming mode to study nearshore hydrogeological processes, we have implemented a new, stationary resistivity cable that consists of 56 electrodes spaced 2 m apart to study in high resolution, the freshwater–saltwater interface. Instead of rapid coverage of long transects, this system permits very detailed examination of the freshwater–saltwater interface over a 100-m lateral

scale, and to ~20 m depth, with better signal-to-noise than the streaming systems. The high resolution offers detailed examination of coastal hydrology and the response of the freshwater–saltwater interface to tidal forcing (Swarzenski *et al.*, 2006b).

METHODS

We surveyed the resistivity of Tampa Bay sediments (Fig. 2) using an Advanced Geosciences Inc. (AGI) Marine SuperSting R8 multi-channel system connected either to: (a) a streaming cable that consisted of two current producing electrodes and eight potential electrodes that acquire data with dipole-dipole geometries, or (b) an external switching box that controlled the flow of current along a 56 electrode, stationary cable. In stationary mode, current potentials are measured in a distributed array, with array geometry set by the user. For every resistivity measurement (~1 per sec), the SuperSting R8 injects an optimized current, reverses the polarity and then re-injects the current again to cancel spontaneous voltages that may occur down-cable. This process is typically repeated, and if the error is less than a pre-determined threshold value (i.e. 7%), then the next reading advances. Replicate measurements provide a means to assess wave-induced artefacts. Resistivity measurements were processed using an inverse modelling routine (AGI EarthImager) that can accommodate water-column salinity and depth observations. The resolution was optimized by using a starting model with the apparent resistivity pseudo-section and a best-fitting layered model was then developed using an iterative least squares smooth model inversion method.

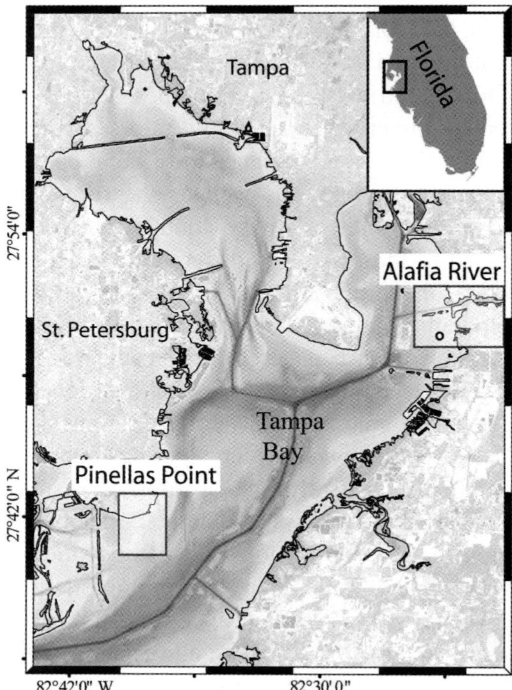

Fig. 2 Tampa Bay, Florida, showing the Pinellas Point and Alafia River study sites. O denotes the offshore well site at Bull Frog Creek used in the formation factor calculations.

RESULTS AND DISCUSSION

Formation resistivity (R_f) values of saturated sediments are derived from both the resistivity of the sediment grains (R_s) and the porewater resistivity (R_p), and are complexly controlled by sediment lithology, porewater salinity, porosity, and temperature. To interpret a resistivity profile obtained through the methods described above, it is useful to better understand the local relationships between the formation resistivity, lithology, and porewater resistivity. The porewater resistivity (ohm m) is easily measured where the porewater can be sampled directly, using a water conductivity meter. The porewater salinity–resistivity relationship can be expressed as $S = 7.042 \times R_p^{-1.0233}$ (at 25°C) where S = salinity; Manheim *et al.*, 2004). The formation resistivity in representative lithologies can be manually measured by extracting a sample of the matrix and using a Wenner-type probe. In saturated sediments the relationship between, R_f and R_p is often summarized in a simplified form of Archie's Law, in which a formation factor, F, is defined as $F = R_f R_p^{-1}$. From the measurements described above one can derive a local formation factor F for the types of sediments present. Assuming locally homogenous and saturated sediments, the formation factor, F, can then be used to interpret formation salinities from the time-series resistivity observations.

From auger cuttings collected during the installation of an offshore well in Tampa Bay (Fig. 2, latitude 27°50.308N; longitude 82°24.127W), we measured R_f, and R_p values from discrete sediment intervals to depths >11 m. In general, porewater salinities decreased in the surface sediments from a value of 20.5 (1 cm) to 17.9 at 90 cm, and then peaked at values in excess of 23 at 603 cm (Fig. 3(a)). Lowest porewater salinities (to 4.8) were observed at depths around 800 to 850 cm. Mean R_f values of 1.80 ± 0.18 ohm m ($n = 17$) at depth <1 m increased to values above 8 at a depth of 845 cm. Such R_f and R_p values yield formation factors, F, that ranged from 3 to above 6 (mean = 4.7 ± 1.1; $n = 13$) (Fig. 3(b)) and that predictably increase as sediments become coarser-grained.

Figure 4 shows two examples of modelled streaming and stationary resistivity profiles from the Tampa Bay region. A streaming resistivity profile (Fig. 4(a)) of the bed sediments from the lower Alafia River was acquired by boat travelling ~14 km upriver from Tampa Bay to evaluate surface water/groundwater (hyporheic) exchange processes prevalent in Florida's coastal rivers. For this line, a 100 m cable consisting of two current producing electrodes and eight potential electrodes spaced 10 m apart was used, and the 8-channel dipole-dipole resistivity data was subsequently merged with streaming GPS, depth and water column salinity. Periodic porewater and sediment resistivity values collected along this transect were used to derive a mean formation factor, $F = 3$, which was used to convert resistivity (ohm m) to salinity values. At the beginning (0 km) of this line, uniform saltwater intrusion can be seen to the cable exploration depth (i.e. ~20 m). Moving upstream, this saline water is gradually replaced by fresher groundwater, and by ~10 km, the resistivity results imply that the underlying groundwater is mostly fresh. The bathymetry of the river bed beyond 10 km also shows marked change in that the river bed transitions from being relatively smooth in the lower reaches of the river to exhibiting karst-type terrain features beyond 10 km.

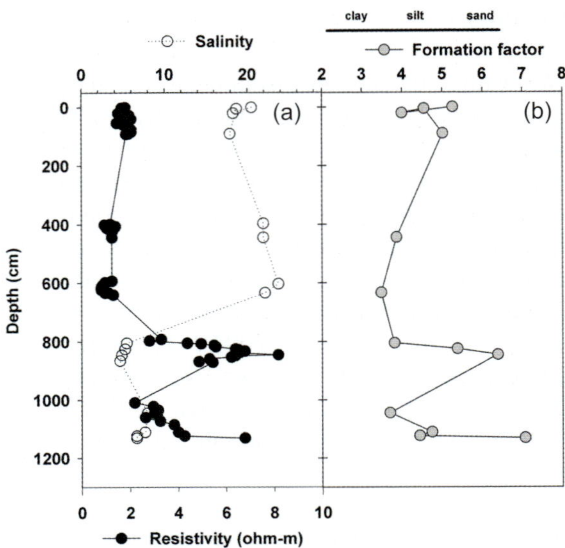

Fig. 3 (a) Seabed resistivity, porewater salinity, and (b) calculated formation factor, F, values derived from auger well cuttings collected during installation of an offshore well at Bull Frog Creek, Tampa Bay.

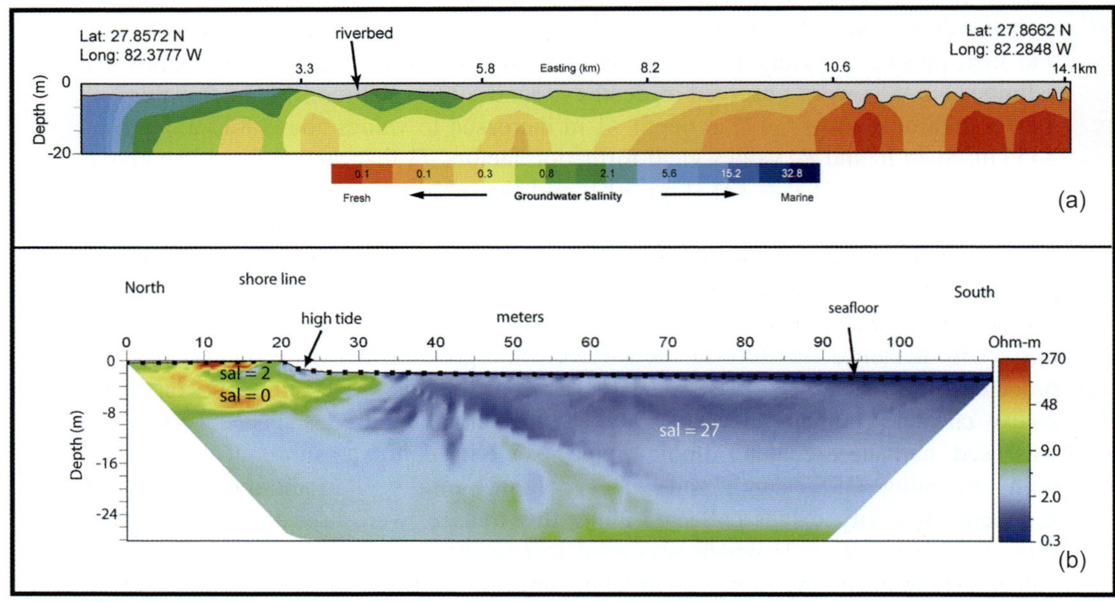

Fig. 4 (a) A 14.1 km long streaming resistivity profile of the lowest reach of the Alafia River, depicting the subsurface freshwater–saltwater interface to a depth of 20 m below the river bed. A formation factor, F, of 3 was used to convert resistivity units to representative groundwater salinity values (adapted from Swarzenski *et al.*, 2007). (b) A 116 m stationary resistivity profile across a beach face at Pinellas Point, showing the nearshore mixing dynamics of the surficial aquifer to 25 m depth as it discharged into bay surface waters. Annotated salinity values were measured using piezometer porewater samples (adapted from Swarzenski, 2007).

A stationary resistivity profile (Fig. 4(b)) was collected across a ~120 m long shore-perpendicular transect at Pinellas Point in southern Tampa Bay to examine submarine groundwater discharge processes. In contrast to the streaming resistivity method, the stationary cable lies fixed on the sea floor and contact of the 56 electrodes with sediment is facilitated using either sand bags or metal spikes. To ground truth the stationary resistivity measurements, we collected porewater samples by drive-point piezometer at discrete horizons for salinity values. At Pinellas Point, stationary resistivity measurements revealed a distinct tongue of freshened water that was being discharged about 30 m down cable. The expression of this water lens ("freshwater tunnelling") is characteristic of density-driven recirculation combined with tidal fluctuations and wave setup (Robinson *et al.*, 2006). By examining the detailed change in resistivity over a tidal cycle one can estimate a first-order SGD rate (Swarzenski *et al.*, 2006b).

CONCLUSION

Multi-channel electrical resistivity is a powerful tool to examine diverse coastal water mass mixing processes, including submarine groundwater discharge and hyporheic exchange within a coastal river. In streaming mode, large coastal areas may be examined rapidly by boat, yet we still have much to learn about the multi-channel system's performance in terms of reproducibility and resolution. In stationary mode, artefacts due to data acquisition or lithology are typically much lower than in streaming mode, and so this approach can provide high resolution information of the freshwater–saltwater interface and the response of this interface to tidal and other forcing factors.

Acknowledgements PWS thanks the US Geological Survey Coastal and Marine Geology Program for continued financial support and Frank Manheim for extending the field of multi-channel electrical resistivity towards coastal hydrological phenomena. Thanks are also due to two anonymous reviews.

REFERENCES

Boyle, E. A., Edmond, J. M. & Sholkovitz, E. R. (1977) The mechanism of iron removal in estuaries. *Geochim Cosmochim. Acta* **41**, 1313–1324.

Burnett, W. C., Bokuniewicz, H., Huettel, M., Moore, W. S. & Taniguchi, M. (2003) Groundwater and porewater inputs to the coastal zone. *Biogeochemistry* **66**, 3–33.

Burnett, W. C., *et al.* (2006) Quantifying submarine groundwater discharge in the coastal zone via multiple methods. *Sci. Total Environ.* doi:10.1016/j.scitotenv.2006.05.009.

Charette, M. A. & Sholkovitz, E. R. (2006) Trace element cycling in a subterranean estuarine: Part 2. The geochemistry of the pore water. *Geochim Cosmochim. Acta* **70**, 811–826.

Manheim, F. T., Krantz, D. E. & Bratton, J. F. (2004) Studying ground water under DELMARVA coastal bays using electrical resistivity. *Ground Water* **42**, 1052–1068.

Michael, H. A., Mulligan, A. E., & Harvey, C. F. (2005) Seasonal oscillations in water exchange between aquifers and the coastal ocean. *Nature* **436**, 1145–1148.

Moore, W. S. (1996) Large groundwater inputs to coastal waters revealed by ^{226}Ra enrichments. *Nature* **380**, 612–614.

Moore W. S. (1999) The subterranean estuary: a reaction zone of ground water and sea water. *Marine Chemistry* **65**, 111–125.

Robinson, C., Li, L. & Barry, D. A. (2006) Effect of tidal forcing on a subterranean estuary. *Adv. Water Resour.* doi:10.1016/j.advwatres.2006.07.006.

Sholkovitz, E. R. (1976) Flocculation of dissolved organic and inorganic matter during mixing of river water and seawater. *Geochim Cosmochim. Acta* **40**, 831–845.

Sholkovitz, E. R. (1977) The flocculation of dissolved Fe, Mn, Al, Cu, Ni, Co, Cd during the estuarine mixing. *Earth and Planetary Science Lett.* **41**, 77–86.

Swarzenski, P. W. (2007) U/Th series radionuclides as tracers of coastal groundwater. *Chemical Reviews* **107**(2), 663–674. doi: 10.1021/cr0503761.

Swarzenski, P. W., Orem, W. G., McPherson, B. F., Baskaran, M. & Wan, Y. (2006a) Biogeochemical transport in the Loxahatchee River estuary: the role of submarine groundwater discharge. *Marine Chemistry* **101**, 248–265. doi:10.1016/j.marchem.2006.03.007.

Swarzenski, P. W., Burnett W. C., Weinstein, Y., Greenwood, W. J., Herut, B., Peterson, R. & Dimova, N. (2006b) Combined time-series resistivity and geochemical tracer techniques to examine submarine groundwater discharge at Dor Beach Israel. *Geophys. Res. Lett.* **33**, L24405, doi:10.1029/2006GL028282.

Swarzenski, P. W., Reich, C., Kroeger, K. & Baskaran, M. (2007) Ra and Rn isotopes as natural tracers of submarine groundwater discharge in Tampa Bay, FL. *Marine Chemistry* **104**, 69–84. (Special Issue, *Biogeochemical Cycles in Tampa Bay, Florida*, ed. by P. W. Swarzenski & M. Baskaran).

3 CHEMICAL APPROACHES

Remaining uncertainties in the use of Rn-222 as a quantitative tracer of submarine groundwater discharge

WILLIAM C. BURNETT[1], ISAAC R. SANTOS[1], YISHAI WEINSTEIN[2], PETER W. SWARZENSKI[3] & BARAK HERUT[4]

1 *Department of Oceanography, Florida State University, Tallahassee, Florida 32306, USA*
wburnett@mailer.fsu.edu

2 *Bar-Ilan University, Ramat-Gan, 52900, Israel*

3 *US Geological Survey, St Petersburg, Florida 33701, USA*

4 *Israel Oceanographic and Limnological Research, Haifa, Israel*

Abstract Research performed in many locations over the past decade has shown that radon is an effective tracer for quantifying submarine groundwater discharge (SGD). The technique works because both fresh and saline groundwaters acquire radon from the subterranean environment and display activities that are typically orders of magnitude greater than those found in coastal seawaters. However, some uncertainties and unanswered problems remain. We focus here on three components of the mass balance, each of which has some unresolved issues: (1) End-member radon – what to do if groundwater Rn measurements are highly variable? (2) Atmospheric evasion – do the standard gas exchange equations work under high-energy coastal mixing scenarios? And (3) "mixing" losses – are there other significant radon losses (e.g. recharge of coastal waters into the aquifer) besides those attributed to mixing with lower-activity waters offshore? We address these issues using data sets collected from several different types of coastal environment.

Key words radon; submarine groundwater discharge; mass balance model

INTRODUCTION

The discharge of fresh and saline groundwater into the coastal zone is now recognized as an important pathway between land and sea for nutrients and other dissolved species (Slomp & Van Cappellen, 2004). Quantifying the amount of this flow is difficult as it typically occurs as a low unit flux process but influencing a large area. One of the more successful assessment techniques has been through the use of natural isotopic tracers such as radon and radium isotopes (Moore, 1996; Burnett *et al.*, 2006). The radon approach works because groundwaters often have ^{222}Rn activities orders of magnitude greater than coastal seawaters, radon is conservative, having a half-life on the same order as many coastal processes, and is relatively easy to measure. In addition, improvements in automated monitoring systems (e.g. Dulaiova *et al.*, 2005) have made continuous measurements of radon at environmental activities a reality.

Estimating groundwater discharges via radon is based on a mass balance approach. Inventories are measured, either as a snapshot or continuously over time (usually over at least one tidal cycle), and these inventories are converted to input fluxes after making allowances for losses due to decay, atmospheric evasion, and net coastal

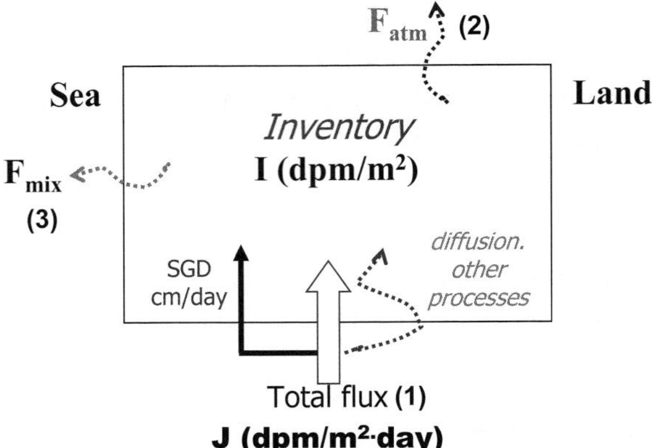

Fig. 1 Diagrammatic view of the radon mass balance with designations of the three fluxes discussed here (arrows). The inventory is measured over time and assessments are made for the atmospheric (F_{atm}) and mixing (F_{mix}) losses, allowing the total flux (J) to be calculated (Burnett & Dulaiova, 2003).

"mixing" terms (Fig. 1). Although changing radon concentrations in coastal waters could be in response to a number of processes (sediment re-suspension, long-shore currents, etc.), advective transport of radon-rich groundwater (pore water) through sediment is often the dominant process. Thus, if one can measure or estimate the radon concentration in these advecting fluids, the ^{222}Rn fluxes may be easily converted to water fluxes.

We examine here the problems and implications of assessing the various portions of the radon mass balance. Namely, we will discuss the issues concerning estimates of: (1) the "groundwater end-member" ^{222}Rn activity; (2) atmospheric evasion losses; and (3) mixing losses of radon – all essential components of the mass balance model.

GROUNDWATER END-MEMBER

Perhaps the most serious issue with the radon mass balance approach is the selection of the appropriate end-member radon activity. The mass balance is constructed in terms of radon fluxes, so in order to convert to a water flux (e.g. $m^3\ m^{-2}\ h^{-1}$), we divide the calculated total radon flux (e.g. dpm $m^{-2}\ h^{-1}$) into the reservoir by the estimated radon-in-water activity (e.g. dpm m^{-3}).

What is the best approach to estimate the average radon in the fluids advecting into the coastal waters? One can rely on measurements made from monitoring wells on shore, piezometers in the coastal zone, fluids sampled from seepage meters offshore, or perform "sediment equilibration experiments" to assess the equilibrium activity of radon in pore solutions (Cable *et al.*, 1996). In areas where seepage through sandy sediments is the main mode of discharge, equilibrating seawater with sediments from the study site should be an excellent manner to estimate the radon activity of advecting fluids. With flow rates typically on the order of a few centimetres per day, there should be ample time for the fluids to equilibrate with the radium in the solid phases of the sediment. We have found that in some cases the ^{222}Rn data from shallow wells matches the sediment equilibration results very well (e.g. Florida Bay; Corbett *et al.*, 2000). In other cases, there may be large discrepancies between these different approaches. For example, an investigation in a harbour in Sicily suggested that much of the water being discharged was from a mixture of shallow low-radon water with much higher activity water from a deeper aquifer (Burnett & Dulaiova, 2006). A mixing model based on radium isotopes allowed us to discern the relative fraction of each component (Moore, 2006).

During a recent investigation of an embayment on the Mediterranean coast of Israel (Dor Beach), measurements were made in several piezometers and shallow wells established in the thin sandy surficial aquifer and the deeper fossilized sandstone aquifer (called "Kurkar") as well as a few seepage meters offshore (Weinstein *et al.*, 2007). The seepage meter data fits on a good mixing line between the Kurkar data and that of the coastal seawater (Fig. 2(a)). This provides evidence that the majority of the groundwater discharge into the embayment is derived from mixtures of water from this deeper aquifer with seawater, and only limited amounts from the sand aquifer.

On the other hand, data from the Florida State University Marine Laboratory (FSUML) on the Gulf of Mexico show that radon in seepage meter chambers does not fall on a good mixing line with many of the monitoring wells sampled on shore (Fig. 2(b)). There is one good radon-salinity mixing line from well B4 through the seepage meter results to the mean value of the coastal seawater (r^2 = 0.81). Sediment equilibration experiments on samples from this site indicated a radon-in-water end-member of 129 dpm/L (n = 6) that overlaps the seepage meter results shown in Fig. 2 (Lambert & Burnett, 2003). However, if one used either well B1 or B5, the end-member activity would have been overestimated by more than an order of magnitude.

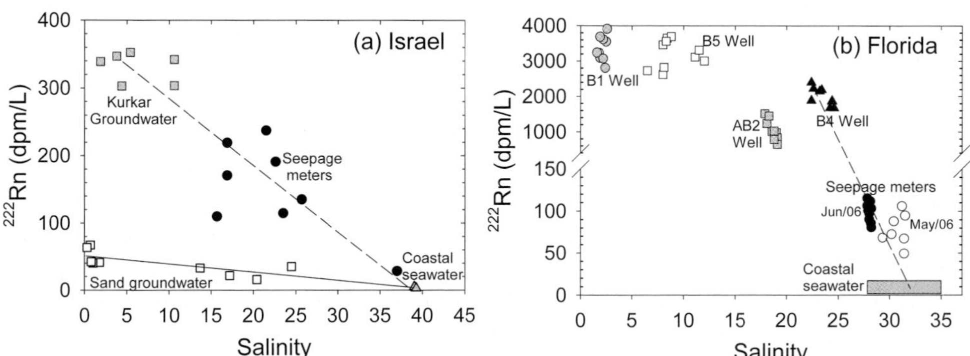

Fig. 2 (a) ^{222}Rn versus salinity from Dor Beach, Israel (Weinstein *et al.*, 2007); and (b) ^{222}Rn *vs* salinity for a site near FSUML, Gulf of Mexico (unpublished). Note that the break in scale in (b) distorts the relationships somewhat – well "B4" fits best on a mixing line with the mean value of coastal seawater and samples collected from seepage meters.

ATMOSPHERIC EVASION

In order to calculate a mass balance, we should account for the loss of radon to the atmosphere. Radon flux (F_{atm}) across the air–sea boundary is governed by molecular diffusion produced by the concentration gradient across the air–water interface and turbulent transfer, and can be calculated from:

$$F_{atm} = k(C_w - \alpha C_{atm}) \quad (1)$$

where C_w and C_{atm} are the radon concentrations in water and air, respectively; α is Ostwald's solubility coefficient; and k is the gas transfer coefficient or "piston velocity". This coefficient is a function of kinematic viscosity, molecular diffusion, and turbulence. The evaluation of k is often based on empirical relationships observed in different environments for different gases. The piston velocity depends on the Schmidt number (Sc, the ratio of kinematic viscosity and molecular diffusion) which offers a way to convert the coefficients from one gas to another. The Schmidt number of each gas is a function of temperature and salinity with k being proportional to $Sc^{-2/3}$ for a smooth liquid and proportional to $Sc^{-1/2}$ for a rough surface (Wanninkhof, 1992).

Based on a number of field studies, empirical equations that relate k to wind speed as a source of turbulence have been proposed. We have elected to use an equation presented in MacIntyre *et al.* (1995) where k represents the piston velocity for a given wind speed normalized to the Schmidt number for CO_2:

$$k(600) = 0.45 u_{10}^{\ 1.6} (Sc/600)^{-2/3} \quad (2)$$

where u_{10} is the wind speed at 10 m height above the water surface and Sc for the dissolved gas of interest is divided by 600 to normalize k to CO_2 at 20°C in freshwater. Turner *et al.* (1996) showed in calculating the gas transfer coefficient for DMS as a function of wind speed, that the (Sc/600) term in equation (2) should be raised to the power of –2/3 for $u_{10} \leq 3.6$ m/s and –1/2 for $u_{10} > 3.6$ m/s. Happell (1995) suggested that k is independent of wind speed below 1.6 m/s.

We were able to test this theoretical approach experimentally by sampling ^{222}Rn and ^{224}Ra on a transect away from the Chao Phraya River estuary (Gulf of Thailand) that has elevated radon and radium activities due to inputs by groundwater discharge and desorption processes (Dulaiova & Burnett, 2006). Since ^{224}Ra stays dissolved in seawater until it decays or it is transported away by mixing but radon also escapes to the atmosphere, we can assess the difference in the $^{222}Rn/^{224}Ra$ ratio as a function of radon evasion.

The isotope distributions were plotted against the corresponding "radium ages" (Moore, 2000a) in the estuary, and fitted by exponential regressions (Fig. 3(a)). As expected, in each case radon has a steeper profile with a more negative slope than ^{224}Ra. The slope of each equation at different points along the transect is equivalent to the rate of loss of the isotope from the water column over time (dpm m^{-3} day^{-1}) and may be calculated as the first derivatives of the trend lines. The difference in the slopes of the radon and radium curves at corresponding points will provide the estimates of radon loss by atmospheric evasion (dpm m^{-3} day^{-1}). The difference of the ^{224}Ra–^{222}Rn slopes at each radium sampling point is than multiplied by the water depth (there was no stratification when these samples were collected) to derive an estimated radon flux from the water column (dpm m^{-2} h^{-1}). The result is an exponential function with the apparent radium age as a variable. Because we know the relationship between distance and the radium age we can also plot atmospheric evasion against distance (Fig. 3(b)). The resulting radon fluxes from these estimates were 10 to 200 dpm m^{-2} h^{-1}. The higher

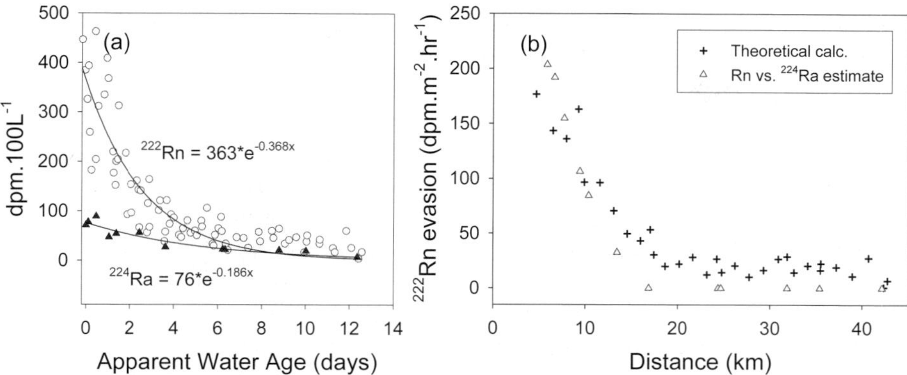

Fig. 3 (a) Radon and ^{224}Ra transects away from the Chao Phraya River mouth (Gulf of Thailand). Activities of ^{222}Rn (circles) and ^{224}Ra (triangles) *vs* apparent water age as calculated from the $^{224}Ra/^{223}Ra$ ratios (Moore, 2000a). Since the mixing processes and half-lives of ^{222}Rn and ^{224}Ra are nearly the same, the difference in slopes is due to radon evasion. (b) Atmospheric evasion based on the experimental data (triangles) and calculated from the gas exchange equations (crosses).

fluxes were near the river mouth (higher activities) and they decrease as the water radon concentrations drop offshore. While this experimental approach becomes imprecise where radon and radium concentrations get close to their supported values offshore, the agreement between the theoretical and experimental values are generally excellent.

We also had an occasion to experimentally test the gas exchange equations during a period of alternating high winds (~10 m/s) during the afternoons and very calm (~0 m/s) winds in the evening in a protected small boat harbour in Donnalucata, Sicily (Burnett & Dulaiova, 2006). In that case, the calculated fluxes during the high wind periods corresponded very closely to the observed change in inventories. The protected nature of the harbour and its very low tides may have simplified the system sufficiently that these estimates were better than would be expected in more "open" environments.

While the gas exchange equations do appear to work reasonably well in many conditions, it is likely that they do not predict the correct evasion rates under more extreme circumstances. For example, storms in most coastal areas will be associated with considerable turbulence that is not predictable by the standard equations. Breaking waves, bubble formation and bursting, tidal surges, etc., would likely all contribute to gas exchange in complex ways.

We show two examples of recent experiments that captured the change in radon inventories during storm events (Fig. 4). The first example, from Dor Beach (Israel) shows a very large drop in radon inventories during the storm – in some cases leaving no detectable ^{222}Rn in the water column. While the wind speeds during this storm were not extremely high with velocities generally less than 10 m/s, there was an increase in water level (~29 cm rise in the high tide with respect to the pre-storm conditions). The storm surge running up on the beach likely contributed to unaccounted-for radon loss. Perhaps more importantly, the trend in maximum wave height (H_{max}, the maximum height of individual waves relative to mean sea level over 1-hour observations) coincides almost exactly with the depletion in radon inventory. Assuming an average depth in the bay of 1.0–1.5 m, we expect to have breaking waves at 0.6–0.8 times

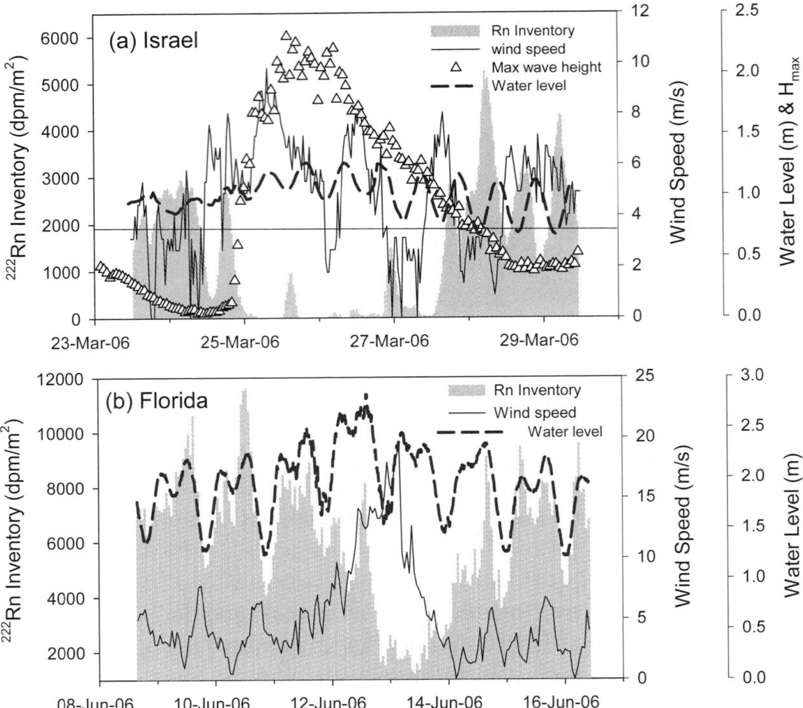

Fig. 4 Radon inventories, wind speeds, and water levels over time from: (a) Dor Beach Israel; and (b) FSUML, Gulf of Mexico. Both examples display dramatic reductions in ^{222}Rn inventories resulting from storm events. Maximum wave heights (H_{max}) are also shown for Dor. Note that the drop in radon inventories corresponds closely to the period when H_{max} exceeds ~0.7 m height (indicated by horizontal line) in (a).

water depth or about when $H_{max} \geq 0.7$ m. Waves would be especially effective in an area surrounded by rocks as Dor, as this would produce sea spray that would enhance gas exchange.

The gas exchange equations would obviously break down during periods of zero radon inventories as the calculation is based on a concentration gradient between the water and air phases. Based on pre-storm conditions, we estimate that it would have been necessary to have prolonged atmospheric evasion rates of approx. 500 dpm m^{-2} h^{-1} to deplete the inventory, yet the gas exchange equations only predicted an average of 24 dpm m^{-2} h^{-1} and a maximum of 270 dpm m^{-2} h^{-1}during the storm period.

In the case of Florida (Fig. 4(b)), the storm (tropical storm "Alberto") had much higher winds (up to ~20 m/s) and a significant storm surge (~0.5 m maximum) compared to the case in Israel. In this case, the gas exchange equations underestimated the observed drop in inventories by ~40–80%. The underestimate in this case may simply be due to the losses via "mixing" with lower activity waters offshore, i.e. we cannot independently test the theoretical approach for atmospheric loss as we do not have an independent estimate for the mixing losses. We do note that the average atmospheric evasion fluxes during the storm (160 dpm m^{-2} h^{-1}) are essentially the same as those estimated during pre-storm conditions (150 dpm m^{-2} h^{-1}). Apparently, the large drop in ^{222}Rn inventories compensated for the increased flux that would be expected at higher wind speeds.

MIXING LOSSES

We evaluate what we have called "net" radon fluxes by taking the change in inventories over time after correcting for atmospheric evasion and normalizing to a common tidal height. Invariably, we observe both positive and negative fluxes and these are often systematically related to the tidal record. We interpret the positive fluxes as due to input of radon-rich groundwater while the negative fluxes are the result of mixing with lower activity waters offshore. Note in the plot from an experiment on Shelter Island, New York (Fig. 5) that the negative fluxes are mostly associated with the incoming tide while the positive fluxes begin showing up on the outgoing tide and reach a maximum at low tide. If we assume that groundwater flow is low or absent during the rising tides, we can estimate the minimum flux due to mixing based on the most negative values of these fluxes (indicated in Fig. 5 by the dashed line). Of course, it is certainly possible that flow continues even at high tide and so these evaluations would be lower estimates. In the case of Shelter Island, seepage meter measurements support the low to no-flow assumption (Paulsen *et al.*, 2001). While the record is not always so clear, this illustrates a fairly typical example of how the mixing term can be handled.

One can also assess mixing losses of radon by application of the mixing model presented by Moore (2000b) using short-lived radium isotopes as tracers. One can calculate a mixing coefficient, k_h, which would then be multiplied by the mixed layer thickness and the linear radon gradient obtained along the same transect. The final step is to then convert this offshore flux to a seabed flux that is used for the mass balance model. This is done by multiplying the offshore flux by the ratio of the cross-sectional area (layer thickness by width) by the estimated seepage area (seabed area). We have had occasion to use this approach during four experiments and the results have overlapped the radon-based estimates each time (Table 1). We should caution, however, that the calculation may be quite sensitive to some parameters. For example,

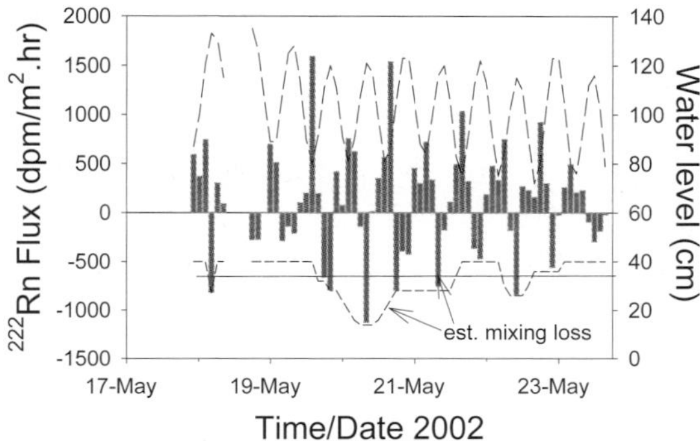

Fig. 5 Calculated "net" radon fluxes (bars), water level (top dashed curve), and mixing losses estimated from inspection of the radon fluxes (dashed line) and via use of short-lived radium isotopes (straight horizontal line; Dulaiova *et al.*, 2006).

the Ra isotope-based value (90 dpm m^{-2} h^{-1}) calculated for the Kona experiment assumed that the mixed layer thickness was only 0.5 m on average. If we had used a 1-m thick layer, the result would have been much higher at 375 dpm m^{-2} h^{-1}. So, while the results are encouraging, the possibility of having higher mixing rates is clearly present.

We have also considered the question of how recharge of coastal waters into the aquifer may represent part of what we have been calling the "mixing" loss. In most cases, we think that this is likely a minor component of the overall balance. Consider the case of Dor Beach (Israel). We use the average measured radon activity in the bay waters during the March 2006 sampling (2200 dpm/m^3) and assume that the recharge rates are comparable to the discharge rates (~10 cm/day). Based on these assumptions, we calculate that the radon loss by this process would only represent ~9 dpm m^{-2} h^{-1} or only about 2% of the estimated mixing loss of 450 dpm m^{-2} h^{-1}. While some other environments or situations would be different, it is likely that recharge does not typically represent an important part of the radon mass balance.

How important are the atmospheric and mixing losses relative to the total mass balance of ^{222}Rn? In most cases, the atmospheric evasion represents a minor loss compared to mixing losses with offshore waters. We compiled data from five different coastal environments showing how the absolute fluxes compare to the measured inventories, mixing losses, and to each other (Table 1). Note that of the examples shown, the evasion rates are highest at FSUML (Florida) and Dor Beach (Israel) while very low in the embayments of Shelter Island (New York) and Ubatuba (Brazil) as well as in Kona (Hawaii) where the setting was also relatively protected. As a percent of the total estimated flux into each system (not shown in Table 1, but the total flux is equal to the sum of the atmospheric plus mixing losses), the atmospheric evasion is highest at ~18–25% of the total at the Florida and Israel sites and ~4–7% of the total flux at the other locations.

Table 1 Radon average inventories, calculated atmospheric fluxes, and mixing estimates based on inspection of the radon fluxes and independent estimates based on short-lived radium isotopes. NA = not available.

Location/Date	Description	^{222}Rn inventories (dpm/m^2)	Atm fluxes (dpm/m^2 h)	Mixing Est from ^{222}Rn (dpm/m^2 h)	Mixing est Ra isotopes (dpm/m^2 h)	Reference
FSUML, Gulf of Mexico/Oct01	Coastal plain; silt/sand overlying limestone	7580	150	690 ± 460	390 ± 140	Burnett & Dulaiova (2003)
Shelter Island, NY/May02	Glacial till aquifer; shallow enclosed embayment	4130	30	670 ± 200	730 ± 260	Dulaiova *et al.* (2006)
Ubatuba, Brazil/Nov03	Fractured rock aquifer; small tropical embayment	7150	40	1060 ± 420	1220 ± 440	Burnett *et al.* (2007)
Kona coast, Hawaii/Feb06	Volcanic aquifer; springs discharging directly to ocean	1510	5	70 ± 70	90 ± 30	Peterson *et al.* (2007)
Dor Beach, Israel/Mar06 before storm	Coastal plain embay- ment; sandy aquifer overlying consol- idated calcaranitic sandstones called "Kurkar"	1820	100	330 ± 120	NA	Weinstein *et al.* (2006)
Dor Beach, Israel/Mar06 after storm		390	100	450 ± 150	NA	Swarzenski *et al.* (2006)

CONCLUSIONS

Radon is a useful tool for studying coastal processes. Because of its very high activity in most groundwaters and the development of improved technologies for measurement, one can now investigate the spatial distribution and temporal dynamics of groundwater discharges as well as make discharge estimates based on a mass balance approach. In most situations, the groundwater end-member concentration represents the largest uncertainty for calculating discharge via a radon mass balance. We recommend that investigations include multiple assessments of this value in order to evaluate the range for any particular situation. End-member estimates may be assessed via sediment equilibration techniques as well as sampling from monitoring wells, piezometers, and seepage meters. Atmospheric losses can be reliably evaluated under "normal" conditions using standard gas exchange equations. While there is some debate concerning the best approach for determining piston velocities (k), our experience has been that the loss via atmospheric evasion is typically less than ~25% of the overall mass balance. Thus, precise determinations, while always desirable, are not critical. Mixing losses are typically more important than atmospheric fluxes but conservative assessments can easily be made by inspection of the radon "net" fluxes. Comparisons of these fluxes to estimates based on short-lived radium isotopes have been favourable.

Acknowledgements We thank the graduate students from FSU (R. Peterson, N. Dimova and B. Mwashote) for their assistance in the field. We thank Eng. Dov Rosen for providing sea state and wind data from Hadera Sea Level Observing Station (IOLR) in Israel. The authors gratefully acknowledge financial support from the Chemical Oceanography Program of the National Science Foundation (OCE04-51379 and OCE05-20723) and the US-Israel Bi-National Science Foundation (BSF2002-381). I.R. Santos acknowledges support from a fellowship from the Brazilian government (CAPES/Fulbright 2150/04-2).

REFERENCES

Burnett, W. C. & Dulaiova, H. (2003) Estimating the dynamics of groundwater input into the coastal zone via continuous radon-222 measurements. *J. Environ. Radioact.* **69**(1-2), 21–35.

Burnett, W. C. & Dulaiova, H. (2006) Radon as a tracer of submarine groundwater discharge into a boat basin in Donnalucata, Sicily. *Cont. Shelf Res.* **26**(7), 862–873.

Burnett, W. C., Aggarwal, P. K., Aureli, A., Bokuniewicz, H., Cable, J. E., Charette, M. A., Kontar, E., Krupa, S., Kulkarni, K. M., Loveless, A., Moore, W. S., Oberdorfer, J. A., Oliveira, J., Ozyurt, I. N., Povinec, P., Privitera, A. M. G., Rajar, R., Ramessur, R. T., Schollten, J., Stieglitz, T., Taniguchi, M. & Turner, J. V. (2006) Quantifying submarine groundwater discharge in the coastal zone via multiple methods. *Sci. Total Environ.* **367**(2-3), 498–543.

Burnett W. C., Peterson R., Moore W. S. & Oliveira J. (2007) Radon and radium isotopes as tracers of submarine groundwater discharge - results from the Ubatuba, Brazil SGD assessment intercomparison. *Estuar. Coast. Shelf Sci.* (submitted).

Cable, J. E., Burnett, W. C., Chanton, J. P. & Weatherly, G. (1996) Modeling groundwater flow into the ocean based on ^{222}Rn. *Earth Planet. Sci. Lett.* **144**, 591–604.

Corbett, D. R., Dillon, K., Burnett, W. C. & Chanton, J. P. (2000) Estimating the groundwater contribution into Florida Bay via natural tracers ^{222}Rn and CH_4. *Limnol. Oceanogr.* **45** 1546–1557.

Dulaiova, H. & Burnett, W. C. (2006) Radon loss across the water-air interface (Gulf of Thailand) estimated experimentally from ^{222}Rn-^{224}Ra. *Geophys. Res. Lett.* **33** L05606, doi:10.1029/2005GL025023.

Dulaiova, H., Burnett, W. C., Chanton, J. P., Moore, W. S., Bokuniewicz, H. J., Charette, M. A. & Sholkovitz, E. (2006) Assessment of groundwater discharges into West Neck Bay, New York, via natural tracers. *Cont. Shelf Res.* **26**(16), 1971–1983.

Dulaiova, H., Peterson, R., Burnett, W. & Lane-Smith, D. (2005) A multi-detector continuous monitor for assessment of ^{222}Rn in the coastal ocean. *J. Radioanal. Nucl. Chem.* **263**(2), 361–365.

Happell, J. D., Chanton, J. P. & Showers, W. J. (1995) Methane transfer across the air-water interface in stagnant wooded swamps of Florida: evaluation of mass-transfer coefficients and isotopic fractionation. *Limnol. Oceanogr.* **40**(2), 290–298.

Lambert, M. & Burnett, W. C. (2003) Submarine groundwater discharge estimates at a Florida coastal site based on continuous radon measurements. *Biogeochem.* **66**, 55–73.

MacIntyre, S. R., Wanninkhof, G. B. & Chanton, J. P. (1995) Trace gas exchange across the air-sea interface in freshwater and coastal marine environments. In: *Biogenic Trace Gases: Measuring Emissions from Soil and Water* (ed. by P. A. Matson & R. C. Harris), 52–97. Blackwell Science, Malden, Massachusetts, USA.

Moore, W. S. (1996) Large groundwater inputs to coastal environments revealed by ^{226}Ra enrichments. *Nature* **380**, 612–614.

Moore, W. S. (2000a) Ages of continental shelf waters determined from ^{223}Ra and ^{224}Ra. *J. Geophys. Res.* **105**(C9), 117–122.

Moore, W. S. (2000b) Determining coastal mixing rates using radium isotopes. *Continental Shelf Res.* **20**, 1995–2007.

Moore, W. S. (2006) Radium isotopes as tracers of submarine groundwater discharge in Sicily. *Cont. Shelf Res.* **26**(7), 852–861.

Paulsen, R., Smith, C. F., O'Rourke, D. & Wong, T. F. (2001) Development and evaluation of an ultrasonic groundwater seepage meter. *Ground Water* **39**(6), 904–911.

Peterson, R. N., Burnett, W. C., Glenn, C. R. & Johnson, A. J. (2007) A box model to quantify groundwater discharge along the Kona coast of Hawaii using natural tracers. In: *A New Focus on Groundwater–Seawater Interactions* (ed. by W. Sandford, C. Langevin, M. Polemio & P. Povinec) (Proc. Symposium HS1001 at IUGG2007, Perugia, July 2007). IAHS Publ. 312. IAHS Press, Wallingford, UK (this volume).

Slomp, C. P. & Van Cappellen, P. (2004) Nutrient inputs to the coastal ocean through submarine groundwater discharge: controls and potential impact. *J. Hydrol.* **295**(1-4), 64–86.

Swarzenski, P. W., Burnett, W. C., Greenwood, W. J., Herut, B., Peterson, R., Dimova, N., Shalem, Y., Yechieli, Y. & Weinstein, Y. (2006) Combined time-series resistivity and geochemical tracer techniques to examine submarine groundwater discharge at Dor Beach, Israel. *Geophys. Res. Lett.* **33**, L24405, doi:10.1029/2006GL028282.

Turner, S. M., Malin, G. Nightingale, P. D. & Liss, P. S. (1996) Seasonal variation of dimethyl sulphide in the North Sea and an assessment of fluxes to the atmosphere. *Marine Chemistry* **54**, 245–262.

Wanninkhof, G. B. (1992) Relationship between wind speed and gas exchange over the ocean. *J. Geophys. Res.* **97**, 7373–7382.

Weinstein, Y., Less, G., Kafri, U. & Herut, B. (2006) Submarine groundwater discharge in the southeastern Mediterranean (Israel), preliminary results. *Radioactivity in the Environment* **8**, 360–372.

Weinstein, Y., Burnett, W. C., Swarzenski, P. W, Shalem, Y., Yechieli, Y. & Herut, B. (2007) The role of coastal aquifer heterogeneity in determining fresh groundwater discharge and seawater recycling: an example from the Carmel coast, Israel. *J. Geophys. Res.* (submitted).

In situ underwater gamma-ray spectrometry as a tool to study groundwater–seawater interactions

PAVEL P. POVINEC
Faculty of Mathematics, Physics and Informatics, Comenius University, SK-84248 Bratislava, Slovakia
povinec@fmph.uniba.sk

Abstract A new technology based on *in situ* underwater gamma-ray spectrometry of radon daughter products in water has been applied for groundwater–seawater interaction studies in the coastal regions of SE Sicily (offshore Donnalucata) and SE Brazil (offshore Ubatuba). The continuous monitoring carried out at the Donnalucata (and Ubatuba) site have revealed an inverse correlation between the ^{222}Rn concentration *versus* the tides and salinity, as ^{222}Rn concentrations in seawater varied from 2 kBq m^{-3} (1 kBq m^{-3}) during high tide to 5 kBq m^{-3} (5 kBq m^{-3}) during low tide. The observed variations in ^{222}Rn concentrations are likely caused by sea level changes, as tidal effects induce variations of hydraulic gradient, which can increase ^{222}Rn concentrations during a falling tide, while during a high tide, ^{222}Rn concentrations decrease.

Key words groundwater–seawater interaction; radon; radon decay products; seawater; submarine groundwater discharge; underwater gamma-spectrometry; SE Brazil; SE Sicily

INTRODUCTION

One of the frequently studied coastal processes is groundwater–seawater interaction (GSI) because of its importance for the protection of coastal zones against contamination from land-based sources, as well as for the management of freshwater resources in coastal areas (Burnett *et al.*, 2006). Several isotope techniques for GSI studies have been developed using stable (^{2}H, ^{18}O, $^{87/86}$Sr, etc.) and radioactive (^{3}H, ^{14}C, Ra isotopes, ^{222}Rn, etc.) isotopes (Burnett *et al.*, 2006; Povinec *et al.*, 2006a). New technologies developed in recent years are based on analysis of radon or its daughter products emitting alpha-rays (Burnett *et al.*, 2001; Burnett & Dulaiova, 2003, 2006; de Oliveira *et al.*, 2003), or gamma-rays (Povinec *et al.*, 2001; Levy-Palomo *et al.*, 2004; Povinec, 2004, 2005; Povinec *et al.*, 2006b,c).

Radon is a conservative radioactive tracer and because its concentration in groundwater is much higher than in seawater, it is an ideal tracer for studying GSI. ^{222}Rn is a decay product of ^{226}Ra (half life = 1.6 ky) in the ^{238}U natural decay chain and its short half life (3.82 d) makes it a suitable tracer for studying dynamic coastal systems. ^{222}Rn daughters are short lived radionuclides such as ^{214}Pb, ^{214}Bi, etc., which further decay by alpha and beta decays to ^{210}Pb (22.2 y) and ^{210}Po (138 d), and finally to stable ^{206}Pb. In the ^{232}Th decay chain there is another radon isotope, ^{220}Rn (thoron), with a very short half life (55.6 s). While ^{228}Ra (^{228}Ac) has been (as a part of the radium quartet, together with ^{226}Ra, ^{223}Ra and ^{224}Ra) used very often as a tracer of coastal processes (Moore, 2006), ^{220}Rn is still waiting for its applications in oceanography. Especially in coastal areas rich in thorium rocks, such as observed along the south-eastern Brazilian coast, ^{220}Rn may be a useful tracer of rapid coastal processes.

A coordinated research project (CRP) on "Nuclear and Isotopic Techniques for the Characterization of Submarine Groundwater Discharge (SGD) in Coastal Zones" has been jointly organized by the IAEA's Marine Environment Laboratories (IAEA-MEL, Monaco) and the Isotope Hydrology Section (Vienna), with the aim to develop new isotope techniques for studying SGD. The CRP has been carried out in cooperation with UNESCO's Intergovernmental Oceanographic Commission (IOC), the International Hydrological Programme (IHP), and several laboratories (Povinec *et al.*, 2006a). In the framework of the CRP, two expeditions were carried out to the Ionian Sea (offshore Sicily), and one to offshore Ubatuba (Sao Paulo region, Brazil). It has been a great challenge to investigate SGD using underwater gamma-ray spectrometry in such different geological and hydrological environments.

METHODS

An underwater gamma-ray spectrometer (consisting of a 5 cm diameter and 15 cm long NaI(Tl) scintillation detector) previously developed for seabed mapping and stationary monitoring of radionuclides in seawater (Osvath & Povinec, 2001) was used in the GSI studies. Additional sensors for monitoring of temperature, water depth and wave impacts were located in front of the NaI(Tl) detector. The detector unit was connected via a 70 m long, double armoured steel coaxial cable to a PC with processing electronics and a multichannel analyser. The PC and an auxiliary low voltage power supply were located on a ship, or in a car when operating close to the coast.

The corresponding ^{214}Bi peaks (representing a decay product of ^{222}Rn) used in spectra evaluations were either at 609, 1120 or 1765 keV, depending on background conditions. The data acquisition system evaluates gamma-ray spectra every minute. Later, the obtained spectra are integrated to one hour intervals and the activity concentration of ^{214}Bi (and ^{222}Rn after calibration) in seawater is calculated. The system is fully automatic and can operate without any surveillance. The mean input flux of ^{222}Rn from ^{226}Ra present in sediments was estimated from measurement of ^{226}Ra activities of sediment samples. A detail description of the system and its calibration can be found in the papers of Povinec *et al.* (2006a,b).

RESULTS AND DISCUSSION

Study area offshore SE Sicily

In situ underwater gamma-ray spectrometry measurements were carried out from 16 to 25 March 2002 in the Donnalucata boat basin, where several sites, situated close to manual and automatic seepage meter posts, were occupied (Povinec *et al.*, 2006b). Seepage rates of up to ~30 cm day^{-1} were reported by Taniguchi *et al.* (2006).

Time series of ^{222}Rn in seawater, salinity and tide were recorded at a site close to the coast. Results presented in Fig. 1 document that after the maximum tide the ^{222}Rn activity concentration of seawater was at minimum (down to 2.3 kBq m^{-3}), and after

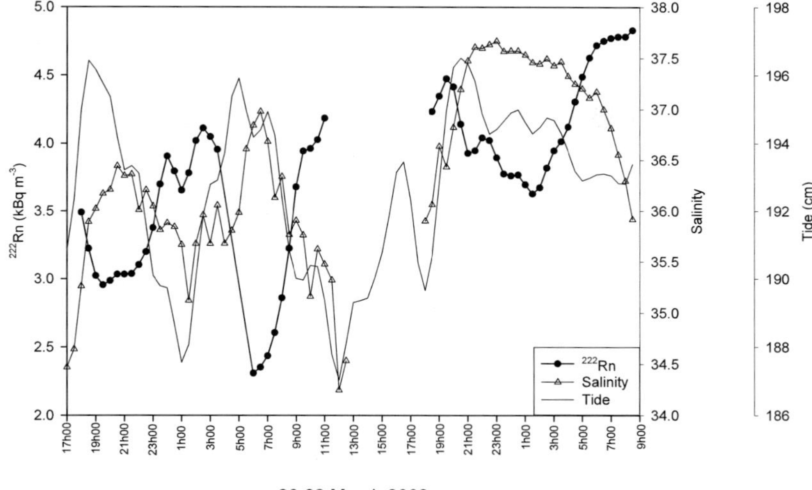

Fig. 1 Time series of radon concentration in seawater in the Donnalucata boat basin (Sicily) *vs* salinity and tide.

the minimum tide the ^{222}Rn activity concentration was at maximum (up to 4.8 kBq m^{-3}) with a delay of about one hour. A shift of approximately two hours was observed between the maximum tide and the salinity maximum.

Study area offshore SE Brazil

Time series of ^{222}Rn activity concentration in seawater, salinity and tide recorded from 22 to 26 November 2003 in Flamengo Bay (Ubatuba area) are shown in Fig. 2. The ^{222}Rn activity concentration in seawater varied between 1.0 and 5.2 kBq m^{-3}, while the tide varied between 4.4 and 5.6 m. The usual inverse relationship between the ^{222}Rn concentration in seawater and tide/salinity was not observed during 22 November, despite large variations in water levels. The observed changes in salinity during this time were, however, also smaller than during 25 and 26 November. The inverse relationship between the ^{222}Rn activity concentration in seawater and tide/salinity was, however, established from 23 to 25 November, when a few hours shift between the tide minimum/maximum and the ^{222}Rn maximum/minimum activity concentration was again observed.

Comparison of Sicilian and Brazilian results

In contrast to the Sicililian sites, which represent classic karstic terrain with low background radioactivity (^{238}U ~10 Bq kg^{-1} and ^{232}Th ~5 Bq kg^{-1}), the Brazilian site is in a tropical coastal area characterised by granite rocks where the concentrations of ^{238}U and ^{232}Th in the rock samples were higher by a factor of 5 and 10, respectively.

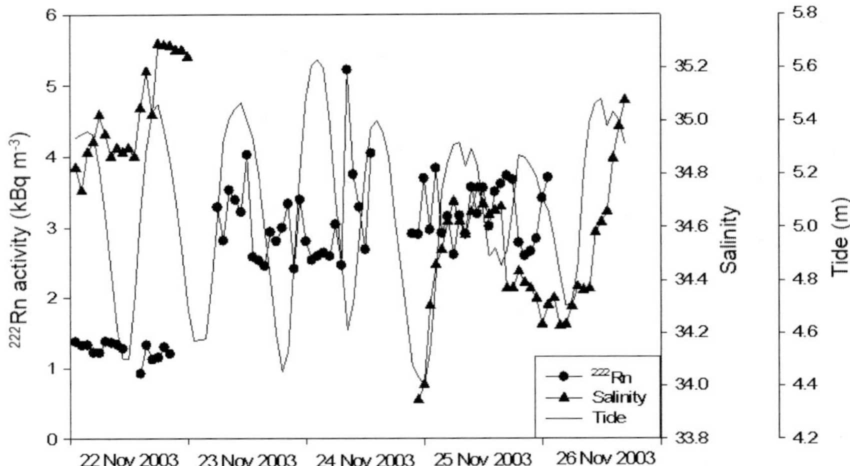

Fig. 2 Time series of radon concentration in seawater in Flamengo Bay (Brazil) *vs* salinity and tide.

The results obtained for Sicilian and Brazilian waters confirm the inverse relationship between tide and the ^{222}Rn concentration. During a falling tide, the observed ^{222}Rn concentration increases, while during a high tide the ^{222}Rn concentration decreases. While the ^{222}Rn concentration in Sicilian waters followed the tide with a delay of only one hour, in Brazilian waters a delay of several hours was observed. This is likely caused by the different oceanographic and hydrogeological conditions at both sites. The Sicilian coast is characterised by carbonate rocks with cracks, which facilitate groundwater transport to the sea, while along the Brazilian coast, granite rocks have lower transport capabilities. In spite of different geological/ hydrological settings, the ^{222}Rn activity concentrations in seawater at Sicilian and Brazilian sites were very similar, between 2.3 and 4.8 kBq^{-3}, and 1.0 and 5.2 kBq m^{-3}, respectively. Due to a factor of 5 higher ^{238}U concentration measured in granite than in carbonate rocks, higher ^{222}Rn activity concentrations along the Brazilian coast could have been expected. However, in Flamengo Bay, similar to that observed in the Donnalucata boat basin, the SGD may be represented by a mixture of re-circulated groundwater and seawater, having a lower ^{222}Rn concentration.

The inverse relationship between the ^{222}Rn activity concentration and tide in the Donnalucata boat basin was also reported by Burnett & Dulaiova (2006), who analysed ^{222}Rn by alpha-ray spectrometry of its daughter products. The temporal changes in ^{222}Rn concentration measured on 22 March 2002 (at a different point in the Donnalucata boat basin) were from 1.6 to 3.0 kBq m^{-3}, comparable with results presented in Fig. 1. They showed that the observed variations in ^{222}Rn activity concentrations can be related to SGD fluxes, and thus can be used for characterisation of SGD.

CONCLUSIONS

In situ underwater gamma-ray spectrometry measurements carried out during two expeditions in coastal waters offshore SE Sicily and SE Brazil showed the ability of the method to monitor SGD and to study its temporal and spatial variations. This new method of SGD investigations represents a robust technique that can be applied for long-term, continuous monitoring of radon in seawater and/or groundwater.

Time series measurements of ^{222}Rn activity concentrations generally confirmed an inverse correlation between the ^{222}Rn activity concentration and tide/salinity, caused by sea level variations as tide effects induce variations of hydraulic gradients, which increase ^{222}Rn concentrations during decreasing sea level, and opposite, during high tides ^{222}Rn activity concentrations are decreasing.

Such large changes in SGD, observed in a relatively small area, document again why the isotopic characterisation of SGD is important for estimation of real groundwater fluxes to the sea.

Acknowledgements The author thanks J.-F. Comanducci and I. Levy-Palomo of IAEA-MEL for assistance during the deployment and data evaluation from the underwater gamma-ray spectrometer. The members of the IAEA-UNESCO team who participated in the expeditions to Sicily and to Brazil are acknowledged for fruitful collaboration. A. Privitera and the University of Palermo (Italy), and J. de Oliveira and Instituto de Pesquisas Energeticas e Nucleares at Sao Paulo (Brazil), are acknowledged for logistical support during the field-work on SGD.

REFERENCES

Burnett, W. C. & Dulaiova, H. (2003) Estimating the dynamics of groundwater input into the coastal zone via continuous radon-222 measurements. *J. Environ. Radioactivity* **69**, 21–35.

Burnett, W. C. & Dulaiova, H. (2006) Radon as a tracer of submarine groundwater discharge into a boat basin in Donnalucata, Sicily. *Continental Shelf Research* **26**, 862–873.

Burnett, W. C., Kim, G. & Lane-Smith, D. (2001) A continuous radon monitor for assessment of radon in coastal ocean waters. *J. Radioanal. Nuclear Chem.* **249**, 167–172.

Burnett, W. C., Aggarwal, P. K., Aureli, A., Bokuniewicz, H., Cable, J. E., Charette, M. A., Kontar, E., Krupa, S., Kulkarni, K. M., Loveless, A., Moore, W. S., Oberdorfer, J. A., Oliveira,. J., Ozyurt, N., Povinec, P. P., Privitera, A. M. G., Rajar, R., Ramessur, R. T., Scholten, J., Stieglitz, T., Taniguchi, M. & Turner, J. V. (2006) Quantifying submarine groundwater discharge in the coastal zone via multiple methods. *Sci. Total Environ.* **367**, 498–543.

De Oliveira, J., Burnett, W. C., Mazzilli, B. P., Braga, E. S., Farias, L. A., Christoff, J. & Furtado, V. V. (2003) Reconnaissance of submarine groundwater discharge at Ubatuba coast, Brazil, using ^{222}Rn as a natural tracer. *J. Environ. Radioactivity* **69**, 37–52.

Levy-Palomo, I., Comanducci, J.-F. & Povinec, P. P. (2004) Investigation of submarine groundwater discharge in Sicilian and Brazilian coastal waters using underwater gamma-spectrometer. *Aquatic Forum'2004*, 228–229. IAEA-Cn-118, IAEA, Vienna, Austria.

Moore, W. S. (2006) Radium isotopes as tracers of submarine groundwater discharge in Sicily. *Continental Shelf Research* **26**, 852–861.

Osvath, I. & Povinec, P. P. (2001) Seabed γ-ray spectrometry: applications at IAEA-MEL. *J. Environ. Radioactivity* **53**, 335–349.

Povinec, P. P. (2004) Developments in analytical technologies for marine radionuclide studies. In: *Marine Radioactivity* (ed. by H. D. Livingston), 237–294. Elsevier, Amsterdam, The Netherlands.

Povinec, P. P. (2005) Ultra-sensitive radionuclide spectrometry: radiometrics and mass spectrometry synergy. *J. Radioanal. Nuclear Chem.* **263**, 413–417.

Povinec P. P, Osvath I., Mulsow S. & Comanducci J.-F. (2001) La surveillance *in situ* de la radioactivite marine au large de Monaco. In: *Rapport du 36^e Congres de la CIESM*, 36, 155. CIESM, Monaco.

Povinec, P. P., Aggarwal, P. K. Aureli, A., Burnett, W. C., Kontar, E. A., Kulkarni, K. M., Moore, W. S., Rajar, R., Taniguchi, M., Comanducci, J.-F., Cusimano, G., Dulaiova, H., Gatto, L., Groening, M., Hauser, S., Levy- Palomo, I., Oregioni, B., Ozorovich, Y. R., Privitera, A. M. G. & Schiavo, M. A. (2006a) Characterisation of submarine groundwater discharge offshore south-eastern Sicily. *J. Environ. Radioactivity* **89**, 81–101.

Povinec, P. P., Comanducci, J. F., Levy-Palomo, I. & Oregioni, B. (2006b) Monitoring of submarine groundwater discharge along the Donnalucata coast in the south-eastern Sicily using underwater gamma-ray spectrometry. *Continental Shelf Research* **26**, 874–884.

Povinec, P. P., Levy-Palomo, I., Comanducci, J.-F., de Oliveira, J., Oregioni, B. & Privitera, A. M. G. (2006c) Submarine groundwater discharge investigations in Sicilian and Brazilian coastal waters using an underwater gamma-ray spectrometer. In: *Radionuclides in the Environment* (ed. by P. P. Povinec & J. A. Sanchez-Cabeza), 373–381. Elsevier, Amsterdam, The Netherlands.

Taniguchi, M., Burnett, W. C., Dulaiova, H., Kontar, E. A., Povinec, P. P. & Moore, W. S. (2006) Submarine groundwater discharge measured by seepage meters in Sicilian coastal waters. *Continental Shelf Research* **26**, 835–842.

Temporal variability of submarine groundwater discharge: assessments via radon and seep meters, the southern Carmel Coast, Israel

YISHAI WEINSTEIN[1], **YEHUDA SHALEM**[1,2], **WILLIAM C. BURNETT**[3], **PETER W. SWARZENSKI**[4] & **BARAK HERUT**[2]

1 *Department of Geography and Environment, Bar-Ilan University, Ramat-Gan 52900, Israel*
weinsty@mail.biu.ac.il

2 *Israel Oceanographic and Limnological Research, Haifa 31080, Israel*

3 *Department of Oceanography, Florida State University, Tallahassee, Florida 32306, USA*

4 *US Geological Survey, St Petersburg, Florida 33701, USA*

Abstract Seep meter data from Dor Bay, Israel, showed a steady decrease in submarine groundwater discharge (SGD) rates between March and July 2006 (averages of 34, 10.4 and 1.5 cm d^{-1} in March, May and July, respectively), while estimates based on radon time series showed remarkably uniform averages (8 cm d^{-1}). The May seep meter data show a rough positive correlation with sea level, unlike the negative correlation shown by the Rn-calculated rates. Smaller-size meters, deployed in July adjacent to the regular-size ones, showed significantly higher rates (10 cm d^{-1}), which negatively correlated with salinity. It is suggested that the decreased rates documented by the seep meters are the result of an increased shallow seawater recharge in the bay (due to decreasing hydraulic gradients). This is not captured by the radon, since recharging water is radon-poor. The positive correlation of discharge with sea level is due to increased seawater recycling in times of high sea stand.

Key words submarine groundwater discharge; radon; seep meter; seawater recycling; electrical resistivity

INTRODUCTION

Submarine groundwater discharge (SGD) is now well-established as a major process in coastal areas and as an important factor in coastal water mass balances (Moore, 1996). It has been shown that, in most cases, the discharging water is a mixture of saline and fresh groundwater, and it is usually assumed that the saline component is recycled seawater (e.g. Michael *et al*., 2005). The mechanisms, locations and scales involved in this recycling are less clear. Most relevant publications highlight the role of tidal or seasonal forcing (e.g. Li *et al*., 1999; Michael *et al*., 2005; Prieto & Destouni, 2005), and to less extent that of wave-induced recycling (e.g. Li *et al*., 1999; Robinson *et al*., 2006), though its existence is commonly acknowledged (e.g. Taniguchi *et al*., 2006a,b).

The various methodologies used to assess SGD do not always refer to the same component. While hydrological balances usually estimate the discharge of fresh groundwater, radium isotopes are usually used to study recycling of seawater (e.g. Moore, 1996). On the other hand, direct measurements by seep meters, as well as estimates based on ^{222}Rn time series (Burnett & Dulaiova, 2003) are measuring total SGD (fresh + recycled seawater, e.g. Mulligan & Charette, 2006).

In this paper we compare seep meter measurements and radon estimates taken in three campaigns during March–May 2006 in the Dor Bay, northern Israel. We relate the differences between measurements and patterns to discharge–recharge interplay in the bay.

GEOGRAPHICAL AND HYDROGEOLOGICAL BACKGROUND

Dor Bay is located about 30 km south of Haifa, Israel, west of Mount Carmel (inset in Fig. 1). The bay was formed by a submerged calcareous sandstone ridge (locally known as Kurkar), which closes the embayment on its northern and western sides (Fig. 1). Its average dimensions are 100 × 150 m, and its average depth is about 1 m.

The coastal aquifer in the Dor area is composed of two units: the Pleistocene Kurkar sandstone and the overlying Holocene quartz sands (Michelson, 1970; Sivan

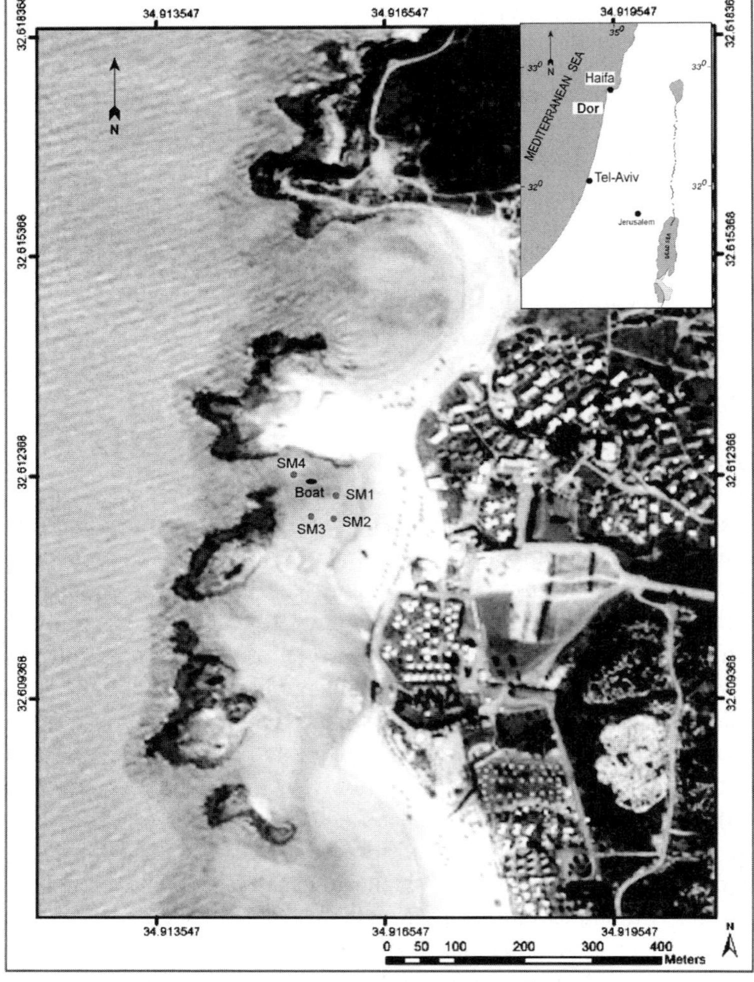

Fig. 1 An air photo of the Dor area, showing Dor Bay and other embayments and the locations of seep meters and the boat, where radon was measured.

et al., 2003). A variably thick clay unit is usually found as a confining layer between the two units (Sivan *et al.*, 2003). In the embayment, the Kurkar is exposed at the seafloor in about one third of the area (Fig. 1), while all the rest is covered by variably-thick loose sand. It is unknown, whether and to what extent the sand in the bay is underlain by clays.

METHODOLOGY

We report here on seep meter measurements taken in Dor Bay during three field campaigns in 2006. We used two different Lee-type seep meters. The first is a dome-shaped plastic meter (hereafter: plastic SM), with a diameter of 80 cm (bottom area of 5.027 cm^2) and 60 cm tall. The second type was smaller, made of metal (hereafter: small SM), 40–60 cm in diameter (1257–2827 cm^2, respectively) and 40 cm tall. All meters follow basic principles of manual seep meters, including a sharp bottom edge that penetrates the sediment to about 10 cm and a threaded pipe nipple at the top to which a 4-L flexible collection bag is attached. The collection bag was pre-filled with 0.5 L water to reduce artifacts due to bag expansion. Measurements started 24 hours after deployment, in order to allow meter equilibration.

Meters were deployed in four sandy sediment sites in Dor Bay (SM1–4, Fig. 1) at distances that varied between 15 and 60 m from the low-tide line. The measurements in March 2006 included just two plastic SMs in SM1 and SM2 and just for a few hours (5 and 2 h, respectively). In May, plastic SMs were deployed in all four sites, and in July two meters (one of each type) were deployed at SM1–3 (Fig. 1). Depths of the bay at deployment sites varied between approx. 1 m (low tide) at SM2, 1.5 m at SM1 and SM3 and approx. 1.7 m at SM4.

RESULTS

The SGD rates measured in Dor Bay are shown in Figs 2–4 and in Table 1. Advection rates measured by the plastic SMs were highest in March 2006, when the few measurements taken indicated averages of 18.1 ± 1.9 and 49.7 ± 4.8 cm d^{-1} in SM1 and SM2, respectively. In May, discharge was significantly lower, with typical advection rates of 0–20 cm d^{-1} and averages of 7.1–12.6 cm d^{-1}, and in July rates were the lowest (Fig. 4), with rates not exceeding 6.4 cm d^{-1}, frequently showing a zero discharge, and averages of 0.2–2.4 cm d^{-1} (Table 1).

Table 1 Average advection rates measured by seep meters and estimated by ^{222}Rn time series in Dor Bay during 2006 (Burnett *et al.*, 2007; Weinstein *et al.*, 2007).

	March	May	July
Large seep meters (cm d^{-1})	33.9 [b]	10.4 ± 5.0 [c]	1.5 ± 1.9
Small seep meters (cm d^{-1})			10.0 ± 7.7
^{222}Rn (calculated)[a] (cm d^{-1})	7.7 ± 6.0	8.4 ± 7.2	8.1 ± 7.2

[a] Calculated from radon time series assuming that radon activities in the groundwater end member is 235 dpm L^{-1} (the weighted average of radon discharging to the Bay, Weinstein *et al.*, 2007).

[b] March average was calculated from the results of two meters that worked for just a few (2–5) hours.

[c] May averages were calculated for the last two days (28–30 May), when discharge was more steady.

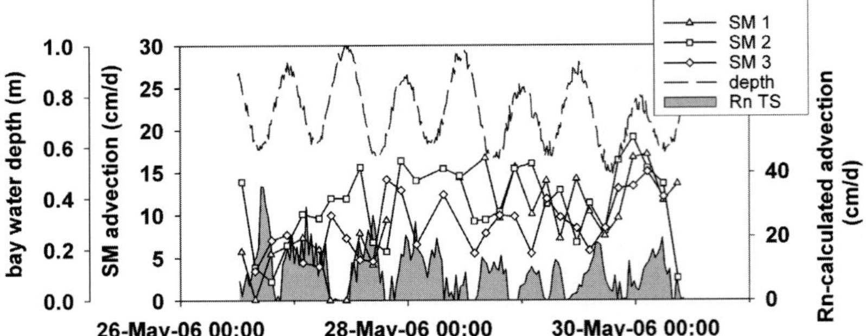

Fig. 2 Seep meter discharge rates measured in May 2006, compared with advection rates estimated by radon time series (Weinstein *et al.*, 2007).

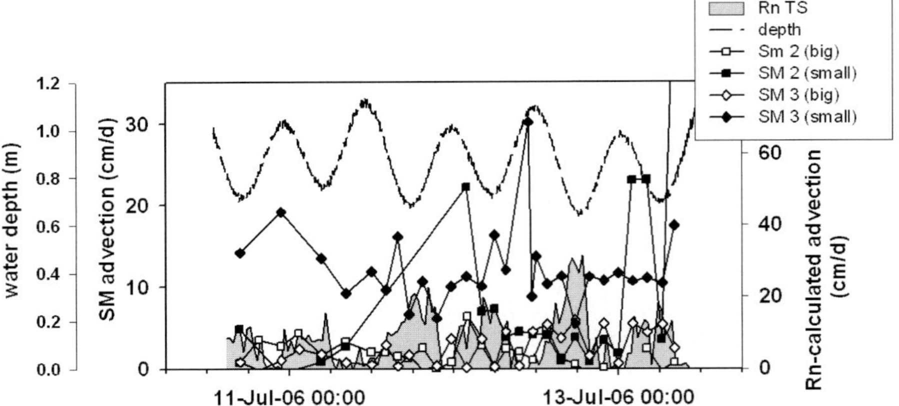

Fig. 3 Seep meter discharge rates (two types of meters, see text) measured in July 2006, compared with advection rates estimated by radon time series (Weinstein *et al.*, 2007).

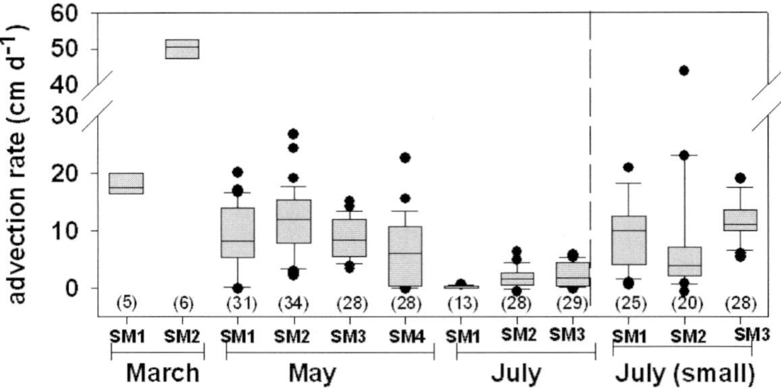

Fig. 4 Box plot of discharge rates measured by the seep meters. July (small) refers to the small SMs (see text). Each box defines the interquartile range; the line inside is the median; the bars are for the 90th and 10th percentiles and the dots for outliers. In parantheses are the numbers of measurements.

Table 2 Average advection rates and discharge measured in July by the two meter types (plas = plastic SM).

	SM1 plas	SM1 small	SM2 plas	SM2 small	SM3 plas	SM3 small
Advection (cm d^{-1})	0.2 ± 0.3	9.7 ± 6.8	1.8 ± 1.7	8.3 ± 11.1	2.4 ± 2.1	12.1 ± 4.7
Discharge (ml m^{-1})	0.8 ± 0.9	19.1 ± 13.3	6.4 ± 5.8	7.2 ± 9.7	8.5 ± 7.2	23.8 ± 9.3

In May, the close-to-shore meters (SM1 and SM2, 15–25 m from the low-tide line) showed the highest rates (averages of 12.6 and 12.1 cm d^{-1}, respectively, compared with 6.8 and 8.8 cm d^{-1} in SM3 and SM4, at 50 and 60 m from shore, respectively). In July, the trend changed and the highest rates were measured in the offshore SM3 site (2.4, compared with 0.2 and 1.8 cm d^{-1} in SM1 and SM2, respectively, Table 2 and Fig. 4).

Advection rates measured in July by the small SMs were significantly higher than those measured simultaneously by the plastic SMs (average rates of 8.3–12.1 cm d^{-1}, Table 2) but very similar to those measured in May by the plastic SMs in the same sites (all-meter average of 10.0 and 10.4 cm d^{-1}, respectively, Fig. 4). A key observation is that even the volumetric discharge was higher through the small SMs than through the plastic meters (by a factor of 2–3 in SM1 and SM3, Table 2), despite the significantly larger area covered by the latter. Advection rate variability was very high, especially in the close-to-shore sites (1–31 and 1–44 cm d^{-1} in SM1 and SM2, respectively, and 6–30 cm d^{-1} in SM3). As in the plastic SMs, the highest rates were measured in the furthest-from-shore meter, SM3 (Table 2).

Rates measured by the plastic SMs showed a tidal pattern during the last two days of the May campaign, which, at least in SM1 and SM2, seemed to correlate positively with Bay water level (Fig. 2). Tidal or any other periodic pattern was not observed in July (Fig. 3).

Salinity was measured in water discharging from the meters after being flushed by advecting water (not less than two days after deployment). Due to the low rates observed in the plastic SMs, salinities in July were measured just in the small SMs. Salinities in the seep meters ranged between 15 and 30. The lowest salinity was observed in SM2, and it increased slightly from March to July (15.2–18.9). However, in July, the lowest average salinity was observed in the furthest-from-shore SM3 (average of 23.4, compared with 24.8 in SM2), coincident with the highest-observed discharge rates during this period (Table 2 and Fig. 4). Water in the closest-to-shore SM1 was the most saline in July (average of 28.3). Unfortunately, due to the large size of the plastic SMs, we do not have a similar follow-up on salinities during the May campaign. Salinities do not show a tidal pattern, but in SM1 and SM3 they do show a general pattern of negative correlation with discharge rates (Fig. 5). In SM2, the correlation is less clear due to a group of samples with low salinities at low discharge rates (circled in Fig. 5).

DISCUSSION

Direct measurements of discharge by seep meters and estimates based on ^{222}Rn time series (Burnett & Dulaiova, 2003) are both documenting total SGD (fresh + recycled

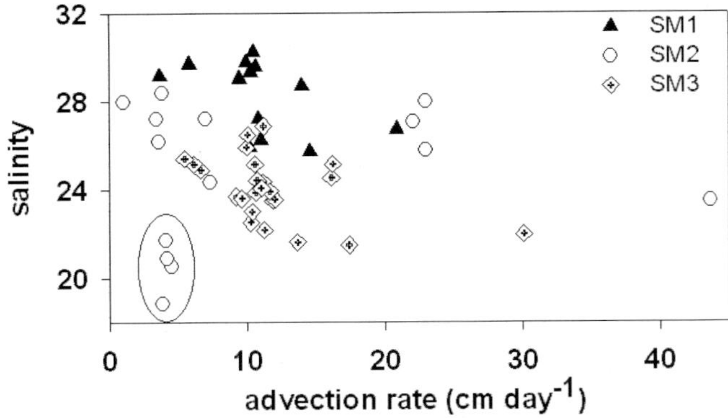

Fig. 5 Salinity *vs* discharge rates measured by the small SMs in July.

seawater, e.g. Mulligan & Charette, 2006). However, unlike the radon mass balance, which yields information on a relatively large area, the seep meters give very local information, and thus often show a patchy pattern with high variability from meter to meter. Nevertheless, often there is a close agreement between the average of meter results and radon mass balance calculations (e.g. Lambert & Burnett, 2003; Mulligan & Charette, 2006), and even if the two methodologies do not quantitatively agree, they tend to show similar patterns (e.g. negative correlation with sea level, Burnett & Dulaiova, 2006; Dulaiova *et al.*, 2006; Taniguchi *et al.*, 2006a). This is not the case in Dor Bay. While calculations based on radon time series found remarkably constant advection rates from the rainy into the dry season (averages of 7.7–8.4 cm d^{-1} in March to July 2006, Table 1 and Figs 2–3; Swarzenski *et al.*, 2006; Burnett *et al.*, 2007; Weinstein *et al.*, 2007), the seep meter data of the plastic SM defines a strong trend of decreasing rates from March to July 2006 (Fig. 4). In March, average rates measured by the meters (though a very small data set) were 2–6 times the average rates calculated from the radon time series, while in July the plastic SM average rates were just 20% of the rate calculated via radon. In May, the two methods were in a good agreement (10.4 and 8.4 cm d^{-1}, Table 1). This suggests that the seep meters and the Rn in Dor documented partly decoupled components of SGD. The positive correlation of seep meter discharge rates with bay water level in May and the absence of tidal patterns in July (in both types of meters), as opposed to the usually very clear negative correlation shown by Rn-calculated rates (e.g. Figs 2–3), also suggest that the two data sets did not document the same discharge process. We note that the accuracy of seep meter measurements has been questioned by several authors (e.g. Shinn *et al.*, 2002). However, erroneous measurements cannot explain the overall pattern of decreasing discharge from March to May. Thus, we prefer to seek a hydrological mechanism.

The radon measurements were taken in the Kurkar area of the northern part of the bay (Fig. 1), while all seep meters were deployed on sand-covered bay bottom. Thus, it could be argued that, while discharge was uniform in the Kurkar areas, it was decreasing in the sandy areas during the dry season (May–July). However, the observation of high discharge rates in all three small SMs placed adjacent to the plastic ones in July implies that discharge did occur in the sand-covered areas, and for some reason was not documented by the larger plastic SMs. Moreover, high radon activities were found in July in all three SMs (115–237 dpm/L, compared to 110–220 dpm/L in March and May; Weinstein *et al*., 2007). Since high radon water is supplied from the underlying Kurkar (Weinstein *et al*., 2007) and since decline of discharge would result in abrupt decrease in radon activity (by dilution and decay), these high activities clearly imply that sand bottom discharge rates in July were similar to those operating in previous months. We suggest that the observations from Dor Bay could be explained by the co-operation of discharge and recharge in the Bay, which is discussed below.

Weinstein *et al*. (2007) showed that radon activities and salinity of the water discharging in the seep meters sit on a mixing line between radon-rich Kurkar groundwater and radon-poor seawater. Weinstein *et al*. suggested that the radon is mainly supplied to the bay by discharge of fresh groundwater from the Kurkar, which, when it discharges through a sandy bay bottom, mixes to various proportions with recycled seawater. Based on this, as well as on hydrological measurements and on multi-electrode resistivity images of onshore and submarine sediments (Swarzenski *et al*., 2006), Weinstein *et al*. (2007) further suggested that the patterns of discharge and seawater recycling in Dor are strongly dependent on the local hydrogeology. While seawater recycling is a major process in the shallow superficial sands, it hardly occurs in the Kurkar due to its relatively high hydraulic heads.

Following the above, it is suggested that the disagreement between Rn-calculated rates and measurement by seep meters are related to the discharge–recharge interplay. We suggest that the abrupt decrease in rates measured by the plastic SMs in July was mainly caused by increased seawater recharge in the bay. While any recharge of seawater into the sediment in the deployment site results in a lower discharge rate measured by the meter, the effect on the Rn inventory (and thus on the calculated advection rate) is negligible, since the recharging bay water is relatively radon-poor. The relatively similar minimum salinities measured in all three campaigns (in SM2, 15–19) and the fact that most salinity measurements were lower than 25 imply that the proportions of recycled seawater in the discharging water did not change much. What could change was the location of seawater recharge, namely: while in May, recharge occurred further away from shore and discharge was the main process in the bay and in seep meters deployment sites, in July the recycling was more local and recharge was more frequent in the deployment sites. This could possibly happen due to a landward shift of the freshwater–saline groundwater interface and its intersection with bay floor (e.g. Michael *et al*., 2005), a result of inland decreasing hydraulic heads in the dry period and of the relatively high sea levels in July (the average level was 12.2 cm higher than in May; IOLR data from the offshore Hadera station).

We note that the seawater recycling we envisage is a shallow process, which at least during March was relevant just to the sand unit (Swarzenski *et al*., 2006), and which operates on a local, high spatial and temporal resolution. This kind of recycling is probably induced by wave setup, which could be enhanced by the very irregular bathymetry of the Dor Bay. This could also be the reason for the large difference in advection rates measured by the two types of meters in July (Table 2 and Fig. 4). The plastic SMs, with relatively large bottom area (approx. 5000 cm^2), possibly documented both recharge and discharge, while the small meters mainly focused on discharge spots.

The rough positive correlation with bay water level in May (Fig. 2), unlike the negative correlation shown by the Rn estimates, could also be the result of increased seawater recycling in the bay during high stands of the sea, which on one hand increased the saline discharge, but on the other hand further diluted the radon in the discharging freshwater. In July, a tidal pattern is not observed, probably because both recharge and discharge occur in the vicinity of the deployment sites. Thus, the possible signal of increased saline discharge during high tide is counterbalanced by an increase in recharge. Unfortunately, we do not have salinity time series from May to support this hypothesis.

The interpretation of low advection rates as actually higher recharge events also explains the negative correlation with salinity observed in the small SMs in July (mainly in SM1 and SM3, Fig. 5). The increased volume of seawater in the sediment due to higher recharge (= low advection rates) caused an increase in salinity, and *vice versa* during times of low recharge (= high advection). The lowest salinities (on average) and highest rates were measured in July in the offshore meter SM3, while the highest salinities were documented in SM1. A similar pattern of lower salinity of SGD in offshore sites was shown by Taniguchi *et al.* (2006b) for the Yatsushiro Sea, Japan. In the Dor Bay, this probably means that recharge was higher at the close-to-shore northeastern corner of the Bay than in its central part (see Fig. 1), which could be the result of stronger wave action close to shore. This is also reflected by the permanent occurrence of saline water (17–25) at the water table in the nearby onshore sands up to at least 10 m from high tide coastline (Weinstein *et al.*, 2007).

The very high discharge rates measured in SM2 in March 2006 (50 cm d^{-1}) were measured two days after a strong winter storm. This storm introduced large volumes of seawater into the onshore sand (Burnett *et al.*, 2007), which could be the source of the high discharge. However, salinity in discharging water was kept low (15) and radon activity was relatively high (110–190 dpm L^{-1}), just slightly lower than the activities measured in May (170–219 dpm L^{-1}, Weinstein *et al.*, 2007), which cast doubt on the feasibility of this scenario. The discharge/recharge regime during winter storms should be studied further.

CONCLUSIONS

We suggest that the observed decrease in discharge rates measured by seep meters in Dor Bay could be caused by an increase in local, shallow seawater recharge in the bay. This process was not evident in radon time series due to the low radon in the recharging water and was apparently "ignored" by small seep meters, which showed high discharge rates. Increased recharge could also explain the positive correlation of SGD with sea level. This suggestion should be further studied by automated meters, preferably with small bottom area.

Acknowledgements We wish to thank Eng. Dov Rosen for providing sea state from Hadera Sea Level Observing Station (IOLR); Y. Gertner (IOLR), R. Peterson (FSU), and J. Coddington and H. Lutzki (HU), for their helpful assistance in the field. This work was funded through a US-Israel BSF grant (no. 2002381).

REFERENCES

Burnett, W. C. & Dulaiova, H. (2003) Estimating the dynamics of groundwater input into the coastal zone via continuous radon-222 measurements. *J. Environ. Radio.* **69**, 21–35.

Burnett, W. C. & Dulaiova, H. (2006) Radon as a tracer of submarine groundwater discharge into a boat basin in Donnalucata, Sicily. *Cont. Shelf Res.* **26**, 862–873.

Burnett, W. C., Santos, I., Weinstein, Y., Swarzenski, P. W. & Herut, B. (2007) Remaining uncertainties in the use of Rn-222 as a quantitative tracer of submarine groundwater discharge. In: *A New Focus on Groundwater–Seawater Interactions* (ed. by W. Sanford, C. Langevin, M. Polemio & P. Povinec), 109–118. IAHS Publ. 312. IAHS Press, Wallingford, UK (*this volume*).

Dulaiova, H., Burnett, W. C., Chanton, J. P., Moore, W. S., Bokuniewicz, H. J., Charette, M. A. & Sholkovitz, E. (2006) Assessment of groundwater discharges into West Neck Bay, New York, via natural tracers. *Cont. Shelf Res.* **26**, 1971–1983.

Lambert, M. J. & Burnett, W. C. (2003) Submarine groundwater discharge estimates at a Florida coastal site based on continuous radon measurement. *Biogeochemistry* **66**, 55–73.

Li, L., Barry, D. A., Stagnitti, F. & Parlange, J. -Y. (1999) Submarine groundwater discharge and associated chemical input to a coastal sea. *Water Resour. Res.* **35**, 3253–3259.

Michael, H. A., Mulligan, A. E. & Harvey, C. F. (2005) Seasonal oscillations in water exchange between aquifers and the coastal ocean. *Nature* **436**, 1145–1148.

Michelson, H. (1970) The geology of the Carmel coast. MS Thesis, The Hebrew University of Jerusalem, Tahal Rep., HG/70/ 025 (in Hebrew).

Moore, W. S. (1996) Large groundwater inputs to coastal waters revealed by ^{226}Ra enrichments. *Nature* **380**, 612–614.

Mulligan, A. E. & Charette, A. C. (2006) Intercomparison of submarine groundwater discharge estimates from a sandy unconfined aquifer. *J. Hydrol.* **327**, 411–425.

Prieto, C. & Destouni, G. (2005) Quantifying hydrological and tidal influences on groundwater discharges into coastal waters. *Water Resour. Res.* **41**, W12427, doi:10.1029/2004WR003920.

Robinson, C., Gibbes, B. & Li, L. (2006) Driving mechanisms for groundwater flow and salt transport in a subterranean estuary. *Geophys. Res. Lett.* **33**, L03402, doi:10.1029/2005GL025247.

Shinn, E. A., Reich, C. D. & Hickey, T. D. (2002), Seepage meters and Bernoulli's revenge. *Estuaries* **25**(1), 126–132.

Sivan, D., Eliyahu, D. & Raban, A. (2003) Late Pleistocene to Holocene wetlands now covered by sand, along the Carmel Coast, Israel, and their relation to human settlement: an example from Dor. *J. Coast. Res.* **20**, 97–110.

Swarzenski, P. W., Burnett, W. C., Greenwood, W. J., Herut, B., Peterson, R., Dimova, N., Shalem, Y., Yechieli, Y. & Weinstein Y. (2006) Combined time-series resistivity and geochemical tracer techniques to examine submarine groundwater discharge at Dor Beach Israel. *Geophys. Res. Lett.* **33**, L24405, doi:10.1029/2006GL028282.

Taniguchi, M., Burnett, W. C., Dulaiova, H., Kontar, E. A., Povinec, P. P. & Moore W. S. (2006a) Submarine groundwater discharge measured by seepage meters in sicilian coastal waters. *Cont. Shelf Res.* **26**, 835–842.

Taniguchi, M., Ishitobi, T., Shimada, J. & Takamoto, N. (2006b) Evaluation of spatial distribution of submarine groundwater discharge. *Geophys. Res. Lett.* **33**, L06605, doi:10.1029/2005GL025288.

Weinstein, Y., Burnett, W. C., Swarzenski, P. W, Shalem, Y., Yechieli, Y., & Herut, B. (2007) The role of coastal aquifer heterogeneity in determining fresh groundwater discharge and seawater recycling: an example from the Carmel coast. *Israel J. Geophys. Res.* (accepted).

Seasonal changes in the radium-226 distribution on the southeastern USA continental shelf: implications for changing submarine groundwater discharge

WILLARD S. MOORE
Dept Geological Science, University of South Carolina, Columbia, USA
moore@geol.sc.edu

Abstract Enrichments of radium isotopes in coastal waters have served as indicators of submarine groundwater discharge (SGD). Because coastal waters exchange with the open ocean on a time scale of weeks to months, seasonal patterns of radium isotope distribution may be used to indicate changes in SGD through the year. Here I report the seasonal distributions of ^{226}Ra measured in surface waters of the continental shelf of southeastern USA. The study area encompassed most of the South Atlantic Bight. Activities of ^{226}Ra were highest off the coast of Georgia. In summer, these high activities extended throughout the study area; but during spring and winter they decreased markedly off the coast of South Carolina. The primary source of excess ^{226}Ra (that is activities in excess of open ocean values) is SGD. Because the activities of ^{226}Ra in SGD vary little with season, the lower excess activities off South Carolina imply lower rates of SGD during the spring and winter.

Key words submarine groundwater discharge; radium; coastal ocean; nutrients

INTRODUCTION

We now recognize that submarine groundwater discharge (SGD) is an important component of the hydrological cycle because it transfers nutrients, metals and carbon to the coastal ocean (Valiela *et al.*, 1990; Paeril, 1997; Cai *et al.*, 2003; Windom *et al.*, 2006). Recent studies have found that SGD provides a major source of nutrients to salt marshes, estuaries and other communities on the continental shelf. For example, it is estimated that the fluxes of nitrogen and phosphorus to the shelf of South Carolina and Georgia, USA, from submarine groundwater discharge exceeded fluxes from local rivers (Krest *et al.*, 2000; Moore *et al.*, 2002). Because nutrient concentrations in coastal groundwater may be several orders of magnitude greater than in surface waters, groundwater input may be a significant factor in the eutrophication of coastal waters.

Material fluxes from the coastal to open ocean are difficult to quantify because small scale temporal and spatial variability make these systems exceedingly complex. Chemical tracers offer promise but few techniques have been developed and tested to study this complex region. Moore (2000a,b) developed new methods based on four radium isotopes, which enable oceanographers to quantify SGD and determine fluxes of dissolved components across the continental shelf. These early studies were in July 1994, primarily along the coast of South Carolina (Moore, 1996). In this paper I investigate the distributions of ^{226}Ra across the continental shelf of the South Atlantic

Bight from Cape Fear, North Carolina, to St Augustine, Florida. The present study includes measurements during September–October 1998, April 1999, and February 2000.

EXPERIMENTAL METHODS

Samples were collected and processed using standard methods (Moore, 1976). Radium isotope measurements followed Moore (1984).

RESULTS

The results are presented in Figs 1–4, with different time periods shown on different figures. The first cruise (PW) was divided into two legs: leg 1, 6–20 September 1998 (Fig. 1), and leg 2, 22 September–3 October 1998 (Fig. 2). Cruise AW was from 9–18 April 1999 (Fig. 3); cruise FW was from 8–17 February 2000 (Fig. 4).

Surface waters in the open Atlantic Ocean contain 7–8 dpm ^{226}Ra/100L with little spatial variation (Broecker *et al.*, 1976; Key *et al.*, 1985). Higher activities measured in coastal waters indicate a local source. These sources include riverine input (both dissolved and desorbed from particles), regeneration and release from marine sediments, erosion of old terrestrial sediments, and submarine groundwater discharge (SGD).

Moore (1987) reported much higher near-shore ^{226}Ra and ^{228}Ra activities for samples collected in February and April compared to September–October. The station closest to shore had activities almost double the values measured at this station during the other cruises. For samples collected in July 1994, Moore (1996) reported ^{226}Ra activities on the South Carolina inner shelf in the range 15–28 dpm/100L. Activities of 10–15 dpm/100L were present between the inner shelf and the shelf break. These activities are similar to values measured in September–October 1982 (Moore, 1987) and in September 1998, leg (Fig. 1). Moore (1996) demonstrated that such high activities could not be supported by riverine or sedimentary inputs; they must result primarily from SGD.

The present study extends the temporal ^{226}Ra data set to autumn, winter and spring and the spatial data from the coast of central North Carolina to the northern Florida coast, about 80 km south of Jacksonville. The pattern during 6–20 September 1998 from Cape Fear to Crescent Beach, Florida, (Fig. 1) was similar to the July distribution along the South Carolina coast reported by Moore (1996) with activities >20 dpm/100L on the inner shelf and >9 dpm/100L over the remainder of the shelf. By late September the pattern changed significantly. The 9 dpm/100L contour was much closer to shore and near-shore activities in Long Bay (between Winyah Bay and Cape Fear) decreased from 13–18 dpm/100L to 7–8 dpm/100L (Fig. 2). These lower values are equal to the average ^{226}Ra activity in the open North Atlantic, thus there was essentially no enrichment of ^{226}Ra in near-shore Long Bay during the late September sampling period. The same was true in April 1999 when activities in the entire northern half of the study area again were not significantly enriched over the open North Atlantic with the exception of a few samples on the inner shelf (Fig. 3). The distribution off southern Georgia and northern Florida during this period were similar to the distribution in

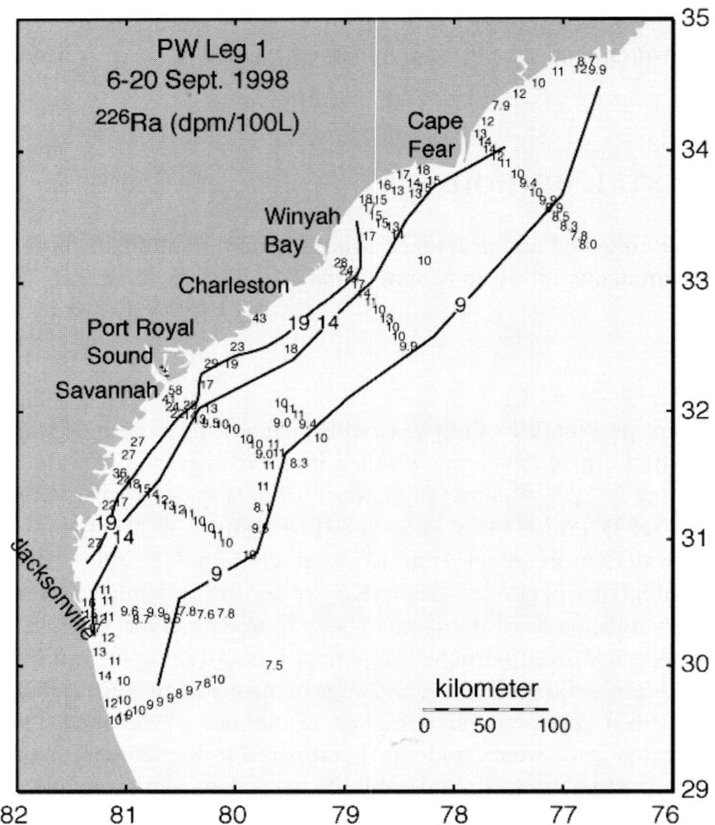

Fig. 1 The seasonal distribution of ^{226}Ra in South Atlantic Bight surface waters during early September was similar to prior studies in the summer with high enrichments throughout the inner shelf.

September–October 1998. Samples from February 2000 followed a pattern similar to April 1999, a slight enrichment in the northern half of the study area and higher enrichments in the southern half (Fig. 4).

DISCUSSION

Spatial distribution

The Georgia Bight between Port Royal Sound and Jacksonville clearly has higher activities of ^{226}Ra compared to Long Bay and Onslow Bay and Crescent Beach. Although this region has higher river runoff than the rest of the SAB, Moore & Shaw (2007) concluded that activities of ^{226}Ra in the estuaries of the major rivers of the SAB could not be explained by input from the rivers. They considered these river mouths to be "marsh-dominated", where chemical fluxes are strongly augmented by interactions with marsh pore waters.

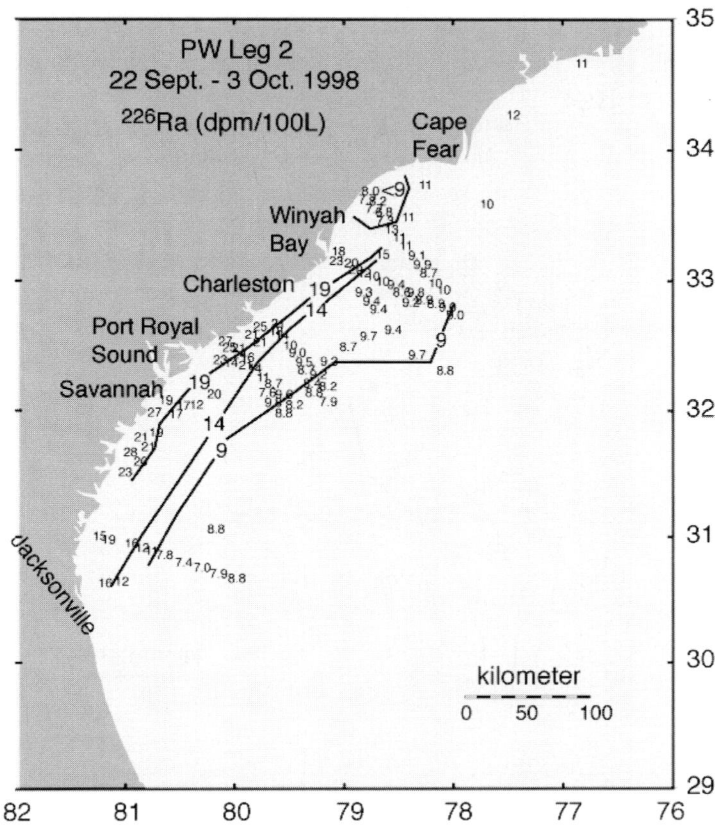

Fig. 2 By late September the high ^{226}Ra activities measured in Long Bay (between Cape Fear and Winyah Bay) had diminished to open ocean values. High activities were present off the coast of Georgia.

The tide range increases from the coast of North Carolina (average ~1 m) to the Georgia Bight, where it reaches an average of ~2 m. This leads to a much greater exchange of salty water across the estuaries and coastal marshes of the Georgia Bight compared to Long Bay, Onslow Bay, and the northern coast of Florida. Studies in these salt marshes (Rama & Moore, 1996; Krest *et al*, 2000; Crotwell & Moore, 2003; Moore *et al*., 2006; Moore & Shaw, 2007) reveal that circulation of sea water through coastal aquifers underlying the salt marshes strongly enriches the resulting SGD in radium as well as nutrients and carbon. Near shore SGD certainly is an important source of radium isotopes to the coastal waters.

Coastal marshes are not the only source of excess radium to the SAB. Moore *et al.* (2002) and Moore & Wilson (2005) demonstrated that SGD leaking from limestone on the inner shelf is another source of radium and nutrients. These studies showed that tidal pumping, storms, and buoyancy are important factors in supplying radium to overlying waters. Thus, higher tidal ranges in the Georgia Bight may increase offshore additions of radium.

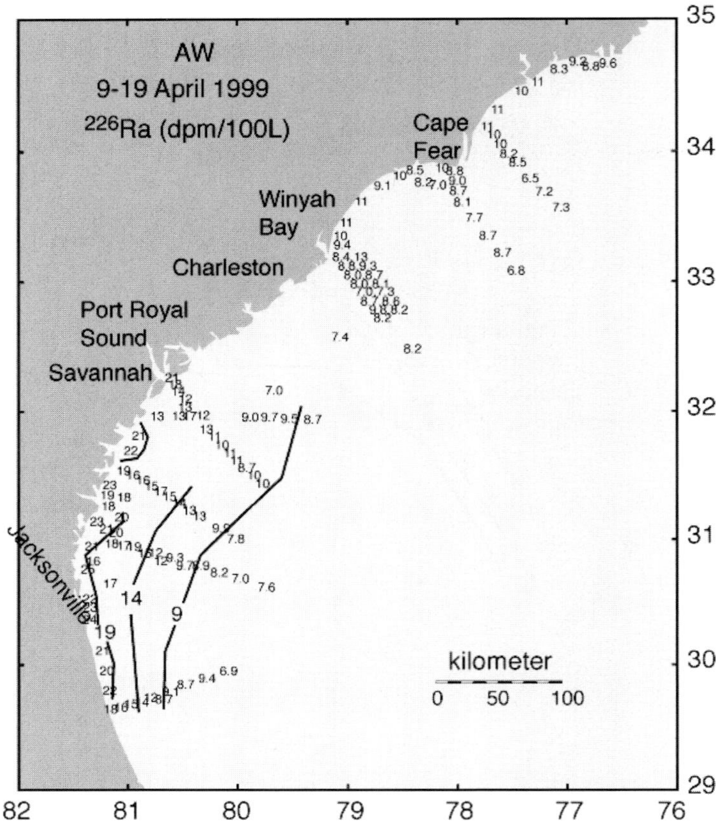

Fig. 3 In April the ^{226}Ra distribution in South Atlantic Bight surface waters was similar to late September.

Temporal changes

The most dramatic temporal changes in ^{226}Ra activity occurred between legs 1 (6–20 September) and 2 (22 September–3 October) of the 1998 cruise. In less than four weeks the inshore surface water in Long Bay (<39 km from shore) decreased in ^{226}Ra by a factor of 2 (Figs 1 and 2). During the second leg of this cruise, offshore activities of ^{226}Ra in this region were similar to those measured on leg 1. The leg 2 samples with low activity were from stations where the pycnocline had disappeared. However, the decrease in the surface activity cannot be explained by an increased mixing depth (i.e. dilution by mixing with bottom water) because deep samples from this region collected on leg 1 had activities similar to those measured in surface waters. The low leg 2 values represent a reduction in the total inventory of ^{226}Ra in Long Bay. The lower ^{226}Ra values cannot be explained by simple intrusion of low ^{226}Ra waters from offshore. The salinity of the inshore waters was lower than open ocean water and only slightly higher than salinity during leg 1. The low activity samples are lower in ^{226}Ra than samples of similar salinity collected on leg 1. There was a significant change in temperature, with leg 2 samples being up to 5° colder.

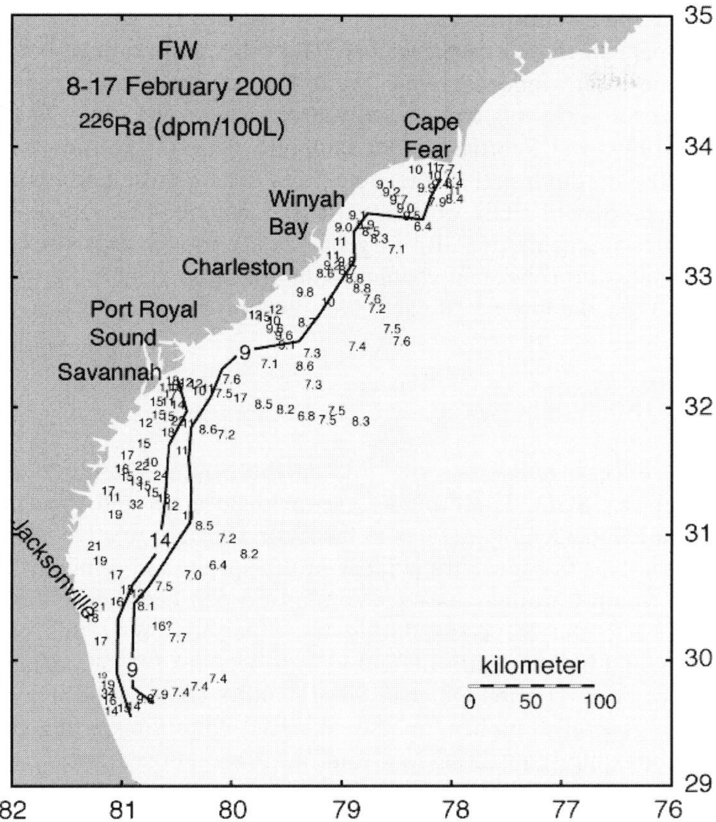

Fig. 4 A pattern similar to late September and April was present in February, with low ^{226}Ra activities in Long Bay and higher activities off the coast of Georgia.

Two processes are required to explain the data. First there must have been an exchange event that flushed the inshore water with water from the open ocean. Secondly, there must have been a reduction or cessation in SGD between the flushing event and the collection of the leg 2 samples. During this interval the salinity of the water derived from the offshore intrusion was measurably diluted by surface runoff that contained low ^{226}Ra activity. Although there are no significant rivers entering this portion of Long Bay, there are inlets to the Intracoastal Waterway that could supply the freshened water. In addition longshore drift of water leaving Winyah Bay may enter Long Bay.

In April 1999 the ^{226}Ra activity throughout Long Bay and extending to the region off Winyah Bay was considerably lower than in September–October 1998. Few values were greater than typical open ocean activities of 8 dpm/100L. Nevertheless, the salinity during each sampling was significantly lower than open ocean values. This means that the residence time of water in the area was long enough to reflect dilution with river water. The Winyah Bay transect was sampled at the start and end of the cruise; results from each transect were similar. These results imply that there must have been a reduction in SGD lasting at least a month before the cruise.

These new data reinforce the hypothesis that SGD is the primary source of the ^{226}Ra enrichment in these coastal waters. The other source functions should operate throughout the year. Indeed, rivers were flowing well and frequent storms re-suspended coastal sediments and eroded subareal sediments during the late September 1998, April 1999, and February 2000 sampling periods. The lower ^{226}Ra activities measured in the northern part of the study area during these periods indicate either a substantial reduction of SGD or a much lower activity of ^{226}Ra in the discharging fluids. Our measurements of ^{226}Ra in groundwater monitoring wells along the coast and offshore show no consistent changes with season (Moore & Wilson, 2005; Moore *et al.*, 2006). Thus, the lower ^{226}Ra activities must be due to reduction of SGD.

CONCLUSIONS

The offshore excess inventories of ^{226}Ra in the South Atlantic Bight are supplied almost entirely by SGD. These inputs vary with location and time of year. Radium enrichments are considerably greater in the Georgia Bight relative to the remainder of the study area. This is especially evident in the spring and winter when activities of ^{226}Ra off the South Carolina coast are similar to open ocean activities, implying low inputs of SGD to the SC coast during these periods. Why this pattern occurs is unknown. It could be a terrestrial effect as lags between seasons of high precipitation and SGD have been reported (Michael *et al.*, 2005). Or it could be a marine effect as sea level is generally higher in the summer (http://ibis.grdl.noaa.gov/SAT/hist/tp_products/topex.html), resulting in greater infiltration of sea water into aquifers.

Acknowledgements Many people contributed to the collection and analyses of these samples. Financial support was provided by the US National Science Foundation.

REFERENCES

Broecker, W. S., Goddard, J. & Sarmiento, J. (1976) The distribution of ^{226}Ra in the Atlantic Ocean. *Earth Planet. Sci. Lett.* **32**, 220–235.

Cai, W.-J., Wang., Y., Krest, J. & Moore, W. S. (2003) The geochemistry of dissolved inorganic carbon in a surficial groundwater aquifer in North Inlet, South Carolina, and the carbon fluxes to the coastal ocean. *Geochim. Cosmochim. Acta* **67**, 631–639.

Crotwell, A. M. & Moore, W. S. (2003) Nutrient and radium fluxes from submarine groundwater discharge to Port Royal Sound, South Carolina. *Aquatic Geochem.* **9**, 191–208.

Key, R. M., Sarmiento, J. L. & Moore, W. S. (1985) Transient tracers in the ocean, North Atlantic study final data report for ^{228}Ra and ^{226}Ra. *Ocean Tracer Lab Technical Report #92-2*, Dept Geol. Geophys., Princeton Univ., Princeton, New Jersey, USA.

Krest, J. M., Moore, W. S., Gardner, L. R. & Morris, J. (2000) Marsh nutrient export supplied by groundwater discharge: evidence from Ra measurements. *Global Biogeochem. Cycles* **14**, 167–176.

Michael, H. A., Mulligan, A. E. & Harvey, C. F. (2005) Seasonal oscillations in water exchange between aquifers and the coastal ocean. *Nature* **436**, 1145–1148.

Moore, W. S. (1976) Sampling ^{228}Ra in the deep ocean. *Deep-Sea Res. Oceanogr. Abstr.* **23**, 647–651.

Moore, W. S. (1984) Radium isotope measurements using germanium detectors. *Nucl. Instrum. Methods Phys. Res., Sect. B* **223**, 407–411.

Moore, W. S. (1987) ^{228}Ra in the South Atlantic Bight. *J. Geophys. Res.* **92**, 5177–5190.

Moore, W. S. (1996) Large groundwater inputs to coastal waters revealed by ^{226}Ra enrichments. *Nature* **380**, 612–614.

Moore, W. S. (2000a) Determining coastal mixing rates using radium isotopes. *Cont. Shelf. Res.* **20**, 1993–2007.

Moore, W. S. (2000b) Ages of continental shelf waters determined from ^{223}Ra and ^{224}Ra. *J. Geophys. Res.* **105**, 22117–22122.

Moore, W. S. & Shaw, T. J. (2007) Fluxes and behavior of radium isotopes, barium, and uranium in Southeastern US rivers and estuaries. *Marine Chem.* (in press)

Moore, W. S. & Wilson, A. M. (2005) Advective flow through the upper continental shelf driven by storms, buoyancy, and submarine groundwater discharge. *Earth Planet. Sci. Lett.* **235**, 564–576.

Moore, W. S., Krest, J., Taylor, G., Roggenstein, E., Joye, S. & Lee, R. (2002) Thermal evidence of water exchange through a coastal aquifer: implications for nutrient fluxes. *Geophys. Res. Lett.* **29**, doi:10.1029/2002GL014923.

Moore, W. S., Blanton, J. O. & Joye, S. (2006) Estimates of flushing times, submarine groundwater discharge, and nutrient fluxes to Okatee River, South Carolina. *J. Geophys. Res.* **111**, (C09006) doi:10.1029/2005JC003041.

Paerl, H. W. (1997) Coastal eutrophication and harmful algal blooms: Importance of atmospheric deposition and groundwater as “new” nitrogen and other nutrient sources. *Limnol. Oceanogr.* **42**, 1154–1167.

Rama & Moore, W. S. (1996) Using the radium quartet for evaluating groundwater input and water exchange in salt marshes. *Geochim. Cosmochim. Acta* **60**, 4645–4652.

Valiela, I., Costa, J., Foreman, K., Teal, J. M., Howes, B. & Aubrey, D. (1990) Transport of groundwater-borne nutrients from watersheds and their effects on coastal waters. *Biogeochem.* **10**, 177–197.

Windom, H. L., Niencheski, L. F., Moore, W. S. & Jahnke, R. (2006) Submarine groundwater discharge: a large, previously unrecognized source of dissolved iron to the South Atlantic Ocean. *Marine Chem.* **102**, 252–266.

A box model to quantify groundwater discharge along the Kona coast of Hawaii using natural tracers

RICHARD N. PETERSON[1], WILLIAM C. BURNETT[1], CRAIG R. GLENN[2] & ADAM J. JOHNSON[2]

1 *Dept of Oceanography, Florida State University, Tallahassee, Florida 32306, USA*
peterson@ocean.fsu.edu

2 *Dept of Geology and Geophysics, School of Ocean and Earth Science and Technology, University of Hawaii, Honolulu, Hawaii 96822, USA*

Abstract Major islands such as Hawaii typically exhibit conditions favourable for high submarine groundwater discharges (SGD) to the ocean. Quantitative aerial thermal imaging along the leeward Kona coast of Hawaii reveals plumes of relatively cold groundwater discharging from distinct portals along the coastline. Many of these plumes are thought to represent substantial volumes of groundwater discharge. The goal of our tracer work is to quantify groundwater fluxes to the coastal ocean from some of these plumes as a means of calibrating the aerial imaging. We employed coincident mass balance equations for two tracers (salinity and radon) and water fluxes to develop a mass balance box model for quantifying groundwater discharge. Our results indicate that a small SGD discharge plume emanating from Kahualoa Bay represents water fluxes on the order of thousands of m^3/day to the coastal ocean.

Key words submarine groundwater discharge; radon; Hawaii; mass balance equations

INTRODUCTION

By definition, submarine groundwater discharge (SGD) includes any and all outward movement of water from the aquifer to the overlying water column (Burnett *et al.*, 2003). Terrestrial factors that can enhance SGD include high precipitation rates, relief, and permeability, as well as a lack of a well-developed river system (Zektser, 2000). The islands of Hawaii exhibit all these characteristics, and so are ideal sites for developing SGD assessment tools. The “Big Island” of Hawaii is geologically the youngest of the islands, so all these factors are most pronounced on this island. Several authors have described the high volumes of SGD along the western coast of the Big Island of Hawaii (Fig. 1). While most of these prior studies (Kanehiro & Peterson, 1977; Kay *et al.*, 1977; Bienfang, 1980; Brock, 1980; Dollar & Atkinson, 1992; Oki, 1999) involved calculations of SGD via a water balance approach, an earlier study led by George Wilkins (Univ. Hawaii) produced a videotape in 1992 by flying over the coastline around the Kona coast of Hawaii with a hand-held infrared camera. That investigation revealed surface water temperature anomalies and demonstrated that there are distinct portals along the coastline where relatively cold groundwater discharges out of the aquifer and mixes with the warmer ocean water.

Working with the University of Hawaii’s Airborne Hyperspectral Imager (AHI), we conducted a more advanced and quantitative aerial infrared survey in the same

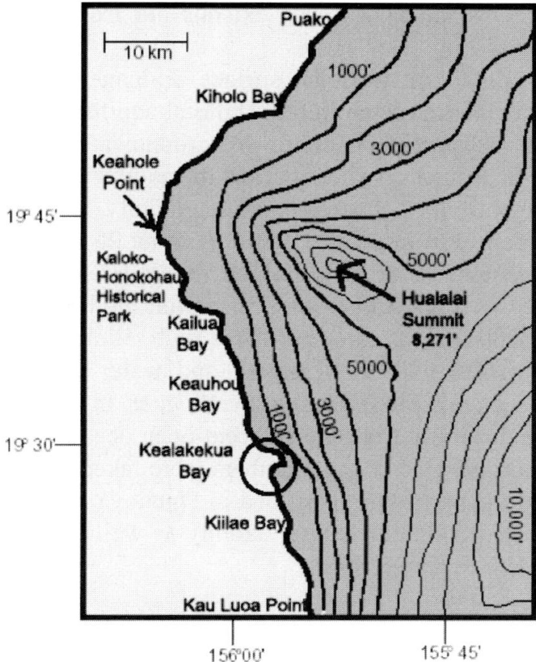

Fig. 1 Map depiction of the study site in the western Kona coast area of Hawaii. Kealakekua Bay is marked by the circle, and Kahualoa Bay is found just south (below) the large bay in the circle. Contours are feet above mean sea level.

area. In order to calibrate the aerial technique, we are employing ground-truthing via radioisotopic methods. Burnett & Dulaiova (2003) described a radon box model approach to evaluate rates of SGD in coastal regions. However, that approach was designed for diffuse seepage flow and thus requires information regarding the area of seepage in order to convert specific discharge (e.g. m^3/m^2 day) to absolute discharge (m^3/day). To reduce that uncertainty, we designed a coincident series of interconnected tracer mass balance equations to determine the groundwater flux. The tracers employed in this model are salinity and radon with a temporal resolution of one hour that corresponds to the integration time of the radon measurements.

STUDY SITE AND EXPERIMENTAL METHODS

The coastal regions on the western, leeward side of Hawaii are typically arid with a mean annual rainfall of only 25 cm. Within 10 km of the coastline, however, the porous mountain slopes receive a mean annual rainfall of 102 cm. These slopes are thus likely areas of groundwater recharge for the region (Kay *et al.*, 1977). The study site discussed here, Kahualoa Bay, is actually a small inlet just south of the much larger Kealakekua Bay, approximately 20 km south of Kailua-Kona village. It measures approximately 130 m long, 20 m wide at the head of the bay, and 110 m wide at its mouth where uninhibited tidal exchange occurs. A gentle sloping bottom, predominantly composed of small lava rocks, extends out from the coastline to an average depth of 2.5 m about 25 m from shore.

Due to the lack of rainfall in the area, surface drainage to this small bay is negligible. In addition, seawater intrusion into the coastal aquifer prevents residents in this area from pumping substantial volumes of groundwater for domestic use. Therefore, no anthropogenic influences should affect the aquifer dynamics at this site.

Continuous measurements of ^{222}Rn, temperature, salinity, and water level were taken over a several-day period from a fixed platform about 25 m from the shoreline. We measured radon concentrations in near surface waters (~0.5 m depth) at 1-hour intervals using a Durridge Co. Rad7 Radon-in-Air Monitor, modified to measure radon in water (Burnett *et al.*, 2001). Water temperatures and salinities were recorded at 1-minute intervals with a YSI 600 XLM probe, and water levels were recorded continuously using an Onset Corp. HOBO water level logger. In addition, end-member concentrations of radon and salinity from the adjacent open ocean and nearby groundwater wells were also recorded. Ocean end-members were taken from the results of a radon survey according to the procedure described in Dulaiova *et al.* (2005). Groundwater samples were collected and analysed for salinity as well as radon according to the procedure reported in Lee & Kim (2006).

Model development

The mass balance box model presented here uses the relatively high radon, low salinity nature of groundwater and the low radon, high salinity conditions of open ocean water to determine the flux of groundwater into and out of the coastal ocean. By continuously monitoring the radon and salt concentrations in the coastal surface waters, one can examine how the groundwater flux changes over time.

We represent (Fig. 2) input fluxes of water to the coastal ocean from open ocean exchange (Q_{IN}) as well as SGD (Q_{SGD}). These flows are balanced by outward coastal water exchange to the open ocean (Q_{OUT}). We ignored any meteorological input or output of water that should be negligible compared to the other fluxes on these time scales (hours to days). The variables marked Rn represent the radon concentrations (dpm/m^3) found in the groundwater (Rn_{SGD}), coastal ocean (Rn_c) and the open ocean (Rn_o). Likewise, the variables marked S represent the salinity (g/kg) of the same waters, and variables marked ρ are corresponding densities (kg/m^3).

The geometry of the groundwater plumes such as the one exiting Kahualoa Bay allows for use of a simple, half-box type approximation of its volume (Fig. 2(b)). Estimating the volume of the coastal ocean at two different time steps allows determination of ΔV_c. In order to simplify the geometry further, we assume that the length and width terms do not change throughout the tidal cycle, i.e. the bay has vertical walls. By convention, we consider fluxes directed offshore from the coastal bay as positive, and landward fluxes as negative.

The model is developed using three simultaneous equations for water, salt, and radon mass balance. Initially, these equations are set up for an entire tidal cycle, assuming steady state conditions with respect to the water and salt balances. Coastal water averages over a 24-hour period for radon, salinity, and density are used for Rn_c, S_c, and ρ_c. We used a modified version of the LOICZ box model approach (Gordon

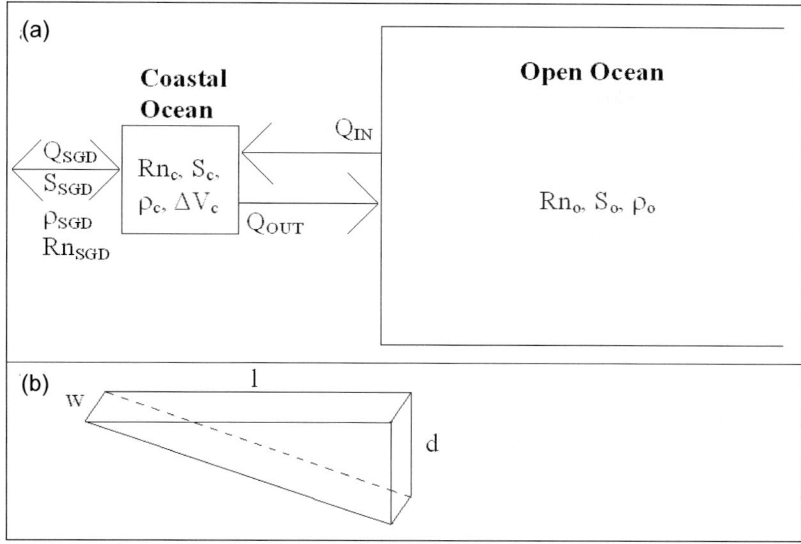

Fig. 2 (a) Diagram of model variables and their interactions; (b) geometry used to simulate the changing volume of the Kahualoa Bay plume. The volume of the tidal wedge is approximated by V = lwd/2.

et al., 1996) to derive the following equations for water balance:

$$\frac{\Delta V}{\Delta t} = 0 = Q_{out} - Q_{in} - Q_{sgd} \tag{1}$$

for salt balance:

$$\frac{lw(S_{c(t+1)}\rho_{c(t+1)}d_{c(t+1)} - S_{c(t)}\rho_{c(t)}d_{c(t)})}{\Delta t} = 0 = S_c\rho_c Q_{out} - S_o\rho_o Q_{in} - S_{sgd}\rho_{sgd}Q_{sgd} \tag{2}$$

and for radon balance:

$$\frac{lw(Rn_{c(t+1)}d_{c(t+1)} - Rn_{c(t)}d_{c(t)})}{\Delta t} = Rn_c Q_{out} - Rn_o Q_{in} - Rn_{sgd}Q_{sgd} \tag{3}$$

Terms marked with the subscript (t) represent the conditions at the beginning of the tidal cycle, and those marked by (t+1) represent the conditions at the end of the tidal cycle. In general, these equations are setup so that the left-hand side equates to a change in total mass (or volume for the case of water) over a set period of time, whereas the right-hand side of the equations represent the contributing effects of input or output fluxes to these mass changes in the groundwater plume. Solving these equations results in average values for water fluxes Q_{IN}, Q_{OUT}, and Q_{SGD} over a complete tidal cycle.

We next obtain better temporal resolution by examining how these fluxes vary within smaller time steps. For 1-hour time steps, the assumption of steady-state conditions for the water and salt balances are no longer valid, but we can examine how the tracer masses change between two time steps. Since the individual water fluxes are known for the entire tidal cycle, it is now necessary to examine their relative change each hour (Q ± ΔQ). A few processes inherently require an inverse relationship between tracer concentration in the end-member and corresponding water flux to achieve an expected response. For these processes, we expect the water flux to decrease if the end-member tracer concentration increases, so we use (Q – ΔQ). The processes that follow this rule are input of radon from groundwater and input of salt from the open ocean.

The resulting equations are the following for water balance:

$$\frac{\Delta V}{\Delta t} = (Q_{out} + \Delta Q_{out}) - (Q_{in} + \Delta Q_{in}) - (Q_{sgd} + \Delta Q_{sgd}) \tag{4}$$

for salt balance:

$$\frac{lw(S_{c(t+1)}\rho_{c(t+1)}d_{c(t+1)} - S_{c(t)}\rho_{c(t)}d_{c(t)})}{\Delta t} = S_c\rho_c(Q_{out} + \Delta Q_{out}) - S_o\rho_o(Q_{in} - \Delta Q_{in}) - S_{sgd}\rho_{sgd}(Q_{sgd} + \Delta Q_{sgd}) \tag{5}$$

and for radon balance:

$$\frac{lw(Rn_{c(t+1)}d_{c(t+1)} - Rn_{c(t)}d_{c(t)})}{\Delta t} = Rn_c(Q_{out} + \Delta Q_{out}) - Rn_o(Q_{in} + \Delta Q_{in}) - Rn_{sgd}(Q_{sgd} - \Delta Q_{sgd}) \tag{6}$$

Atmospheric evasion losses of radon, while minimal (<1% of radon inventory per hour), can be accounted for by adding this loss term into equation (6). On 1-hour time scales, the radon decay losses are considered unimportant. Solving these simultaneous equations for ΔQ_{IN}, ΔQ_{OUT}, and ΔQ_{SGD} allows for the calculation of net water fluxes into and out of a coastal ocean system during each time step analysed.

RESULTS

Several large diameter shallow coastal wells (remnants of Old Hawaiian ponds with brackish water: av. salinity = 6.4) exist in the vicinity of Kahualoa Bay. The average radon concentration of these waters is rather low at 4980 dpm/m^3, as compared to upland wells that often exceed 50 000 dpm/m^3. These wells respond to tidal changes very quickly, some even going completely dry during low tide. The open ocean waters offshore from this area have relatively constant radon activities of about 62 dpm/m^3 and average salinities of 35.5.

During the fieldwork in Kahualoa Bay, the platform where the continuous measurements were made was anchored 25 m from the shoreline. The radon platform is assumed to be in the middle of a concentration gradient between shore and the open ocean, and so twice the distance between shore and the platform is taken to be the length of the study domain.

The variation of radon and salinity at our measurement platform shows that both salinity and radon vary in opposite manners during tidal oscillations (Fig. 3). The radon activities are surprisingly low for an area so influenced by groundwater. This is likely due to the relatively low concentration of radon in the coastal groundwater. Since the aquifer solids are very young volcanic rock, radon's parent, ^{226}Ra, may not have had sufficient time to grow into equilibrium with the parent uranium.

The results of our model indicate that positive groundwater fluxes occur during outgoing tides and last through the low tides (Fig. 4). While the concentrations of salt and radon as shown in Fig. 3 seem to be in phase with the tide, the calculated ground-water flux in Fig. 4 is out of phase. This is because the Q_{SGD} calculation is based

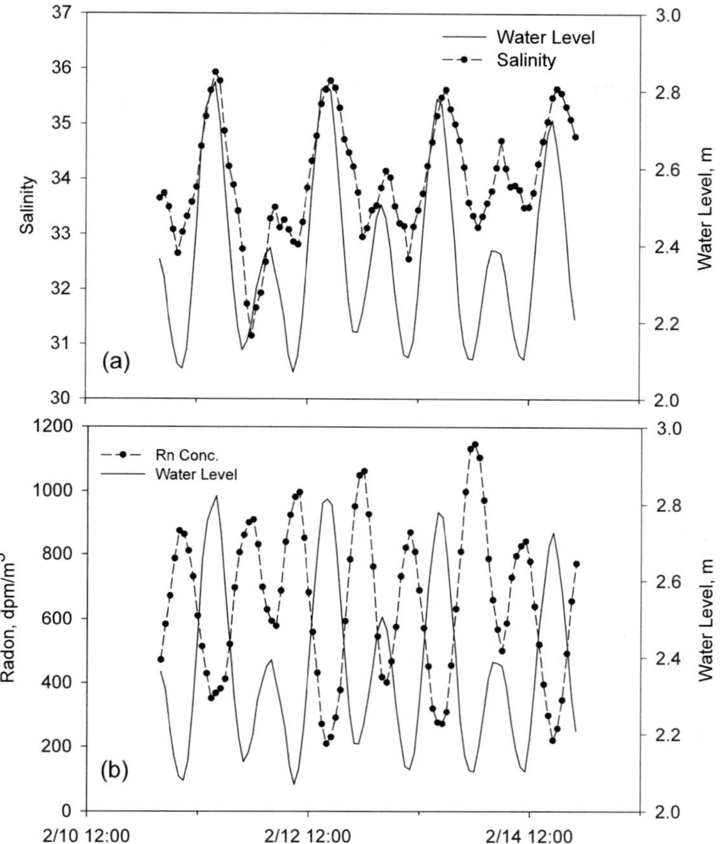

Fig. 3 (a) Salinity; and (b) radon concentration variations together with the water level (solid line) through several days of measurements in Kahualoa Bay.

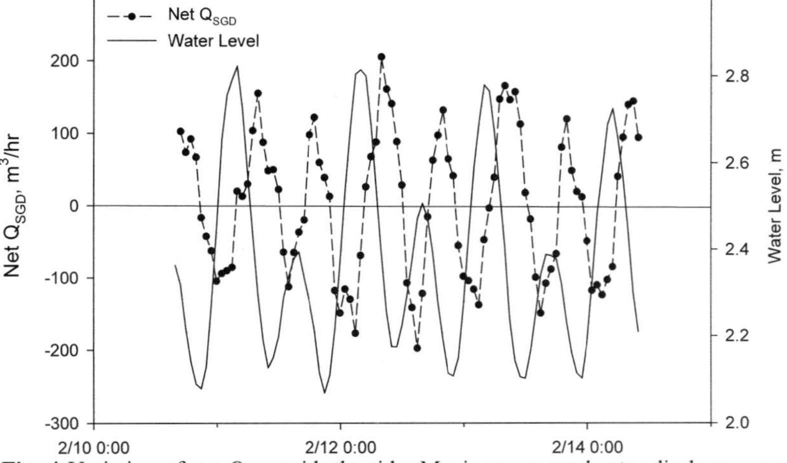

Fig. 4 Variation of net Q_{SGD} with the tide. Maximum groundwater discharge occurs during outgoing tide, and maximum groundwater recharge occurs during the incoming tide.

on the rate of change in the salt and radon masses (inventories), which peak in the middle of the tidal range. Once water levels begin to rise from the incoming tide, the groundwater discharge slows and eventually reverses direction, recharging the aquifer, presumably with saltier, lower radon water. By integrating all positive values of Q_{SGD}, over the course of a complete (24-hour) tidal cycle, the total flux of groundwater per day can be determined. In the case of Kahualoa Bay, we estimate the total groundwater flux at about 1120 m^3/day, with the terrestrial freshwater component of that flux to be 920 m^3/day.

CONCLUSIONS

Our modelled groundwater flux calculations based on natural tracer measurements are comparable to earlier water balance estimates. Such estimates are based on regional water budgets which typically calculate total fresh groundwater discharge from an aquifer system as the total freshwater recharge (precipitation minus evapotranspiration) minus well pumping. Kanehiro & Peterson (1977), for example, estimated average groundwater discharges on the order of 15 000 m^3/d·km for a 26 km length of shoreline north of Kiholo Bay (Fig. 1). While the current model shows that the small, 40-m wide basin we studied discharges 1120 m^3/d, extrapolating to 1 km of shoreline would give the equivalent of 28 000 m^3/d·km. Bienfang (1980) estimated the groundwater discharge of a larger plume issuing from Honokohau Harbor (north of Kailua-Kona) to be between 5600 and 7600 m^3/d. Visual comparisons of the infrared images of these two plumes indicate that the Honokohau Harbor plume is likely on the order of 5–10 times the magnitude of that from Kahualoa Bay. Our modelled result thus appears to be of the correct magnitude.

The model is rather sensitive to changes in some end-members and other parameters. For example, a 5% change in dimensional distance changes the final SGD flux by 5%. Changing the groundwater radon end-member concentration by 5% results in a final SGD flux change of 4%. However, changes within reasonable bounds for S_{SGD}, Rn_o, and S_o have only minor impacts on the final model outcome.

By using simultaneous mass balance equations for water, salt, and radon, it is possible to obtain reasonable results for groundwater fluxes throughout the course of a tidal cycle. In this paper, appropriate trends and values have been obtained for groundwater fluxes into a small coastal bay along the Kona coast of Hawaii. We hope to make additional enhancements to the model in the future including lessening the impact of dimensional length scales and groundwater radon concentration changes on the final result.

Acknowledgements The authors are most indebted to George Wilkins, without whom none of this work would have been possible. The authors also thank Henrieta Dulaiova, Isaac Santos, Axel Schmidt, David Gremminger and Benjamin Sellers for assistance in the field. We also thank Drs Doron Nof and Allan Clarke for their guidance and thoughtful insight in building this model. We also appreciate the insightful comments of two anonymous reviewers. This work was funded by National

Science Foundation Collaborative Research Grants (OCE04-51386 to WCB and OCE04-51379 to CRG). University of Hawaii School of Ocean and Earth Science and Technology (SOEST) Contribution no. 7065.

REFERENCES

Bienfang, P. (1980) Water quality characteristics of Honokohau Harbor: a subtropical embayment affected by groundwater intrusion. *Pacific Sci.* **34**(3), 279–291.

Brock, R. E. (1980) Colonization of marine fishes in a newly created harbor, Honokohau, Hawaii. *Pacific Sci.* **34**(3), 313–326.

Burnett, W. C. & Dulaiova, H. (2003) Estimating the dynamics of groundwater input into the coastal zone via continuous radon-222 measurements. *J. Environ. Radioactivity* **69**(1-2), 21–35.

Burnett, W. C., Kim, G. & Lane-Smith, D. (2001) A continuous radon monitor for assessment of radon in coastal ocean water. *J. Radioanal. Nucl. Chem.* **249**, 167–172.

Burnett, W. C., Cable, J. E. & Corbett, D. R. (2003) Radon tracing of submarine groundwater discharge in coastal environments. In: *Land and Marine Hydrology* (ed. by M. Taniguchi, K. Wang & T. Gamo), 25–42. Elsevier, New York, USA.

Dollar, S. J. & Atkinson, M. J. (1992) Effects of nutrient subsidies from groundwater to nearshore marine ecosystems off the island of Hawaii. *Estuarine, Coastal and Shelf Science* **35**(4), 409–424.

Dulaiova, H., Peterson, R., Burnett, W. C. & Lane-Smith, D. (2005) A multi-detector continuous monitor for assessment of 222Rn in the coastal ocean. *J. Radioanal. Nuc. Chem.* **V263**(2), 361–363.

Gordon, D. C. Jr, Boudreau, P. R., Mann, K. H., Ong, J.-E., Silvert, W. L., Smith, S. V., Wattayakorn, G., Wulff, F. & Yanagi, T. (1996) *LOICZ Biogeochemical Modeling Guidelines.* LOICZ/R&S/95-5, VI LOICZ, Texel, The Netherlands.

Kanehiro, B. Y. & Peterson, F. L. (1977) Groundwater recharge and coastal discharge for the northwest coast of the Island of Hawaii: a computerized water budget approach. University of Hawaii, Water Resources Research Center Technical Report 110.

Kay, A. E., Lau, L. S., Stroup, D., Dollar, S. J., Fellows, D. P. & Young, R. H. F. (1977) Hydrologic and ecologic inventories of the coastal waters of West Hawaii. University of Hawaii, Water Resources Research Center Technical Report 105.

Lee, J.-M. & Kim, G. (2006) A simple and rapid method for analyzing radon in coastal and ground waters using a radon-in-air monitor. *J. Environ. Radioactivity* **89**(3), 219–228.

Oki, D. S. (1999) Geohydrology and numerical simulation of the ground-water flow system of Kona, Island of Hawaii. *US Geological Survey Water-Resources Investigations Report 99-4073.*

Zektser, I. S. (2000) *Groundwater and the Environment: Applications for the Global Community.* Lewis Publishers, Boca Raton, Florida, USA.

Nutrient dynamics with groundwater–seawater interactions in a beach slope of a steep island, western Japan

SHIN-ICHI ONODERA[1], MITSUYO SAITO[2], MASAKI HAYASHI[2] & MISA SAWANO[1]

1 *Graduate School of Integrated Arts and Sciences, Hiroshima University, 1-7-1, Kagamiyama, Higashi-Hiroshima, Hiroshima 7398521, Japan*
sonodera@hiroshima-u.ac.jp

2 *Graduate School of Biosphere Sciences, Hiroshima University, 1-7-1, Kagamiyama, Higashi-Hiroshima, Hiroshima 7398521, Japan*

Abstract To confirm the semi-diurnal and seasonal variation in nutrient flux with the dynamics in groundwater–seawater interaction, we conducted intensive observations at 25 piezometers in a tidal flat over a distance of 100 m across a steep-sloped island. Based on the chloride balance, groundwater was very well-mixed with seawater under the tidal flat. Nitrate-nitrogen (NO_3^--N) concentrations declined from >20 to near 0 mg L^{-1} along the groundwater flowpath from the hillslope to the tidal flat. In addition, nitrate concentrations in pore water of the tidal flat were lower (at 0.1 mg L^{-1}) than in seawater. These results suggest that the reduction process of NO_3^--N occurred in both contaminated groundwater and seawater. Discharge of inorganic nitrogen by groundwater was confirmed offshore. The suspected source was nitrogen mineralization of organic compounds and seawater recirculation. Phosphorus was produced offshore and transported from the land area. Seasonal variations in nutrient dynamics at the tidal flat were also confirmed.

Key words nutrient dynamics; groundwater; recirculated seawater; nitrate contamination

INTRODUCTION

Eutrophication and associated red tides in estuaries and inland seas are one of the critical environmental issues of this century. One of the controlling factors is the inflow of excess nutrients from the land. Recent research has shown that not only river water but also groundwater has a great effect on nutrient discharge from the land to the sea (Zektser & Loaiciga, 1993; Burnett *et al.*, 2001; Slomp & Cappellen, 2004; etc.). Submarine groundwater discharge (SGD) has been reported by many researchers (Kim *et al.*, 2005; Taniguchi *et al.*, 2005; Burnett *et al.*, 2006; etc.) and the dynamics of seawater–groundwater interactions are now better understood. In addition, the nutrient load and its propagation by SGD have been investigated (Slomp & Cappellen, 2004). However, the nutrient discharge has not been well identified in terms of the dynamics and mixing processes of groundwater and seawater.

In the case of nitrogen and phosphorous, inorganic components have very often been input in agricultural and urban lands (Tsurumaki, 1992; Burt *et al.*, 1993; Saito *et al.*, 2005; etc.). However, the nitrate concentration was generally decreased with groundwater flow by denitrification and plant uptake (Howard, 1985; Ishizuka & Onodera, 1997; etc.), and the phosphorous concentration is also affected by the redox condition. Nutrient dynamics are particularly complicated in the hyporheic zone, such as the riverside and seaside where oxic surface water is mixed temporarily with anoxic subsurface water under the ground during water-level changes (Böhlke *et al.*, 1995; Hinkle *et al.*, 2001; Ullman *et al.*, 2003; Phillipe & Hill, 2004; etc.). For the determination of the nutrient dynamics below the beach slope, it is especially necessary to understand the correlation of nutrient concentrations with temporal interactions between groundwater and seawater during tidal fluctuations.

The objective of this research has been to investigate the nutrient and water dynamics by tidal fluctuations and seasonal changes below a beach slope of an island in an inland sea. In addition, we tried to determine the sources of nutrients found on land or in marine sediment, using a hydrochemical approach.

STUDY AREA AND METHODS

The study area is located at Ikuchijima Island at the Seto Inland Sea, in Hiroshima prefecture of western Japan, which is a Japanese national park (Fig. 1). Environmental problems such as eutrophication and red tides were observed in the Sea. The study area is underlain by granites with a maximum altitude of about 450 m. The annual rainfall is about 1100 mm. The monthly rainfall is greatest from June to July, which is the Japanese rainy season. About 50% of the island is covered by orange groves and much fertilizer is applied during a whole year. The annual application amount is about 2400 kg ha^{-1} $year^{-1}$. Consequently, the surface water and groundwater of the island have been significantly contaminated by nitrates.

The observation site is located in the southern part of the island (Fig. 1(c)). This area is one of the steepest topography in the island, and there are no river channels. More than 50% of the slope around the observation site is covered by orange groves. The beach slope is mainly composed of very permeable coarse sands. The saturated hydraulic conductivity is 5×10^{-2} cm s^{-1}. Organic muddy and clayey sand is deposited on the flat area of the lower beach slope (Fig. 1(d)).

To confirm the nutrient and water dynamics below the beach slope, we installed 21 PVC pipes with diameters of 13 mm as piezometers for the manual measurement of water levels at seven points on an observation line over a distance of 100 m. For collecting water samples, 13 mm diameter pipes were installed at the same depths as the piezometers. The installation depths were from 50 cm to 400 cm (open circles in Fig. 1(d)). In addition, a large piezometer with a diameter of 10 cm was installed at a depth of 80 cm at site P1 (Fig. 1) for automatic measurement of water level and electrical conductivity. The altitude at P1 is approximately the mean sea level. At each observation time, we also collected: (1) samples of near-surface pore water at depths of 5 to 10 cm at five points along the beach using a penetration-type seabed pore water sampler, (2) seawater samples around the offshore sites, and (3) groundwater samples at two dug wells and a borehole behind the beach. The observations were carried out intensively from 2 to 5 times during two or three days in a spring tide in August and October in 2005, and in January, April and July in 2006. We purged the pipes for one hour before collecting the water samples.

The collected water samples were analysed for chemical components in the laboratory. The Cl^- concentrations were analysed by ion chromatography after being

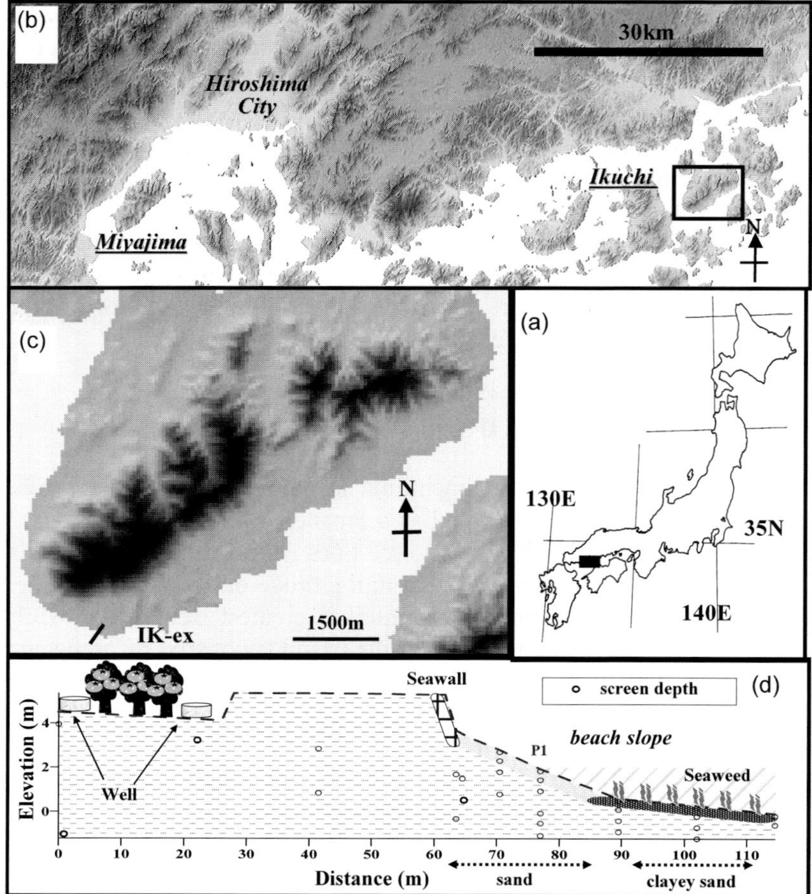

Fig. 1 Location of the observation site at Ikuchijima Island in Hiroshima Prefecture, western Japan; (d) shows the position of screen depth of piezometers and wells on the cross section of the observation beach slope.

filtered using a 0.20 μm cellulose ester filter. In addition, dissolved total nitrogen (DTN), dissolved total phosphorous (DTP), NO_3^--N, NO_2^--N, NH_4^+-N and dissolved silica (DSi) were analysed with a spectrophotometer. Dissolved organic carbon (DOC) was analyzed with a total carbon analyzer.

RESULTS AND DISCUSSION

Variation in subsurface water flow in the beach slope

Figure 2 shows the distribution of water flow direction based on the piezometric potential and seawater contribution ratio at a low (Fig. 2(a)) and high (Fig. 2(b)) tide in August in 2005. The seawater contribution ratio (R_s) is calculated as:

$$R_s = (C_i - C_g)/(C_s - C_g) \quad (1)$$

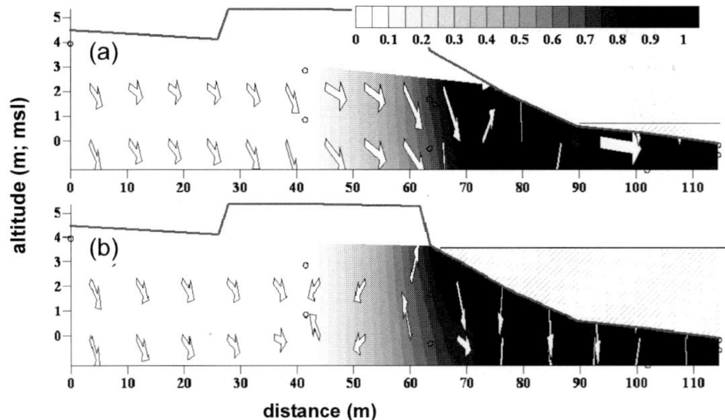

Fig. 2 Distribution of water flow direction and seawater contribution ratio (R_s) in a beach slope at: low (a) and high (b) tide in August 2005.

where C is a concentration and subscripts i, g and s denote a sample of water, groundwater and seawater, respectively. In this study, the Cl^- concentration was applied to calculate R_s, and the concentration in groundwater and seawater which were input to equation (1) were collected at the dug well at the mountain slope foot and offshore at high tide, respectively. In Fig. 2, as the horizontal scale is one fifth of the vertical scale, downward flow directions are emphasized. The flow direction is downward below the beach slope when covered by seawater at high tide (Fig. 2(b)). But, it changes to upward below the exposed beach slope and lateral from the slope to offshore below the seabed that is continuously covered by sea water at low tide (Fig. 2(a)). These results suggest there is infiltration of seawater at the upper beach slope during high tide and discharge of groundwater mixed with seawater at the lower beach slope during low tide. The seawater contribution ratio was more than 80% below the beach slope and 100% farther offshore. This means that the amount of seawater circulation is four times that of groundwater discharge in the beach slope and the discharge water is composed of recirculated seawater from the seabed offshore.

Figure 3 shows the relation between the water level from the bottom (80 cm) of the piezometer and electrical conductivity in the large piezometer at P1 for 15–16 July 2006. The variation pattern was characterized by three types. First, the water level varies by less than 80 cm during low tide with a constant electrical conductivity (EC) of around 47 mS cm^{-1}. The EC value indicates that 10% of terrestrial groundwater is mixed with 90% seawater of 51 mS cm^{-1}. Second, the EC varies between 47 and 51 mS cm^{-1} during mid tide with a constant water level around 80 cm. This water level is closed to the ground surface level, as the ground surface has just been covered by seawater. Third, the water level increases to more than 80 cm during high tide with a constant EC of around 51 mS cm^{-1}. In this stage, the EC value almost equals the value of seawater. These results reveal the temporal fluctuation of the interface between the seawater and brackish groundwater as it relates to the tidal variation. This replacement of brackish groundwater to seawater lasted for 30 minutes according to the observation in the second stage (Fig. 3). If "fingering" of salty water in the high permeability (5×10^{-2} cm sec^{-1}) sediment is assumed, the seawater transport time from the ground

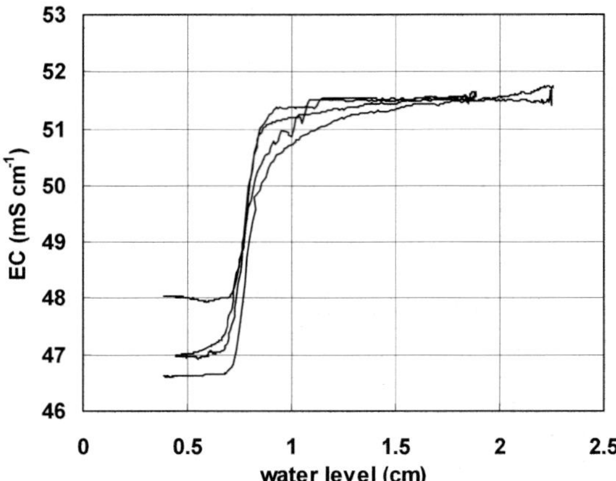

Fig. 3 Relation between water level and electrical conductivity in a large piezometer at the depth of 80 cm at site P1 for 15–16 June 2006.

surface to the depth of 80 cm is estimated to be 30 minutes. This suggests that most of the replaced seawater around the depth of 80 cm at P1 is new water which was input during the recent high tide. On the other hand, the brackish groundwater underlain by the new water (composed of 90% seawater) is relatively old.

Distribution in nutrient content in the beach slope

Figure 4 shows the distribution of nutrients (DSi, NO_3^--N, NH_4^+-N and DTP) below the beach slope at low tide in January 2006. The DSi, NO_3^--N and DTP concentrations decline along the groundwater flowpath from the mountain-foot slope to the beach area. Only the DTP increases below the beach slope. The NO_3^--N concentration in groundwater at the mountain-foot slope is greater than 20 mg L^{-1}. In the agricultural area of this island, nitrate contamination of groundwater is severe. Nevertheless, the concentration in the subsurface water on the beach slope became negligible along the groundwater flowpath. Both the DSi and NO_3^--N concentrations in groundwater at the mountain-foot slope are more than 100 times of those in seawater. One of the expected major decline factors of the concentrations is the dilution by the seawater in the beach slope. However, it is not enough to quantify the nitrate decline process as the average seawater contribution ratio is approximately 90% in the beach slope.

In contrast, the NH_4^+-N and DTP concentrations increase slightly below the beach slope. In the case of NH_4^+-N, the concentration in the groundwater at the mountain-foot slope is close to 0 mg L^{-1} and less than that in seawater. If the nitrate decline is caused by NO_3^--N reduction to NH_4^+-N, the increasing concentration of NH_4^+-N becomes similar to the decline of NO_3^--N. But the increase of NO_3^--N was 100 times that of the NH_4^+-N. This suggests that NO_3^--N reduction to NH_4^+-N is not the major process in the NO_3^--N decline. In the case of the DTP, the concentration in the groundwater is higher than that both in seawater and pore water below the beach slope. It suggests the possibility that the DTP declines along the groundwater flowpath by mineralization of organic matter, and/or deep groundwater mixing.

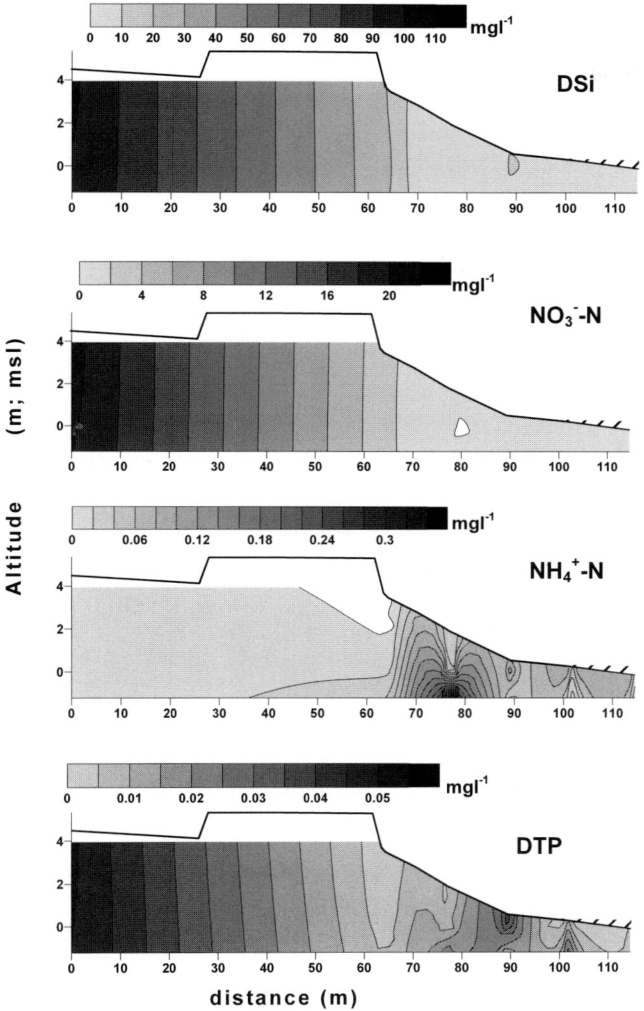

Fig. 4 Distribution of nutrients below the beach slope during a low tide in January 2006.

Dynamics of nutrient in an interaction zone of groundwater and seawater

Figure 5 shows the relations between Cl^- and: (a) NO_3^--N, (b) NH_4^+-N, (c) DSi, and (d) DTP concentrations in January 2006. The broken line in this figure represents the mixing process of groundwater and seawater without chemical reactions. The values used as the end members are the shallow groundwater at the mountain-foot slope as groundwater (GW in Fig. 5), and seawater at high tide offshore as seawater. The area below and above the line in Fig. 5 indicates the disappearance and production process of nutrients, respectively. In the case of DSi, most of the values indicate a mixing process of seawater and groundwater with less dissolution (Fig. 5(c)). In the case of NO_3^--N and NH_4^+-N, the values in the pore water of the beach slope plot below the line (Fig. 5(a), (b)). These show the trends of the nitrogen disappearance in the beach slope.

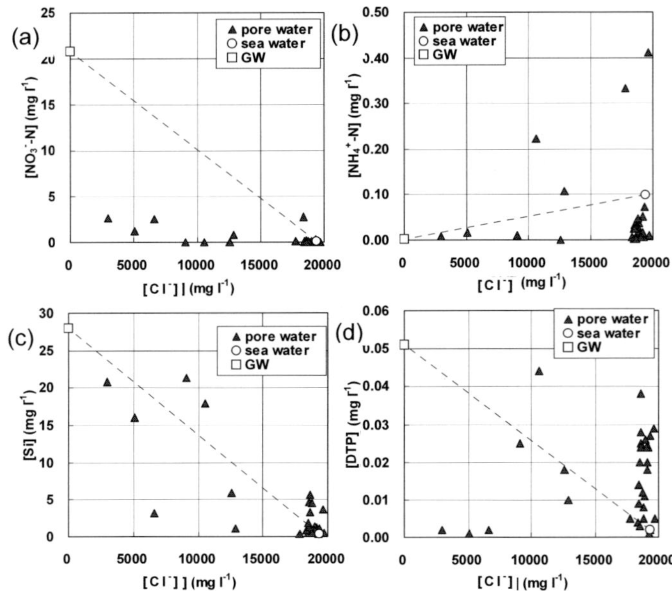

Fig. 5 Relations between Cl^- and (a) NO_3^--N, (b) NH_4^+-N, (c) DSi and (d) DTP concentrations in January 2006.

On the other hand, most of plots between Cl^- and DTP indicate the trend of the production of DTP.

The production concentration (*Pr*) is calculated by the difference of the measured (C_i) and calculated (C_c) concentration:

$$Pr = C_i - C_c \tag{2}$$

The calculated concentration (C_c) is calculated as follows:

$$C_c = C_g (1 - R_s) + C_s R_s \tag{3}$$

Figure 6 shows the distribution of the production concentration of NO_3^--N in the cross section of the beach slope at low tide in January 2006. In equation (2), the negative value denotes the disappearance of a nutrient. The NO_3^--N disappearance is dominant in the groundwater from the mountain-foot slope to behind the beach slope. The concentration of disappearance is 12 mg L^{-1} maximum just behind the beach slope at a point 60 m from the mountain-foot slope. The NO_3^--N disappearance process is typical of denitrification and reduction. Based on the NH_4^+-N concentration, NO_3^--N reduction to NH_4^+-N is negligible. Consequently, denitrification is suggested as the disappearance process. On the other hand, the NO_3^--N production process is dominant in the beach slope. The maximum concentration is 3 mg L^{-1} in the lower beach slope.

Seasonal variation in nutrient dynamics is shown in Fig. 7 as the relation between Cl^- and NO_3^--N in January and July 2006. The NO_3^--N concentration is relatively high in July compared with January. The nitrate production is suggested to be increasing with temperature.

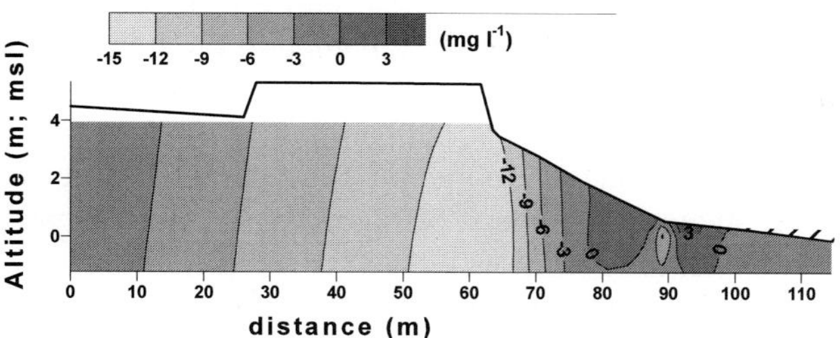

Fig. 6 Distribution of the production concentration of nitrate-nitrogen in the cross section of the beach slope at low tide in January 2006.

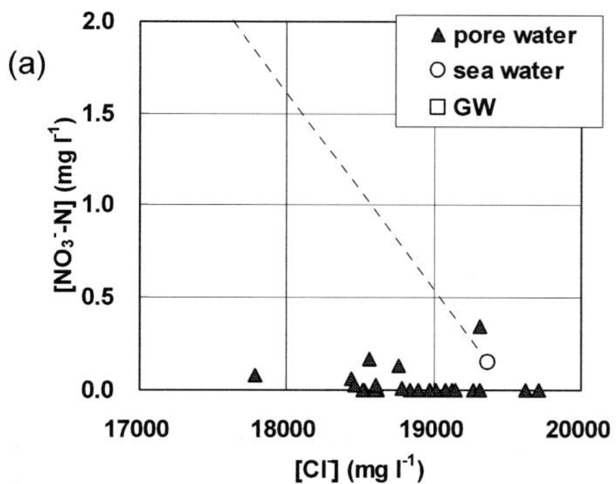

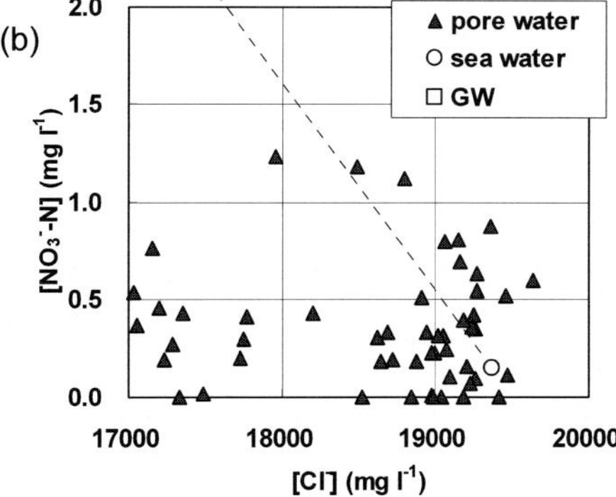

Fig. 7 Relation between Cl^- and NO_3^--N in January (a) and July (b) 2006.

CONCLUDING REMARKS

In this research, water and nutrient dynamics in a tidal flat caused by tidal fluctuation and seasonal changes were confirmed, using intensive observations in a network of piezometers.

1. Based on the chloride balance and piezometric head, groundwater was very well-mixed with seawater under the tidal flat, both temporally and spatially.

2. NO_3^--N concentrations declined from more than 20 mg L^{-1} to nearly 0 mg L^{-1} along the groundwater flowpath from the hillslope to the tidal flat. It is suggested that the nitrate reduction process originated in both the seawater and contaminated groundwater.
3. Based on the nitrogen balance, inorganic nitrogen discharge was confirmed offshore, with the suggested process being nitrogen mineralization of organic compounds and seawater recirculation.
4. The phosphorus was produced offshore as well as transported from the land area.
5. A seasonal variation in nutrient dynamics was corroborated by changes in temperature.

Acknowledgements The authors were appreciating help of Mr T. Takei, T. Shige-eda, T. Mine and Ms M. Kobayashi in our field observation, and Prof. M. Fukuoka for his suggestions. This study was supported by the Ministry of Education and Science in Japan (no. 18201007).

REFERENCES

Böhlke, J. K. & Denver, J. M. (1995) Combined use of groundwater dating, chemical, and isotopic analyses to resolve the history and fate of nitrate contamination in two agricultural watersheds, Atlantic coastal plain, Maryland. *Water Resour. Res.* **31**, 2319–2339.

Burnett, W. C., Taniguchi, M. & Oberdorfer, J. (2001) Measurement and significance of the direct discharge of groundwater into the coastal zone. *J. Sea Res.* **46**, 109–116.

Burnett, W. C., Aggarwal, P. K., Aureli, A., Bokuniewicz, H., Cable, J. E., Charette, M. A., Kontar, E., Krupa, S., Kulkarni, K. M., Loveless, A., Moore, W. S., Oberdorfer, J. A., Oliveira, J., Ozyurt, N., Povinec, P., Privitera, A. M. G., Rajar, R., Ramessur, R. T., Scholten, J., Stieglitz, T., Taniguchi, M. & Turner, J. V. (2006) Quantifying submarine groundwater discharge in the coastal zone via multiple method. *Sci. Tot. Environ.* **367**, 498–543.

Burt, T. P., Heathwaite, A. L. & Trudgill, S. T. (1993) *Nitrate; Processes, Patterns and Management*. John Wiley & Sons, London, UK.

Freeze, R. A. & Cherry, J. A. (1979) *Groundwater*. John Wiley & Sons, London, UK.

Hinkle, S. R., Duff, J. H., Triska, F. J., Laenen, A., Gates, E. B., Benkala, K. E., Wentz, D. A. & Silva, S. R. (2001) Linking hyporheic flow and nitrogen cycling near the Willamette River. *J. Hydrol.*, **244**, 157–180.

Howard, K. W. F. (1985) Denitrification in a major limestone aquifer. *J. Hydrol.* **76**, 265–280.

Ishizuka, S. & Onodera, S. (1997) Determination of denitrification in shallow groundwater on a forested land in Joso Upland, south western Ibaragi, Japan. *Japan J. Soil and Nutrient*, **68**, 1–7 (Japanese with English abstract).

Kim, G., Ryu, J., Yang, H & Yun, S. (2005) Submarine groundwater discharge into the Yellow Sea revealed by ^{228}Ra and ^{226}Ra isotopes: Implications for global silicate fluxes. *Earth Planet. Sci. Lett.*, **237**, 156–166.

Philippe, G. F. V. & Hill, A. R. (2004) Landscape controls on nitrate removal in stream riparian zones. *Water Resour. Res.* **40**, WS03401, doi: 10.1029/2003WR002473.

Saito, M., Onodera, S. & Takei, T. (2005) Nitrate transport process in a small coastal alluvial fan catchment. *Japan J. Limnol.* **66**, 1–10 (Japanese with English abstract).

Slomp, C. P. & Cappellen, P. V. (2004) Nutrient inputs to the coastal ocean through submarine groundwater discharge: controls and potential impact. *J. Hydrol.* **295**, 64–86.

Taniguchi, M., Burnett, W. C., Cable, J. E. & Turner, J. V. (2002) Investigation of submarine groundwater discharge. *Hydrol. Processes* **16**, 2115–2129.

Tsurumaki, M. (1992) Nitrate-nitrogen in shallow groundwater. *Japan J. Groundwater*, **34**, 153–162 (Japanese with English abstract).

Ullman, W. J., Chang, B., Miller, D. C. & Madsen, J. A. (2003) Groundwater mixing, nutrient diagenesis, and discharges across a sandy beachface, Cape Henlopen, Delaware (USA). *Estur. Coast. Shelf Sci.* **57**, 539–552.

Zektser, I. S. & Loaiciga, H. G. (1993) Groundwater fluxes in the global hydrologic cycle: past, present and future. *J. Hydrol.* **144**, 405–427.

Nutrient inputs through submarine groundwater discharge to Ariake Bay, Kyushu Island, Japan

JUN YASUMOTO[1], MAMORU KATSUKI[1], HIDETOMO TAKAOKA[2], YOSHINARI HIROSHIRO[1] & KENJI JINNO[1]

1 *Institute of Environmental Systems, Graduate School of Engineering, Kyushu University, 744 Motooka Nishi-ku, Fukuoka 812-8581, Japan*
yasumoto@civil.kyushu-u.ac.jp

2 *IDEA Consultants Inc., 1-5-12 Higashihama Higashi-ku, Fukuoka 812-0055, Japan*

Abstract Submarine groundwater discharge (SGD) is now recognized as an important pathway between land and sea. This study attempts to estimate the nutrient inputs through SGD to Ariake Bay. SGD rates and its quality along the coast of Ariake Bay in the Oura Region, Japan, were investigated. It was shown that the on-site SGD rate ranges from 0.01 to 20.52 μm/s, and SGD flows through the shallow confined aquifers, which consist of two kinds of rocks: basalt and pyroclastic rocks. The reduction reaction for SGD proceeded just up to denitrification. SGD associated with nutrient loads of N, P and SiO_2 were estimated to be 1.40, 0.07 and 52.78 g m^{-2} d^{-1}, respectively. This study demonstrates that SGD must be considered as a significant source of nutrient input to the coastal sea area in Ariake Bay.

Key words submarine groundwater discharge; nutrient; redox reaction; field survey; seepage meter; Japan

INTRODUCTION

The ecosystem and aquatic environment for human activities and fisheries in semi-enclosed bays as well as coastal areas have been deteriorating in recent years. This study focuses on the environmental reclamation of Ariake Bay, a semi-enclosed inner bay located to the west of Kyushu Island, Japan, which is experiencing such problems. Although the government has implemented several measures to improve the ecosystem and aquatic environment, significant recovery has not yet been achieved in the bay. One of the causes of the deteriorating situation is considered to be an increase in the nutrient flux, due to fertilizers and wastewater, through continuous surface water and groundwater discharges from the residential and agricultural areas into the Ariake Bay catchment.

In recent studies, it has been revealed that nutrient discharge through submarine groundwater discharge (SGD) is not negligible as compared to river discharge (Taniguchi *et al.*, 2002). Nutrient discharge plays a significant role in the nutrient cycle and primary productivity in the coastal ocean (Slomp & Van Cappellen, 2004).

In the case of Ariake Bay, however, there is only a little quantified information of whether SGD is the nutrient source. A detailed investigation of SGD and nutrient transport via SGD is indispensable for a better understanding of the role of SGD and its anthropogenic or natural perturbations. This paper takes a first step at estimating nutrient inputs through the SGD to the coast of the Saga region of Ariake Bay by seepage meter measurements.

STUDY SITE AND METHODS

Geography and geology

The study site is off the Oura coast, Tara-town in Saga, which is located on the west coast of Ariake Bay, Kyushu Island, Japan. The SGD observation points (St.1, St.2, St.3, St.4) and groundwater observation points (St.A, St.B, St.C, St.D) in the land region are as shown in Fig. 1.

The Taradake area, as shown in Fig. 2, has a topography dominated by Tara-dake Mountain, with the highest peak 1075 m in elevation and a base of 25 km in diameter. The volcano-foot alluvial fan with a gentle slope of 3–4° develops to the foot of a mountain in the northeast. In the southern area, there are a lot of small rivers which radiate from the central part of the area.

Aquifers in this study area consist of Pliocene to early Quaternary volcanic rocks which are listed chronologically, divided into six stratigraphic units: pre-Taradake andesite (PTA), Taradake Older basalts (TOB), Koorigawa volcanic rocks (KVR),

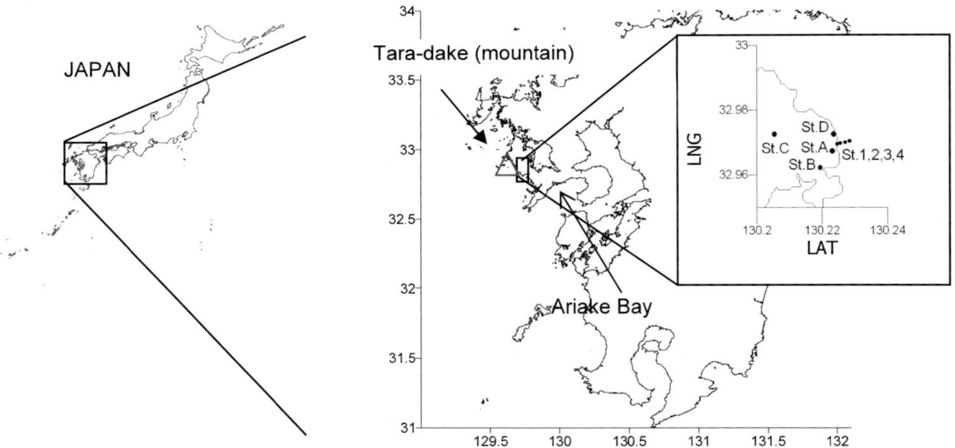

Fig. 1 Location map of Ariake Bay and the study sites.

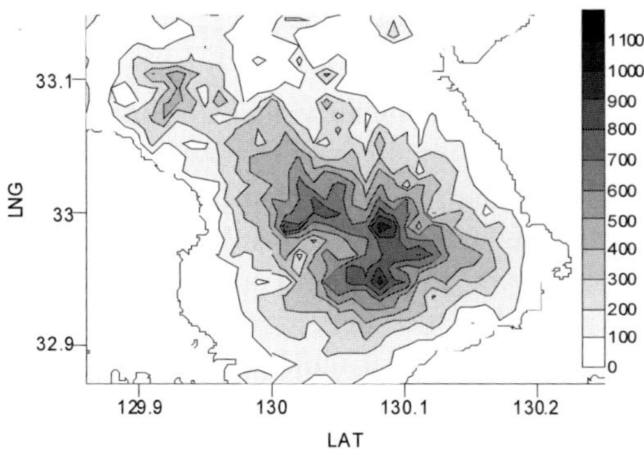

Fig. 2 Topography and geography of the Taradake area.

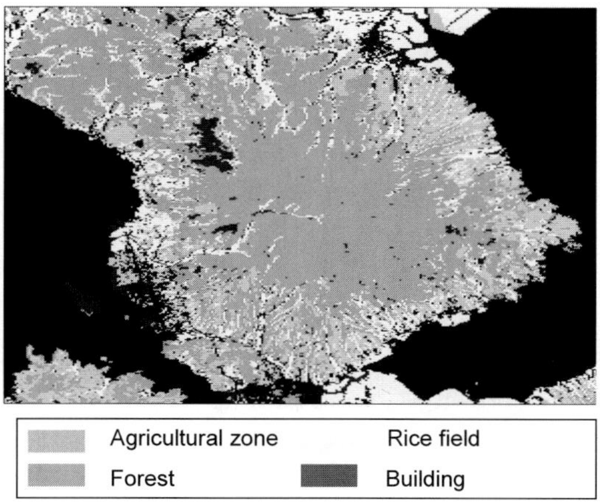

Fig. 3 Use of the land in the Taradake area.

Taradake Older andesites (TOA), Taradake Younger basalts (TYB) and Taradake Younger andesites (TYA). TOB is widely distributed in the base part of Taradake area and is widely exposed to the basement of the mountains except at the southern part. It forms an extensive lava plateau with an area of 15 × 25 km and thickness of 100–250 m. This is one of several large basalt plateaus in the northwestern Kyushu. TOB consists mainly of lava flows with a relatively small amount of pyroclastic rock. The thickness of the lava flow layer is about 30 m. Cracks have developed resulting in high permeability. In contrast, the pyroclastic rock, consisting of volcanic ash and a scoria layer, which consists in many cases of reddish brown clays because of weathering, and has low-permeability.

Figure 3 shows the land use in the Taradake area where there are many agricultural zones at the base of the mountain, which are below 300 m in elevation. The agricultural zones are chiefly used for orange grove cultivation.

The schematic cross-sectional image of SGD in the study area was made based on the geography-topography and geology described above as shown in Fig. 4. It was inferred that the groundwater discharges via a relatively short route through the confined aquifer between the pyroclastic rock with low-permeability and lava flow with high permeability that are formed in the TOB under the sea floor off the coast.

Investigation methods and water analysis items

Ariake Bay is a semi-closed bay, and the average of the tidal change is from 3 to 5 m. Lee-type seepage meters were installed at about 200 m offshore from the coast line. The principles of the Lee-type seepage meter are described in detail by Lee (1977). The seawater depths at high tide are 5.3, 6.4, 6.4, 6.9 m at seepage meters St.1 (20 m offshore), St.2 (35 m offshore), St.3 (100 m offshore) and St.4 (200 m offshore), respectively.

Measurements of the SGD flux and water sampling using the Lee-type seepage meter of 32 cm in diameter were undertaken four times at flood tide (FT), high tide (HT), ebb tide (ET) and low tide (LT) during the one tidal cycle between 20 and 21

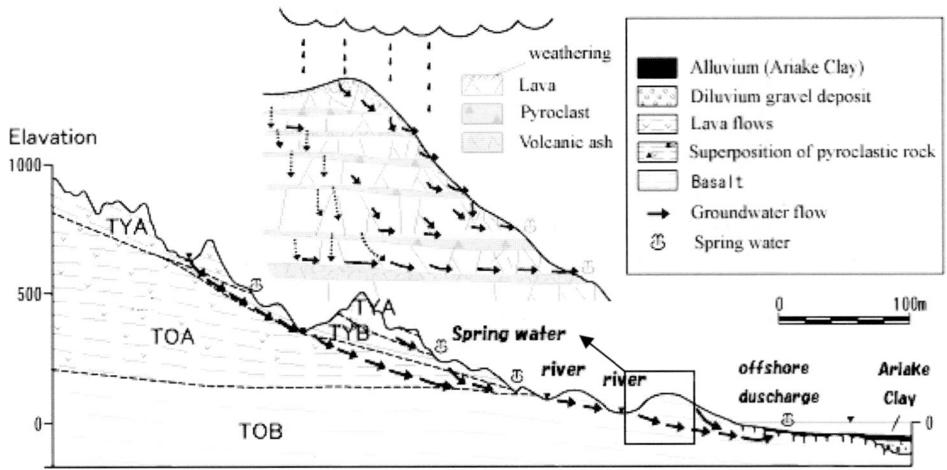

Fig. 4 Schematic cross-sectional image of SGD in the Tara area, Saga, Kyushu Island, Japan.

Table 1 Tidal conditions during the field survey (HT: high tide, LT: low tide).

Date	Sea level: Mean sea level of Tokyo Bay (time, h)	Tide range
20 August 2006	LT −1.22 m (13:05) ~HT 1.4 m (19:49)	2.63 m
21 August 2006	HT 1.76 m (7:31) ~LT −1.48 m (13:57)	3.24 m

August 2006 (Table 1). Moreover, groundwater sampling was done at four points (St.A, St.B, St.C, and St.D) in the coastal region. The sampled water at St.A and St.D represents groundwater discharge from the slope in the vicinity of the observation point of the SGD. St.B is a deep well in the confined aquifer whereas St.D is a disused artesian well about 5 m from ground level.

The chemical components of water samples of SGD, groundwater and seawater were analysed. Water temperature, electrical conductivity (EC), pH, dissolved oxygen (DO), and oxidation–reduction potential (ORP) of water samples were measured in the field using portable meters. Water quality analysis comprised dissolved organic carbon (DOC), total organic carbon (TOC), major ions (cations: Na^+, K^+, Ca^{2+}, Mg^{2+}, Fe^{2+}, Mn^{2+}; anions: Cl^-, HCO_3^-, SO_4^{2-}), nutrients (ammonium-nitrogen: NH_4^--N; nitrate-nitrogen: NO_3^--N; nitrite-nitrogen: NO_2^--N; phosphate: PO_4^{2-}-P), and the ratio of stable isotopes of hydrogen and oxygen (δD:$\delta^{18}O$).

RESULTS AND DISCUSSION

The primary factors controlling the flux of nutrients through coastal aquifers and sediments into coastal waters are the flow paths and rates of the groundwater, since these factors affect the residence time and contact time of the groundwater with the aquifer solids. If the redox potential of groundwater is low then N and P will be strongly affected. Therefore, the analysis of the SGD paths and rates and the evaluation of the nutrient flux affected by the reduction reaction processes are highlighted in the present study.

SGD rates and paths

SGD occurs in response to hydraulic connection and sufficient potential gradient between either the shallow or the deep coastal aquifers and the sea. SGD can be mainly found along the shoreline, or sometimes offshore (Slomp & Van Cappellen, 2004).

The observed SGD rates and EC at each observation point and under each tidal condition are shown in Table 2. From the measurements made by the seepage meter, large SGD rates were observed at St.1 and St.2. Specifically, the observed average SGD rate at St.2 ranged from 24.23 to 15.78 μm/s (average: 25.52 μm/s), while the EC was 32 mS/m, which was close to the freshwater value. On the other hand, the average observed SGD at St.1 was 0.76 μm/s, and the average EC was 2823 mS/m, which was almost half-way between those of freshwater and seawater. There was a small amount of SGD at St.3 and St.4; however, EC was very close to the seawater value, which implies that the measured SGD may contain seawater. This implies that the measured SGD at these points may be contaminated by seawater, suggesting that the SGD might be composed of both fresh groundwater originating from the land region and possible seawater in the aquifer, which has recirculated between the sea and the marine sediment.

Figure 5 shows the distribution of submarine fresh groundwater discharge (SFGD), which is calculated using equation (1) and the measured SGD rates. Equation (1) expresses the mixture ratio of the seawater included in the measured SGD (Taniguchi *et al.*, 2002).

$$q_{SFGD} = \frac{EC_{SGD} - EC_{GW}}{EC_{SW} - EC_{GW}} \cdot q_{SGD} \quad (1)$$

where q_{SGD} is submarine groundwater discharge rate, q_{SFGD} is submarine fresh groundwater discharge rate, EC_{SGD}, EC_{GW}, EC_{SW} are the electric conductivity of SGD, groundwater at the coast region, and seawater at 200 m offshore (St.4), respectively.

The SGD and the SFGD observed at St.2 were almost the same, and the SFGD rate increased at low tide (Table 2). The SFGD component of the measured SGD was 99.6% at St.2, and the evidence of mixing of seawater could be hardly seen. On the

Table 2 The measured SGD rate (μm/s) and EC (mS/cm) (FT: flood tide, HT: high tide, ET: ebb tide, LT: low tide).

Point	Item	Tidal conditions			
		FT	HT	ET	LT
St.1	Rate	1.14	0.36	0.73	0.83
	EC	3155	2815	2695	2625
St.2	Rate	15.78	20.03	22.04	24.23
	EC	27	36	43	21
St.3	Rate	0.18	0.34	0.23	0.31
	EC	3920	3965	4120	4020
St.4	Rate	0.26	0.49	0.13	0.26
	EC	4215	4070	4390	4015

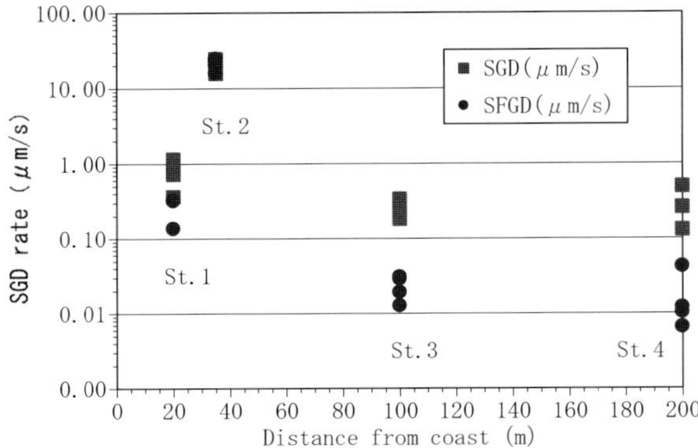

Fig. 5 Distribution of SGD and SFGD rates.

other hand, the SFGD component of the measured SGD at St.1, St.3 and St.4 was 39.4, 8.6 and 5.9%, respectively. The results also suggest that the SGD rate essentially increased from the shoreline to the offshore, but locally considered as a very large value. The SGD through locally developed cracks in the offshore submarine aquiclude from the shallow confined aquifer which consists of the pervious rock (basalt) and aquiclude (pyroclastic rock).

Nutrient fluxes affected by reduction reaction processes

It is important to understand the redox environment of the subsurface in addition to the rates and paths of SGD, since the nutrient transport to the coastal ocean as SGD could be affected by the reduction reaction processes.

Shallow pristine aquifers with short residence times (high recharge and flow rates) and low soil water DOC concentrations may remain largely oxidized. In shallow or deep (intermediate or regional aquifers) with longer residence time and/or higher DOC inputs, dissolved oxygen is usually entirely consumed and organic matter decomposition proceeds via denitrification, Mn and Fe oxide reduction, sulfate reduction and, ultimately, methanogenesis (Lovely & Chapelle, 1995). As a result, the aquifer becomes progressively more reduced toward the direction of groundwater flow (Murphy & Schramke, 1998).

The results of water quality analysis indicate the difference in ORP between the SGD (ORP = 193 mv) and groundwater in the coastal region (ORP = 209 mv), as shown in Table 3. This shows that the SGD is characterized by relatively reductive conditions as compared with the groundwater in the coastal region. Meanwhile, the average NO^{3-}-N concentration in the SGD was 0.75 mg/L, which was low compared to the upstream groundwater (2.44 mg/L). Several studies have shown that a decrease in NO_3^--N concentration in groundwater can be attributed to biochemical denitrification processes (Hill *et al.*, 2000). The reaction formula of this process is:.

$$4NO_3^- + 5C + 3H_2O = 2N_2 + 5HCO_3^- + H^+ \quad (2)$$

Denitrification is the biochemical nitrate reduction, which occurs only in a reduction condition with sufficient organic compounds. Table 3 shows that the HCO_3^-

Table 3 The results of water quality analyses.

Item	Unit	SGD:FT (St.2)	SGD:HT (St.2)	SGD:ET (St.2)	SGD:LT (St.2)	GW (St.A)	GW (St.B)	GW (St.C)	GW (St.D)	SW (St.4)	SW (St.4)
TP	C°	26.9	26.7	26.6	26.8	26.4	26.3	26.4	26.4	26.3	26.4
pH	-	8.2	8.1	8.2	8.1	7.2	8.0	7.4	7.5	7.6	7.7
EC	mS/m	16.9	42.1	22.0	24.7	9.3	11.8	19.4	15.5	3820	3650
DO	mg/L	9.8	9.6	8.7	9.9	9.2	8.8	9.4	9.5	5.9	7.1
ORP	mV	196	192	193	191	213	206	210	208	186	185
DOC	mg/L	0.1	0.2	0.2	0.1	0.2	0.1	0.3	0.1	1.1	1.3
TOC	mg/L	0.2	0.2	0.2	0.2	0.9	0.5	0.9	0.1	1.1	1.3
Na^+	mg/L	17.4	58.3	25.4	30.0	6.4	5.6	13.3	11.7	9310	8240
K^+	mg/L	6.4	5.4	4.0	4.1	1.2	1.6	1.7	1.8	382	358
Ca^{2+}	mg/L	7.6	9.1	8.4	8.8	5.5	11.0	14.8	9.4	466	418
Mg^{2+}	mg/L	4.9	8.5	4.6	5.1	2.4	3.5	4.7	4.7	1040	940
Fe^{2+}	mg/L	0.05	ND	ND	ND	0.09	ND	ND	ND	1.27	1.22
Mn^{2+}	mg/L	0.006	ND	ND	ND	0.027	ND	ND	ND	0.514	0.542
Cl^-	mg/L	18.0	94.2	34.1	42.6	11.6	6.2	21.4	11.2	16700	15800
HCO_3^-	mg/L	52.8	50.5	49.2	49.1	19.1	44.6	31.9	44.1	79.7	80.3
SO_4^{2-}	mg/L	4	16	6	7	2	2	7	6	2500	2300
NH_4^+	mg/L	ND	ND	ND	ND	0.06	ND	ND	0.06	ND	ND
NO_3^-	mg/L	0.75	0.75	0.75	0.75	0.32	1.27	5.31	2.88	0.13	0.10
NO_2^-	mg/L	0.002	0.002	0.002	0.002	0.001	0.002	0.001	0.002	0.026	0.019
TN	mg/L	0.79	0.79	0.79	0.80	0.43	1.31	6.15	3.14	0.69	0.63
PO_4^{2-}	mg/L	0.037	0.041	0.040	0.040	0.016	0.046	0.030	0.036	0.043	0.042
TP	mg/L	0.040	0.043	0.042	0.042	0.062	0.048	0.048	0.038	0.116	0.092
SiO_2	mg/L	31	31	27	29	15	24	46	25	4	4

ND: no detection; SGD: submarine groundwater discharge; GW: groundwater; SW: seawater; FT: flood tide; HT: high tide; ET: ebb tide; LT: low tide; WT: water temperature.

concentrations in the SGD were higher than those of the upstream groundwater, while low TOC was observed in the SGD compared to the shallower groundwater. However, a significant difference in the concentration of $Mn^{2+}$and Fe^{2+} was not seen between SGD and the upstream groundwater. The SGD shows a higher concentration of SO_4^{2-} compared to the upstream groundwater. Similarly, Cl^- concentration in the SGD was higher than in the upstream groundwater. Therefore, it can be assumed that this phenomenon occurred due to the mixing of seawater. The relationship between the SGD and upstream groundwater suggests that the reduction of NO^3-N concentration in the SGD can be attributed to the denitrification process. The denitrification occurred in the SGD, however, it was not observed in the Mn and Fe oxide reduction, and sulfate reduction reaction.

Figure 6 shows the water quality of SGD and upstream groundwater classified by a hexa-diagram. The result shows that the upstream groundwater at St.B is classified into semi-calcium bicarbonate type (Ca-HCO_3), which is generally seen in the deep aquifer. The groundwater at St.A and St.D was classified into non-sodium bicarbonate type (Na-HCO_3). In many cases, the SGD is easily classified into the Na-Cl type, because it often contains seawater and the Cl^- concentration is high. But, the SGD at the flood tide (FT) was classified into Na-HCO_3 type, because of Cl^- concentration of the sample is very low. Then, the SGD water is classified into Na-HCO_3 type, which is similar to the water type of the neighbouring shallow groundwater.

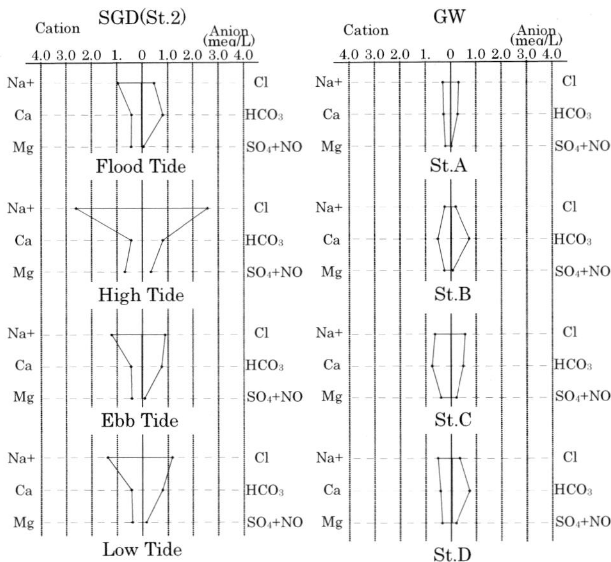

Fig. 6 Water quality classification by hexa-diagram.

Table 4 Nutrient loads at St.2.

Item	Concentration (mg/L)	SGD rate (μm/s)	Loads (g m^{-2} d^{-1})
TN	0.79	20.5	1.40
TP	0.04	20.5	0.07
SiO_2	29.8	205	52.78

Nutrient loads, such as TN, TP, SiO_2, in the SGD estimated at St.2 are presented in Table 4. These loads are larger than those found in previous investigations (Slomp & Van Cappellen, 2004). The average of molar ratio of TN and TP (N/P) in the SGD was 19.2 and this value is larger than the Radfield ratio (N/P = 16), and similar to that of general phytoplankton. As regards dissolved silicate, which is necessary for the proliferation of diatoms, the ratio of TN to DSi in the SGD (0.03) was lower than that in general phytoplankton (1.0). Therefore, the SGD in the study site has comparatively high dissolved DSi, and its value is favourable in terms of the growth of phytoplankton such as diatoms.

Figure 7 shows the relationship between hydrogen and oxygen stable isotopes ($\delta D:\delta^{18}O$) in the SGD of the upstream groundwater and seawater. The observed $\delta^{18}O$ in the sampling SGD indicated that SGD was formed by rain, which had fallen on the land at elevations below 300 m, according to the altitude effect in central Japan (Waseda *et al.*, 1983). Geochemical tracer tests using the hydrogen and oxygen stable isotope ratio ($\delta D:\delta^{18}O$) indicated that SGD would be formed by the infiltrated rainwater at piedmont surface of low altitudes.

The groundwater discharge component of the watershed water budget in the Taradake area seems large, because the area consists of volcanic rocks with high permeability and low river baseflow rate. Therefore, a significant impact of the discharge through the SGD toward the Taradake sea area would be expected.

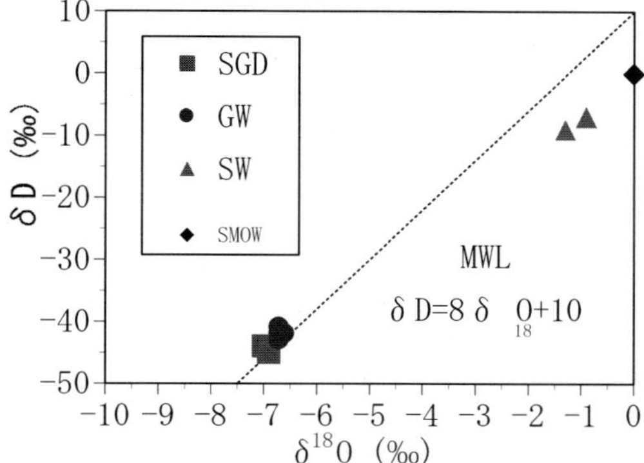

Fig. 7 Relationship between the hydrogen and oxygen stable isotopes (δD:$\delta^{18}O$) in SGD (SMOW: standard mean ocean water, MWL: metric water line, Crage, 1961).

CONCLUSIONS

The present work is intended to take a first step at estimating nutrient inputs through SGD to Ariake Bay by measuring the inputs using a seepage meter. From the field measurements, it was shown that the SGD rate ranged from 0.01 to 20.52 μm/s at St.2. Water quality analyses indicated that SGD water was classified as semi-sodium bicarbonate type, which is similar to the neighbouring shallow groundwater. Moreover, denitrification is likely to be the reduction reaction process for SGD, while $Mn^{2+}$and Fe^{2+} oxide reduction and sulfate reduction are unlikely to proceed. Geochemical tracer tests using the hydrogen and oxygen stable isotope ratio (δD:$\delta^{18}O$) indicated that SGD would be formed by the infiltrated rainwater at the piedmont surface of low altitudes, below 300 m. The SGD is likely to have been formed through locally developed cracks in an offshore submarine aquiclude from a shallow confined aquifer, consisting of pervious rock (basalt) and aquiclude (pyroclastic rock) with shorter residence time. Observed nutrient loads at the St.2 site have high values compared to previous investigations results. Therefore, the SGD found in the area would be a significant source of nutrients to the coastal sea area in Ariake Bay.

Hydrogeological maps and land-use maps in the catchment area of the Ariake Bay are to be compiled in the near future. In addition, a water balance model based on the quasi-3D freshwater–saltwater interface model, incorporating groundwater recharge, will be constructed.

Acknowledgements This research has been supported by Japan Science and Technology Agency.

REFERENCES

Crage, H. (1961) Isotopic variations in meteoric waters. *Science* **133**, 1702–1703.

Hill, A. R., Devito, K. J., Campangnolo, S. & Sanmugadas, K. (2000) Subsurface denitrification in a forest riparian zone: interactions between hydrology and supplies of nitrate and organic carbon. *Biogeochem.* **51**, 193–223.

Lee, D. R. (1977) A device for measuring seepage flux in lake and estuaries. *Limnol. Oceanogr.* **22**, 140–147.

Lovely, D. R. & Chapelle, F. H. (1995) Deep subsurface microbial processes. *Rev. Geophys.* **33**, 365–381.

Murphy, E. M. & Schramke, J. A. (1998) Estimation of microbial respiration rates in groundwater by geochemical modeling constrained with stable isotopes. *Geochim. Cosmochim. Acta* **62**(21/22), 3395–3406.

Slomp, C. P. & Van Cappellen, P. (2004) Nutrient inputs to the coastal ocean through submarine groundwater discharge: controls and potential impact. *J. Hydrol.* **295**, 64–86.

Taniguchi, M., Burnett, W. C., Cable, J. E. & Turner, J. V. (2002) Investigation of submarine groundwater discharge. *Hydrol. Processes* **16**, 2115–2129.

Waseda, A. & Nakai, N. (1983) Isotope composition of natural water in central Japan and Tohoku, Japan. *J. Geochem. Soc. Japan* **17**, 83–91.

Evaluation of fresh groundwater contributions to the nutrient dynamics at shallow subtidal areas adjacent to metro-Bangkok

YU UMEZAWA[1], TOMOTOSHI ISHITOBI[1], SOMPOP RUNGSUPA[2], SINICHI ONODERA[3], TSUTOMU YAMANAKA[4], CHIKAGE YOSIMIZU[5,6], ICHIRO TAYASU[6], TOSHI NAGATA[6], GULLAYA WATTAYAKORN[2,7] & MAKOTO TANIGUCHI[1]

1 *Research Institute for Humanity and Nature, 457-4, Motoyama, Kita-ku, Kyoto 603-8047, Japan*
yu@chikyu.ac.jp

2 *Aquatic Resources Research Institute, Chulalongkorn University, Phyathai Road, Patumwan, Bangkok 10330, Thailand*

3 *Faculty of Integrated Arts and Sciences, Hiroshima University, 1-7-1, Kagamiyama, Higashi-Hiroshima 739-8521, Japan*

4 *Terrestrial Environment Research Center (TERC), University of Tsukuba, 1-1-1, Ten-nou dai, Tsukuba 305-8577, Japan*

5 *CREST, Japan Science and Technology Agency, 4-1-8 Honcho Kawaguchi, Saitama, Japan*

6 *Center for Ecological Research, Kyoto University, 509-3, 2-chome, Hirano, Otsu, Shiga 520-2113, Japan*

7 *Department of Marine Science, Faculty of Science, Chulalongkorn University, Phyathai Road, Patumwan, Bangkok 10330, Thailand*

Abstract Both submarine groundwater discharge (SGD) and the Chao Phraya River are major agents of nutrient supply into the Gulf of Thailand. With the development of the city of Bangkok, however, lowered groundwater levels due to over-pumping suggests a decrease of the fresh groundwater flux into the sea. In this study, time-series resistivity monitoring under the seabed adjacent to Bangkok city did not actually show any evidence of fresh groundwater fluxes, and δD and $\delta^{18}O$ signatures in porewater also followed this phenomenon. Consequently, the observed upward water flux can be mainly attributed to recirculation of the overlying water. $\delta^{15}N$ and $\delta^{18}O$ values in nitrate suggested that nitrate was mainly supplied via the river, and rapidly reduced in the surface suboxic sediment, while re-mineralized ammonium and phosphate were substantially released into the overlying water. River water-derived nutrient could be still important as original sources of organic matter, even at the area where high amounts of SGD is observed.

Key words groundwater; resistivity measurement; stable isotopes; Thailand; Chao Phraya River

INTRODUCTION

Groundwater is a fundamental resource providing reliable and low-cost water for domestic, industrial and agricultural purposes. Thus many Asian cities have depended on groundwater for sustenance and used the resource to facilitate economic activity. Furthermore, fresh groundwater is an important pathway for bringing land-derived nutrients to coastal waters (e.g. Umezawa *et al*., 2002).

During the progress of urbanization and intensive dwelling in city areas, however, discharge of septic and industrial wastes has caused severe groundwater contamination by nutrients such as nitrate, and several toxic metals. These contaminants are a potential threat to human health through drinking water, and also affect the primary production of adjacent coastal ecosystems.

Burnett *et al.* (2007) reported that the seepage of fresh groundwater with high dissolved inorganic nitrogen (DIN) concentrations could have important implications for the magnitude and type of productivity in the coastal waters in Upper Gulf of Thailand, because this area was nitrogen-limited from the viewpoint of the Redfield ratio. A complicating factor, however, is over-pumping of groundwater from deep aquifers lying underneath the cities. In Bangkok, continuous groundwater pumping of over 1.5 million m^3/day during recent decades has caused severe subsidence throughout the city areas (e.g. Phien-wej & Nutalaya, 2006), and has dramatically changed hydraulic potentials in the aquifers. Therefore, intrusion of saltwater into the aquifers along the coastal area is another significant impact (e.g. Das Gupta, 1985).

The objectives of this study were: (a) to confirm whether fresh groundwater fluxes exist along the coastal line, where groundwater pumping has been intensively conducted, and (b) to better understand related nutrients dynamics at shallow inter- and sub-tidal areas. At shallow coastal areas, the term ''groundwater'' includes both fresh-water with land-derived nutrients, and recirculated seawater with marine-borne nutrients, including mineralized ones from sediments. To separate each source of groundwater, conductivity is efficiently used in the case of areas where river input is minor (Taniguchi *et al.*, 2006). In the Gulf of Thailand, however, about half of the river flow to this system was provided through the Chao Phraya River, and so salinity in the water column is very low around the river mouth, especially in the rainy season (Burnett *et al.*, 2007), suggesting that conductivity is unlikely to be an effective tool to separate fresh groundwater from the brackish water. In this study, therefore, we tested the use of stable isotopes in water to separate them, in combination with other physical approaches such as resistivity monitoring. The coupled N and O isotopes of nitrate with/without nitrite were also analysed to identify the source of nitrate and to understand its transformation at the shallow estuary.

MATERIALS AND METHODS

Study site and equipments settings

The Bangkok metropolitan area is located about 25 km to the north of the Gulf of Thailand, on the flood plain of the Chao Phraya River. Dry and wet seasons clearly exist, and river discharge in the wet season amounts to 10 times of that in the dry season (Dulaiova *et al.*, 2006). Although subsidence in the centre of the city area mostly stopped after the regulation of groundwater pumping, the largest subsidence rate is still observed around the industrial southeast suburb areas in 2005.

To monitor fresh groundwater discharge from the shallow coastal areas adjacent to the metro-Bangkok area, a transect line was selected about 10 km eastward from the river mouth (Fig. 1(b)).

Continuous heat-type automated seepage meters (i.e. vented benthic chambers to automatically measure water flow), were deployed on the subtidal seabed at each of four stations (A to D; Fig. 1(c)) from 20 to 23 June 2006. The average depth of the

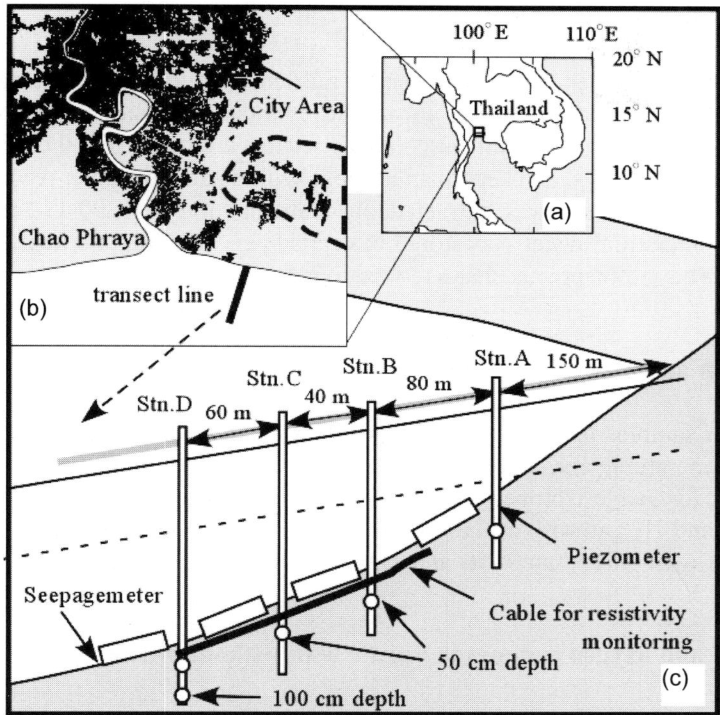

Fig. 1 Location of study area, and setting design of instruments (piezometers and seepage meters) along the transect line. In (b), enclosed areas with a broken line indicate the areas, where subsidence of over 30 mm/year is currently observed, probably due to groundwater pumping (Phien-wej & Nutalaya, 2006).

seepage meter during the observation period was about 1.4, 1.6, 1.7 and 1.9 m at Stations A, B, C and D, respectively. Two set of piezometers (ϕ = 12 mm; vinyl chloride), each of which had an intake with a strainer at 50 cm and 100 cm depth, respectively, were also installed at each station. The water accumulated inside the piezometric tube was discarded using tygon tubing and a plastic syringe 1 hour before the sampling. The porewater sample recharged during the subsequent 1-hour period was collected every 3 hours (low, flood and high tide) on 23 June, and individually stored for the analysis of each chemical component (see below). Seawater overlying the sediment at each station, several river water samples and one offshore water sample (13°19′55″N, 100°49′54″E) were also collected for chemical analysis.

Resistivity under the seabed was measured along the transect line using a Sting R1 IP/Swift (American Geophysical Instrument) every 1 hour from low tide (08:00 h) to high tide (16:00 h) on 22 June. About 14 probes were installed along the 140–m transect (interval length between probes was 10 m). The Schlumberger method and RES2DINV version 3.50 (Geotomo Software) were used for the resistivity analyses. Tidal shift, current speed and direction were continuously monitored using a CTD sensor (HOBO U20-001-01) and an electromagnetic current meter (COMPACT-EM) at Stn D.

Nutrient profiles in sediment

The sediment samples for the analysis of nutrient profiles were collected in duplicate at the same four stations using an acrylic cylinder with silicon stoppers. After the recovery on the boat, the sediment core was sectioned into 1.0 to 2.0 cm intervals, transferred into plastic bags, and kept in an ice box. Within half a day after recovery, interstitial water in the sediment was extracted by centrifugation at 1200 G for 20 min. The extracted interstitial water was stored in capped acrylic tubes and kept frozen for later analysis. The sediment was characterized mostly as clay throughout the transect line.

Hydrogen and oxygen isotopes in water

The water samples for the stable isotope ratios in water, i.e. δD and $\delta^{18}O$, were kept at room temperature, and analysed by mass-spectrometry (Finnigan Mat252). As pretreatment for stable isotopes analysis, water samples were equilibrated with CO_2 gas for $\delta^{18}O$ and H_2 gas with a platinum catalyst for δD. The reproducibility of $\delta^{18}O$ and δD values was better than 0.1‰ and 1.0‰, respectively.

Nitrogen and oxygen isotopes in nitrate with/without nitrite

The water samples for the nutrients and their stable isotopes analyses were processed using 0.45 μM cellulose acetate filter in the field, and kept frozen until the analyses. Nutrient concentrations ([NO_3^-], [NO_2^-], [NH_4^+], and [PO_4^{3-}]) in river water, seawater and porewater were measured colorimetrically with an autoanalyser (AACS III, BRAN+RUEBBE). The stable nitrogen and oxygen isotope ratios, i.e. $\delta^{15}N$ and $\delta^{18}O$, in nitrate + nitrite were determined with the "denitrifier" method of Sigman *et al.* (2001) and Casciotti *et al.* (2002). For the samples that included substantial amounts of nitrite (over 10% of nitrate), $\delta^{15}N$ and $\delta^{18}O$ were determined both for nitrate with and without nitrite, using the mass spectrometer (Finnigan Delta^{plus}XP). For the latter sample, nitrite was excluded in advance using ascorbic acid ($AscH_2$), following Granger *et al.* (2006). Isotope values were calibrated using internationally recognized nitrate standard USGS34 ($\delta^{15}N$ of –1.8‰ and $\delta^{18}O$ of –27.9‰), USGS35 ($\delta^{15}N$ of 3.6‰ and $\delta^{18}O$ of 51.5‰) and laboratory working standard ($\delta^{15}N$ of 1.6‰ and $\delta^{18}O$ of 23.6‰). Based on replicate measurements of standards and some samples, the analytical precision for $\delta^{15}N$ and $\delta^{18}O$ was generally better than ±0.2‰ and ±0.4‰, respectively.

RESULTS

Seepage meter and physical conditions

Specific flow rates obtained every minute at each station by automated seepage meters ranged widely from 1.0 to 110 cm/day (Fig. 2). Although the flow rate showed large variation at any given time, the average flow rate at each location during the monitoring period was 17.7 ± 5.3, 8.8 ± 4.8, 1.2 ± 0.2 and 21.4 ± 14.1 cm/d at Stns A,

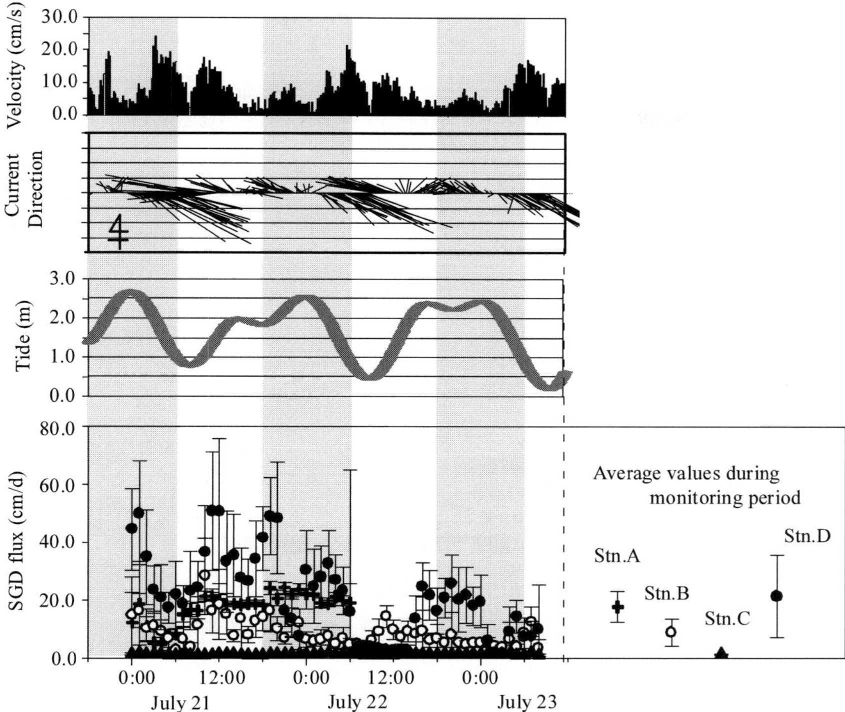

Fig. 2 Time-series data of current velocity and direction at Stn D, and SGD fluxes monitored at Stns A, B, C and D. The SGD data collected every 1 min was compiled into 1-hour average data with S.D. The small flux speed was converted to the daily flux. Average data throughout the monitoring period were separately plotted. Shaded areas mean night time.

B, C and D, respectively. The magnitude of the flux at Stn B was high at low tide, while at Stn D a higher flux was observed randomly at flood and ebb tide.

Higher current velocities, 15 to 20 cm/s, were recorded at flood and ebb tide, while the velocities at low and high tide were a fifth or sixth of the maximum ones.

Resistivity monitoring

Resistivity usually reflects the degree of conductivity and difference of sedimentation and geological characteristics. When the resistivity is continuously monitored at same location, therefore, the change of resistivity could show the change of conductivity (Taniguchi *et al.*, 2006). The distribution of resistivity below 10.0 m depth and onshore areas looked steady condition, while it fluctuated in surface layers according to the tidal shift (Fig. 3).

Hydrogen and oxygen isotopes in water

The δD and $\delta^{18}O$ concentrations showed large variation, but specific values were observed for each water source (Fig. 4). For instance, δD (–7.8‰) and $\delta^{18}O$ (–1.0‰),

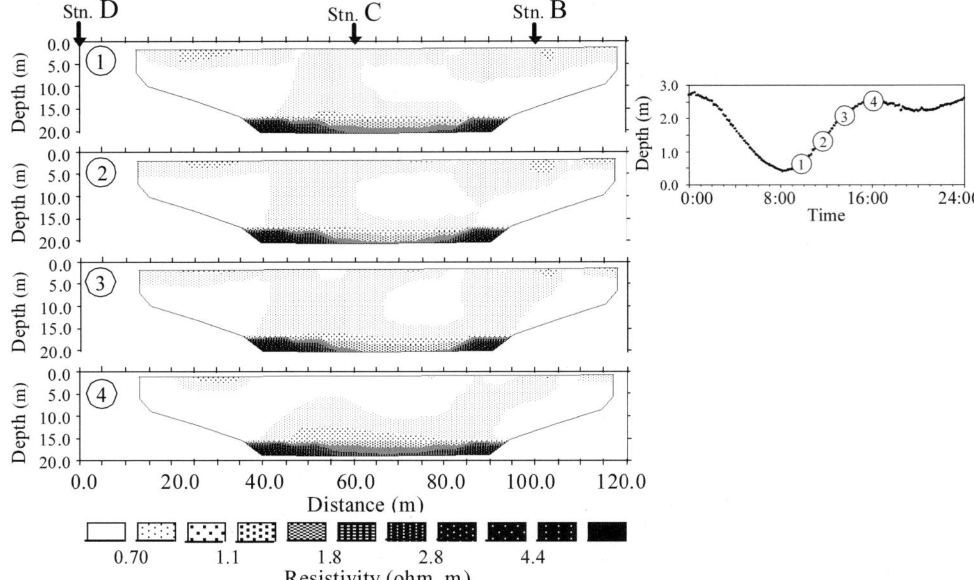

Fig. 3 Resistivity cross-section in a 140 m long transect. The sections 1 to 4 show the situation from low tide to high tide. The *y*-axis shows the depth of the sea bottom.

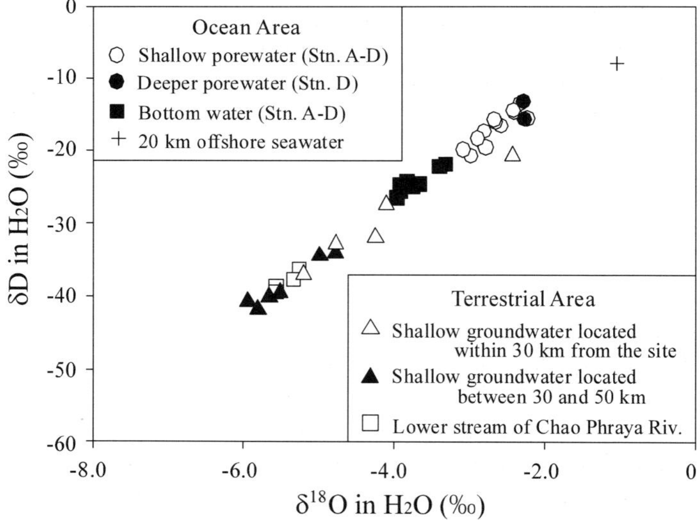

Fig. 4 Variation of δD and $\delta^{18}O$ values in H_2O collected from different water sources around the monitoring station. The data for fresh groundwater in the wells located within 30 or 50 km of the study site were provided by T. Yamanaka (unpubl. data).

for offshore water salinity which was 8.2 psu, were significantly heavier than the others. Values in both fresh groundwater and river water were relatively low. But it was still hard to differentiate groundwater and river water from these signatures alone, because the values in the well groundwater showed variation depending on the locations.

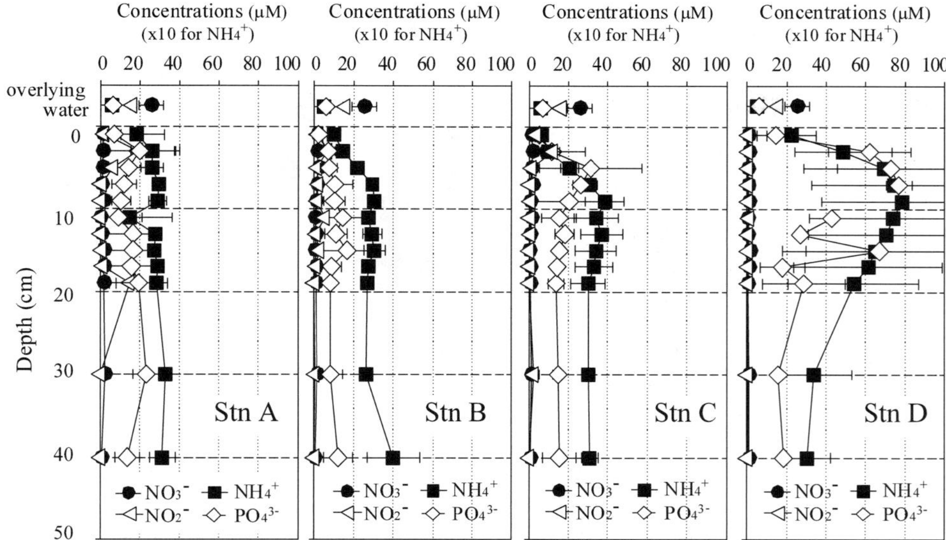

Fig. 5 Vertical profiles of porewater nutrient contents at stations A, B, C and D (from the left). Those in overlying water are also plotted at the top of the profiles using the same symbols.

The values in porewater and overlying bottom water were intermediate between the fresh water and offshore water. However, the characteristics of the offshore water were more similar to that of deeper porewater than the bottom water and shallower porewater. Because these values lined up tightly on the straight line ($y = 7.37x + 2.06$, $r^2 = 0.98$), it was assumed that the porewater and overlying bottom water were formed by simple mixing of fresh ground/river water and offshore water. Some outliers of the groundwater collected from the well might have shown the effect of evaporation or connate water.

Nutrient in sediment and overlying water

Low concentrations of nitrate and nitrite (i.e. 0.8 to 1.5 μM for NO_3^-, 0.8 to 3.5 μM for NO_2^-, Fig. 5) and high concentrations of ammonium and phosphate (i.e., 62.5 to 230 μM for NH_4^+, 2.5 to 15.0 μM for PO_4^{3-}, Fig. 5) were observed in interstitial water of surface sediments at each station, suggesting that suboxic conditions were maintained throughout the shallow subtidal areas. Especially at Stn D, extremely high ammonium and phosphate concentrations were detected at around 5 to 15 cm depth. On the other hand, overlying bottom water, which was collected several times at each location, had relatively steady values independent of location and tidal shift (i.e. 27.5 ± 2.6, 12.0 ± 0.6, 49.2 ± 6.9 and 5.6 ± 0.6 μM or NO_3^-, NO_2^-, NH_4^+ and PO_4^{3+}, respectively; $n = 9$).

Nitrogen and oxygen isotopes in nitrate

$\delta^{15}N$ and $\delta^{18}O$ in nitrate with/without nitrite were analysed for the samples from river water, bottom water, offshore water and porewater, which had enough amounts of

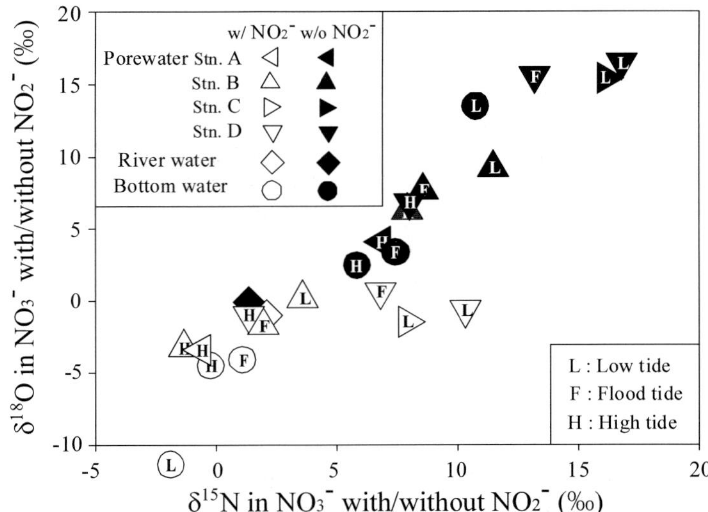

Fig. 6 $\delta^{15}N$ and $\delta^{18}O$ values of nitrate + nitrite (open) and nitrate alone (solid) from river water, bottom water and porewater collected in specific tidal situation at Stn A, B, C and D. The samples for nitrate + nitrite and nitrate analyses were divided from the same sampling bottle.

nitrate for the analyses (i.e. over about 2 μM). The amounts of nitrite in these samples were comparable to nitrate contents, and accounted for 40 to 120% of them.

$\delta^{15}N$ and $\delta^{18}O$ in river water nitrate were around 1.5‰ and –0.5‰, respectively, irrespective of whether it included nitrite or not (Fig. 6). However, those values in nitrate from the other sources became heavier when they were analysed after removing nitrite. In other words, $\delta^{15}N$ and $\delta^{18}O$ in nitrite were significantly lighter than those in nitrate. For example, $\delta^{15}N$ and $\delta^{18}O$ in a bottom water sample shifted to 5.9‰ and 2.5‰ from –0.3‰ and –4.5‰, respectively. As for the samples from piezometers, $\delta^{15}N$ in nitrate + nitrite at C and D were about 5‰ heavier than those at A and B. Furthermore, after the manipulation to remove nitrite, the increase in the ratio of $\delta^{18}O$ to $\delta^{15}N$ (i.e. $\Delta\delta^{18}O : \Delta\delta^{15}N$) was over 2.0 for the samples at C and D, and about 1.0 at A and B.

DISCUSSION

Dynamics of SGD at the subtidal area adjacent to metro-Bangkok

As shown in the resistivity data (Fig. 3), it was unlikely that dynamic mixing between intruded seawater and fresh groundwater occur in the inter- and sub-tidal areas adjacent to the Bangkok metropolitan area. Because the overlying water consists of discharged river water and seawater, depending on the tidal cycles, small change of resistivity in the surface sediment seems to directly reflect the characteristics of the overlying water.

Curious δD and $\delta^{18}O$ signatures of the porewater, which were intermediate between the offshore water and the overlying bottom water, may correspond to such an inactive mixing in the salinity transition zone. Around the shallow areas at Chao

Phraya River mouth, salinity drastically changes between the rainy season and dry season (Burnett *et al.*, 2007). So we assume that the overlying water with high salinity during the dry season still remained in the deeper sea bed, and only the porewater in the surface layer was replaced by the current overlying water with low salinity during our study period, i.e. at the beginning of the rainy season. From the behaviour of radon concentrations around Chao Phraya River, Dulaiova *et al.* (2006) proposed that fresh groundwater flow would be directed down gradient along topographic contours and thus would tend to be concentrated toward the central axis of the river valley or just beyond its mouth. Other than intensive pumping of groundwater, this may be another explanation of the small flux of fresh groundwater at inter- and sub-tidal areas close to metro-Bangkok city.

Consequently, the upward water flow observed by seepage meters can be mostly attributed to the recirculation of the overlying water. Although a major mechanism of seawater intrusion is tidal oscillation in the coastal area (Prieto & Destouni, 2005; Burnett *et al.*, 2007), the correlation between SGD flux and tidal shift was not clear in our transect (Fig. 2). Intensive study is required to investigate the agents enhancing the SGD flux, focusing on other potential sources like surge currents with higher velocity.

Nutrient behaviour around Chao Phraya River mouth

Burnett *et al.* (2007) concluded that DIN supplied by SGD amounted to 40–50% of that delivered by the Chao Phraya River, despite the small flux of freshwater via SGD. As for the interaction between the sediment and water column, however, only a uni-directional flux from the sediment through the dynamic SGD was taken into account in the estimation in their paper. Here, therefore, the potential of nutrient transportation via bidirectional diffusion fluxes was compared with that via advective flux (Table 1). The vertical fluxes of DIN/P by diffusion between sediments and overlying water were calculated using the nutrient concentration in both (Fig. 5) based on the Fick's First Law of diffusion (detail in Table 1). The advective flux was roughly calculated by multiplying seepage flow rate (cm/d) and nutrient concentrations in the surface porewater (0–2 cm depth). Due to the high concentrations of nitrate and nitrite in the overlying water, delivered by river water, these components showed a downward flux into the sediment by diffusion. In contrast, ammonium and phosphate showed upward fluxes into the overlying water, but the total fluxes were considerably smaller compared to the fluxes via advective flow (Table 1).

$\delta^{15}N$ and $\delta^{18}O$ in nitrate became heavier with a ratio increase of 1.0 or 2.0 for $\delta^{18}O$ to $\delta^{15}N$, after removing nitrite. Denitrification in the ocean water preferentially consumes isotopically lighter nitrate, so its occurrence leads to a marked increase in $\delta^{15}N$ and $\delta^{18}O$ of nitrate by 20–30‰ for the isotopic effect at suboxic areas (e.g. Sigman *et al.*, 2003). However, sedimentary denitrification in a variety of environments causes very little net isotope enrichment of oceanic nitrate, probably due to the limited rate of nitrate supply to denitrifying bacteria (e.g. Brandes & Devol, 1997). This apparent difference of isotopic fractionation factors between sediments and open waters could be a potential tool for estimating the locations and condition where denitrification is occurring in this estuary. Because tidal oscillation-derived water flow

Table 1 Estimated nutrient fluxes (mmol m^{-2} d^{-1}) between sediments and overlying waters. Positive values mean the flux is from sediment into the overlying water.

Station	Nitrate		Nitrite		Ammonium		Phosphate	
	Diffn*	Advn	Diffn*	Advn	Diffn*	Advn	Diffn*	Advn
A	–0.2	0.3	–0.1	0.35	1.3	32.0	0.01	2.74
B	–0.2	0.1	–0.1	0.15	0.5	8.6	–0.02	0.44
C	–0.2	0.0	–0.1	0.04	0.1	0.8	0.00	0.13
D	–0.2	0.2	–0.1	0.16	1.8	48.3	0.05	6.98

* Fick's 1st law of diffusion; $J = -\Phi D'(\Delta C/\Delta Z)_{Z=0}$ — (1) where J is the diffusive flux of the nutrient species (μg cm^{-2} s^{-1}), Φ is the porosity of that species and D' is the diffusion coefficient in pore water (cm^{-2} s^{-1}). The porosity (Φ) was estimated following to equation (2): $\Phi = \rho / (\rho + (1 - \omega)/ \omega)$ — (2) where ρ is density of the sediments, and ω is water content ratio. $(\Delta C/\Delta Z)_{Z=0}$ is the gradient of dissolved nutrient concentrations across the sediment–water interface, and represented by equation (3) (Rutgers *et al.*, 1984): $(\Delta C/\Delta Z)_{Z=0} = (C - C_0)/0.5\ L$ — (3) where C is the nutrient concentration in the top of the sedimental segment with thickness L and C_0 is the usual concentrations in the overlying water at each area. L was substituted by 2.0, and the spatially averaged values in overlying water was used as C_0 (i.e. 27.5 μM for NO_3^-, 12.0 μM for NO_2^-, 49.2 μM for NH_4^+ and 5.6 μM for PO_4^{3-}) uniformly at all locations. The molecular diffusion coefficient (Ds) was represented approximately by equation (4) for muddy sediments (Yamamoto *et al.*, 1998): $D' = D^0(1 + \alpha t)\Phi^2$ — (4) where D^0 is D' at 0°C and the value for each species are cited in Yamamoto *et al.* (1998) (i.e. 0.978×10^{-5} $cm^2\,s^{-1}$ for NO_3^-, 0.922×10^{-5} $cm^2\,s^{-1}$ for NO_2^-, 0.98×10^{-5} $cm^2\,s^{-1}$ for NH_4^+ and 0.61×10^{-5} $cm^2\,s^{-1}$ for PO_4^{3-}), α is invariable depending on species (i.e. 0.048 for cation and and 0.040 for anion; Lerman, 1979). Temperature was uniformly substituted by 28.0°C.

and wind-derived shear stress actively disturb the surface layer of the sediment, the nitrate supply rate into the surface sediment could be high. Therefore, continuously diffused nitrate into the suboxic sediment or overlying water may be rapidly reduced with significant isotopic discrimination. On the other hand, further investigation is needed to identify the source of nitrate with heavier $\delta^{15}N$, which was distinctly different from the river-borne nitrate (Fig. 6). Determination of $\delta^{15}N$ values in organic nitrogen and/or ammonium in the sediment may demonstrate the potential for nitrification in local areas.

As also suggested in Burnett *et al.* (2007), re-circulated seawater can be directly important to nutrient dynamics in the Upper Gulf of Thailand, even in circumstances where fresh groundwater discharge is not quantitatively important. However, high concentrations of ammonium and phosphate in surface suboxic sediment can depend on the supply of allochthonous and autochthonous organic matter. The discharges of nitrogen (N) and phosphorus (P) into the Chao Phraya River from the Bangkok city area are estimated to account for 97% and 41% of the total loss, respectively (Færge *et al.*, 2001). Therefore, river water-derived nutrient could still be important as original N and P sources, even if high amounts of SGD with nutrients is observed at these areas.

Acknowledgements We thank the staff of the Southeast Asian Fisheries Development Center (SEAFDEC) for providing accommodation and laboratory space. T. Hosono (RIHN) and V. Monyrath (Chiba Univ.) helped us process the samples. This study was conducted in the project "Human Impacts on Urban Subsurface Environment" at the Research Institute for Humanity and Nature, and partly supported by CREST (JST).

REFERENCES

Brandes, J. A. & Devol, A. H. (1997) Isotopic fractionation of oxygen and nitrogen in coastal marine sediments. *Geochim. Cosmochim. Acta* **61**, 1793–1801.

Burnett, W., Wattayakorn, G., Taniguchi, M., Dulaiova, H., Sojisuporn, P., Rungsupa, S. & Ishitobi T. (2007) Groundwater-derived nutrient inputs to the Upper Gulf of Thailand. *Continental Shelf Res.* **27**, 176–190.

Casciotti, K. L., Sigman, D. M., Hastings, M. G., Bohlke, J. K. & Hilkert, A. (2002) Measurement of the oxygen isotopic composition of nitrate in seawater and freshwater using the denitrifier method. *Anal. Chem.* **74**(19), 4905–4912.

Das Gupta, A. (1985) Simulated salt-water movement in the Nakhon-Luang Aquifer, Bangkok, Thailand. *Ground Water* **23**, 512–522.

Dulaiova, H., Burnett, W. C., Wattayakorn, G. & Sojisuporn, P. (2006) Are groundwater inputs into river-dominated areas important? The Chao Phraya River, Gulf of Thailand. *Limnol. Oceanogr.* **51**, 2232–2247.

Færge, J., Magid, J. & Penning de Vries, F. W. T. (2001) Urban nutrient balance for Bangkok. *Ecological Modelling* **139**, 63–74.

Granger, J., Sigman, D. M., Prokopenko, M. G., Lehmann, M. F. & Tortell, P. D. (2006) A method for nitrite removal in nitrate N and O isotope analyses. *Limnol. Oceanogr. Meth.* **4**, 205–212.

Lerman, A. (1979) *Geochemical Processes Water and Sediment Environment*, John Wiley & Sons, New York, USA.

Phien-wej, N., Giao, P. H. & Nutalaya, P. (2006) Land subsidence in Bangkok, Thailand. *Eng. Geol.* **82**, 187–201.

Prieto, C. & Destouni, G. (2005) Quantifying hydrological and tidal influences on groundwater discharges into coastal waters. *Water Resour. Res.* **41**, W12427, doi:10.1029/2004WR003920.

Rutgers van der Loeff, M. M., Anderson, L. G., Hall, P. O. J., Iverfeldt, A. , Josefson, A. B., Sundby, B. & Westerlund, S. F. G. (1984) The asphyxiation technique: An approach to distinguishing between molecular diffusion and biologically mediated transport at the sediment-water interface. *Limnol. Oceanog.* **29**(4), 675–686.

Sigman, D. M., Casciotti, K. L., Andreani, M., Barford, C., Galanter, M. & Bohlke, J. K. (2001) A bacterial method for the nitrogen isotopic analysis of nitrate in seawater and freshwater. *Anal. Chem.* **73**(17), 4145–4153.

Sigman, D. M., Robinson, R., Knapp, A. N., van Geen, A., McCorkle, D. C., Brandes, J. A. & Thunell, R. C. (2003) Distinguishing between water column and sedimentary denitrification in the Santa Barbara Basin using the stable isotopes of nitrate. *Geochem. Geophys. Geosyst.* **4**(5), 1040.

Taniguchi, M., Ishitobi, T. & Shimada, J. (2006) Dynamics of submarine groundwater discharge and freshwater-seawater interface. *J. Geophys. Res.* **111**, C01008, doi:10.1029/2005JC002924.

Umezawa, Y., Miyajima, T., Kayanne, H. & Koike, I. (2002) Significance of groundwater nitrogen discharge into coral reefs at Ishigaki Island, southwest of Japan. *Coral Reefs* **21**, 346–356.

Yamamoto, T., Matsuda, O., Hashimoto, T., Imose, H. & Kitamura, T. (1998) Estimation of benthic fluxes of dissolved inorganic nitrogen and phosphorus from sediments of the Seto Inland Sea. *Research on Ocean* **7**, 151–158 (in Japanese with English abstract).

A New Focus on Groundwater–Seawater Interactions
(Proceedings of Symposium HS1001 at IUGG2007, Perugia, July 2007). IAHS Publ. 312, 2007.

Influence of groundwater discharge through a coastal sandy barrier in southern Brazil on seawater metal chemistry

HERBERT WINDOM[1], WILLARD MOORE[2] & FELIPE NIENCHESKI[3]

1 *Skidaway Institute of Oceanography, 10 Ocean Science Circle, Savannah, Georgia 31406, USA*
herb.windom@skio.usg.edu

2 *Department of Geological Sciences, University of South Carolina, Columbia, South Carolina 29208, USA*

3 *Department of Chemistry, Fundação Universidade Federal do Rio Grande, Brazil*

Abstract Sandy barriers developed during the Holocene transgressing sea are common features of many coastal regions throughout the world. We present here the results of the study of groundwater–surface water interactions associated with a 600-km barrier which created the Mirim-Patos Lagoon system, the largest in South America. Results show that the composition of the groundwater discharge to the ocean from these permeable sands differs significantly in metal concentrations from those of surface freshwater–sea-water mixtures, primarily as a result of redox processes. Estimates of the volume of freshwater transport and seawater cycling through the sands, and metal concentrations in surface waters and in groundwaters, were used to estimate metal fluxes between compartments of this system and indicate that iron, manganese, cobalt and perhaps vanadium, cadmium, copper and zinc are being enriched in adjacent coastal waters, whereas uranium and perhaps molybdenum are being depleted.

Key words metals; coastal ocean; groundwater discharge; Brazil

INTRODUCTION

There is a growing body of evidence indicating the importance of groundwater pathways for the transport of solutes in permeable sediments and aquifers at the land-sea boundary. Inputs may consist of freshwater, seawater-freshwater mixtures and/or recirculated seawater; these are collectively referred to as submarine groundwater discharge (SGD). Whether SGD originates in freshwater aquifers or includes or is dominated by infiltrated seawater, its composition can be quite different from surface waters and its effect on the coastal ocean can be significant. Groundwater discharge from coastal aquifers can transport material to the ocean from land, in addition to rivers, and often at much higher concentrations.

We have been involved in a study of groundwater–surface water interactions associated with Patos Lagoon (Fig. 1), on the extreme southern coast of Brazil, in an attempt to understand better the exchange processes involving groundwater pathways in permeable sediments. The Patos Lagoon system is the largest in South America and is relatively pristine, with little human development surrounding it, especially on the barrier spit separating it from the sea. Given the geological and hydrological characteristics of the Patos Lagoon system, we hypothesized that a relatively large

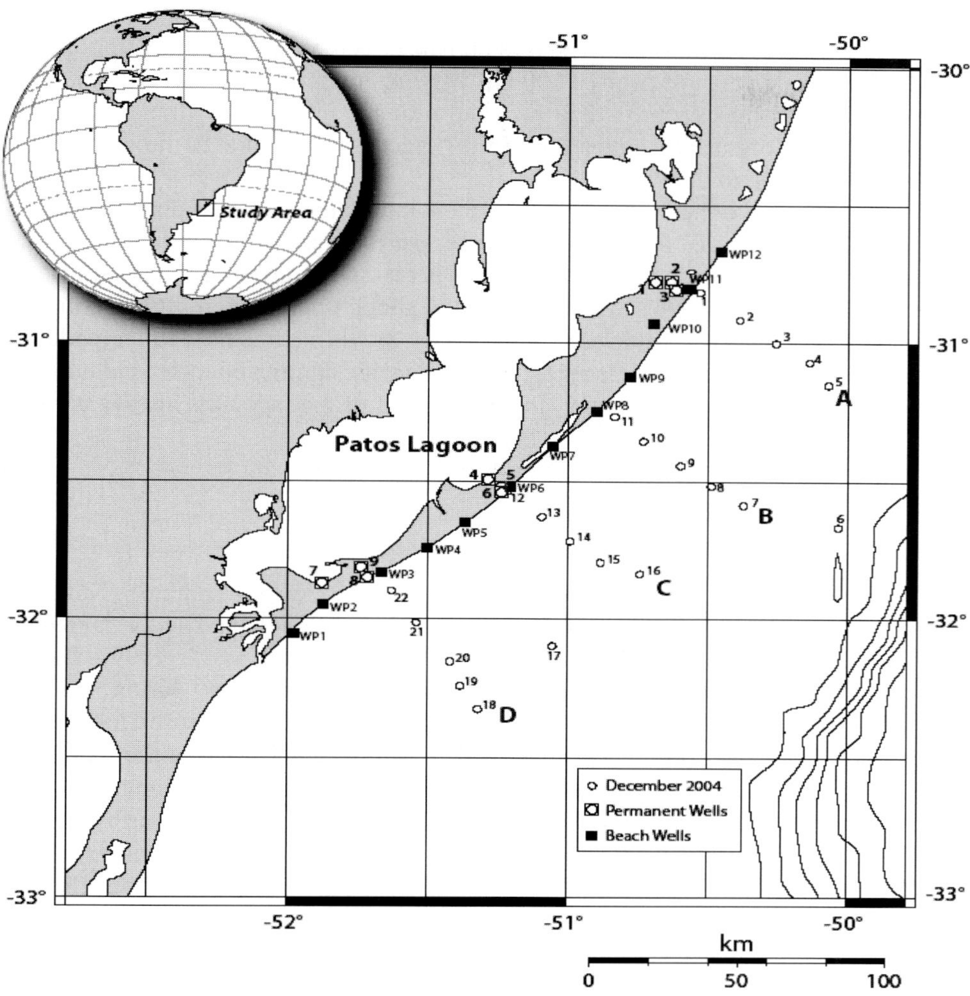

Fig. 1 Patos-Mirim Lagoon system. The Mirim Lagoon is partially shown in the lower left corner of the figure. Its only outlet to the ocean is through the Patos Lagoon. Map of the study area shows locations of permanent wells, beach groundwater and surf zone stations and offshore hydrographic and water sampling stations. The latter are located along four transects referred to in the text, starting from the north, as Transects A, B, C and D.

amount of freshwater flow must occur through the barrier, where it mixes with seawater within the permeable sediments underlying the barrier spit. Here we report the results of trace metal (Fe, Mn, Co, U, V, Mo, Cu, Cd, and Zn) analyses of groundwater recovered from permanent wells installed within the lagoon barrier and from temporary well points deployed along the adjacent beach face. These results are compared to metal concentrations measured in surface waters in the region to assess the impact of SGD on them. Constrained by previously reported Ra measurements (Windom *et al.*, 2006), a simple conceptual model is used to estimate metal fluxes in this groundwater–surface water system.

The Patos-Mirim Lagoon system represents the major coastal expression of a Holocene Post-Glacial Marine Transgression barrier complex (Dillenburg *et al.*, 2002; Fig. 1). It covers an area of over 10 000 km^2. The major surface contact with the sea is a relatively small inlet (approx. 800 m wide) near the city of Rio Grande. This also provides the major surface outlet for Mirim Lagoon, which drains through the Sao Goncalo Channel, into the Patos Lagoon.

High discharge to the lagoon occurs in the winter (June–August) during which the entire lagoon system is fresh. During these periods, circulation within the Patos Lagoon is driven by fresh water discharge whereas wind drives circulation at other times (Möller *et al.*, 1996). Tides (generally about 0.5 m) exert little control on circulation or water level within the lagoon except at its entrance near Rio Grande. But winds and freshwater discharge can result in seasonally significant up-set and down-set of water levels. Generally, however, the water level of the lagoon is always 0.5 m or more above sea level over most of the lagoon.

METHODS

Field sampling campaigns were conducted during the Austral summer (November 2003 and December 2004) when freshwater discharge to the Patos Lagoon is minimal and the salinity of the outflow at its mouth is high (Windom *et al.*, 1999). Data on wind direction and velocity for one week prior and the week during our field campaigns were obtained from a meteorological station at the Fundação Universidade Federal do Rio Grande (FURG) near the mouth of the lagoon. During the November, 2003 campaign mean wind direction was onshore for 4 days with maximum velocities of 6 m s^{-1}, from the south 2 days at less than 2 m s^{-1}, and from the north 9 days at approx. 3 m s^{-1}. During the December 2004 campaign, mean wind direction was onshore 4 days at 2–4 m s^{-1} and 10 days from the north at 2–5 m s^{-1}. The Rio de la Plata discharge is also lowest during the Austral summer and the southward along-shore wind stress further minimizes the intrusion of any freshwater from this source into the study area under these conditions (Piola *et al.*, 2005). This also applies to the discharge from Patos Lagoon.

Sampling methods have been described elsewhere (Windom *et al.*, 2006) and included the following types of samples:

- Permanent wells across three transects of the Patos Lagoon barrier (PLB) with each transect consisting of three sets of wells, one near the lagoon, one in the interior and one near the beach front. Each set generally included three wells screened nominally at 5, 10 and 15 m (Fig. 1).
- Beach groundwater using a drive-point peizometer system (Charette & Sholkovitz, 2006) to sample along the beach of the PLB down to a maximum depth of 8 m, but most samples were collected at shallower depths of a metre or two. Sampling sites were spaced at ~20 km intervals on the southernmost 240 km of the PLB ocean beach (Fig. 1).
- Surf zone samples collected at the same sites as the beach groundwater samples were taken.
- Offshore surface seawater samples collected in December 2004 along four transects perpendicular to shore using the research vessel Atlantico Sul (Fig. 1).

– Surface water samples collected along a salinity gradient starting in Patos Lagoon and extending through its inlet to offshore just south of Transect D (Fig. 1) during June 2001 when freshwater discharge is high and surface water metal variations should be influenced more by river–seawater mixing end members.

Details of metal analyses have been previously described (Windom & Niencheski, 2003; Windom *et al.*, 2006) and involved ultra clean techniques to minimize contamination. Detections limits are generally at least an order of magnitude below the lowest concentration reported below and precision of analysis at concentrations typical of seawater concentrations are generally better than ±10%.

RESULTS

The mean and median metal concentrations in the various compartments of this groundwater–surface water system (Table 1) indicate considerable groundwater enrichment of Fe, Mn and Co above typical levels in surface waters whereas the

Table 1 Summary of average and median metal concentrations in waters of the study area collected in November 2003 and December 2004 compared to average ocean values taken from Bruland (1983).

	Salinity	Fe (μM)	Mn (nM)	Co (nM)	U (nM)	V (nM)	Mo (nM)	Cu (nM)	Cd (pM)	Zn (nM)
Permanent wells										
2003										
Average	0.55	12.7	4351	21	0.27	12.0	6.3	1.7	163	15
Median	0.1	1.7	2781	4	0.13	7.0	2.2	1.2	92	10
2004										
Average	0.2	21.2	2041	24	0.14	9.2	3.5	1.7	57	19
Median	0.1	4.4	1461	4	0.11	5.5	1.2	1.4	42	16
Beach wells										
2003										
Average	11.1	27.8	4324	2.4	1.47	10.6	41.1	1.1	59	11
Median	8.3	22.6	2146	1.9	0.21	6.05	38.2	0.7	41	10
2004										
Average	22.9	24.1	4395	4.5	3.00	9.5	45.2	2.3	54	8
Median	24.5	20.8	3733	1.8	2.47	7.5	54.1	2.3	46	8
Surf zone										
2003										
Average	30.5	4.2	101.3	1.0	10.0	34.8	68.4	5.0	153	9.7
Median	30.3	1.2	60.5	0.8	11.7	35.4	74.0	5.5	143	7.3
2004										
Average	33.6	1.0	27.5	1.0	11.9	39.6	70.2	7.4	175	7.6
Median	34.1	1.0	26.6	1.0	11.6	41.1	67.5	7.4	143	6.7
Shelf										
Average	35.5	0.06	2.3	0.4	13.5	25.9	91.8	2.2	41.8	0.6
Median	35.5	0.04	1.9	0.4	13.3	26.6	93.5	2.1	42.6	0.6
Ocean										
Average		0.001	0.5	0.02	13.8	30	110	4.0	700	6
Surf. depleted		0.0001	0.2	0.01		20		0.5	1	0.05

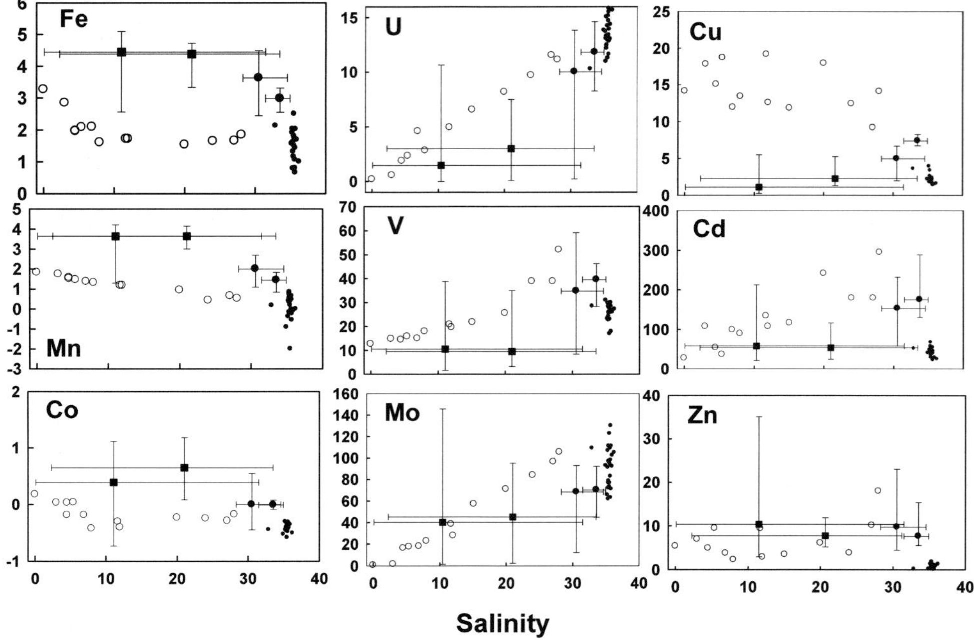

Fig. 2 Mean and range of metal concentrations (in nM except for Cd which is in pM) in beach groundwater (filled squares), and surf zone samples (filled circles, left is 2003 and right is 2004) compared to those in samples collected along a salinity transect in June 2001 (open circles) and shelf water samples collected in December 2004 (filled circles).

concentrations of the other metals, with the exception of Cu, are similar to uncontaminated surface freshwater levels or equivalent surface freshwater–sea water mixtures. Copper concentrations in groundwaters are almost a factor of ten lower than natural surface freshwater values.

Mean concentrations and ranges of metal concentrations in beach groundwater and in samples from the adjacent surf zone for the summer 2003 and 2004 sampling campaigns are shown in Fig. 2 in relation to concentrations observed for surface water samples collected along a salinity gradient during the winter 2001, and for surface shelf water samples collected from offshore transects during December 2004. Results indicate that groundwater concentrations of Fe, Mn and Co (Fig. 2) are virtually always greater than equivalent surface freshwater–sea water mixture. Groundwater U, V, Cu and Cd concentrations (Fig. 2) are generally always less than equivalent surface waters, while Mo and Zn concentrations in ground waters overlap those of surface waters.

DISCUSSION

We use the results presented above, along with the estimates of submarine groundwater discharge (SGD) from Windom *et al.* (2006), based on radium isotopic tracers, to estimate metal fluxes between compartments of this coastal groundwater–

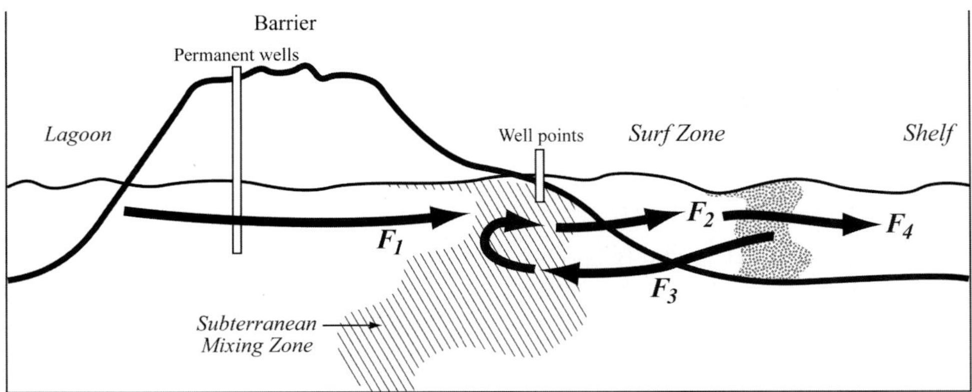

Fig. 3 Conceptual model of groundwater–surface water system.

surface water system. Our conceptual model of how this system operates is as follows is shown in Fig. 3. This is similar to the subterranean estuary studied by Charette & Sholkovitz (2006). By knowing the SGD water flux we can estimate the relative fluxes shown in the conceptual model shown in Fig. 3 starting with the SGdD flux (F_2) which was estimated by Windom *et al.* (2006) using radium isotope tracers as summarized below and based on groundwater and surface water data collected during the December 2004 campaign:

Windom *et al.* (2006) used radium isotopes as tracers to estimate the SGD flux following the method used by Moore (2003). For this estimate they assumed that radium additions to coastal surface waters occurs only in the near-shore zone due to SGD, net cross-shelf advection is zero, cross-shelf mixing is constant and the system is steady state. Given these assumptions, the cross shelf radium isotope gradients in the upper stratified layer of transects A and B must be balanced by the near-shore SGD flux of radium and radioactive decay (there are no additional inputs from the bottom). Using a model based on these assumptions and the average ^{228}Ra concentration of beach groundwater, Windom *et al.* (2006) estimated the SGD to be 8.5×10^7 m^3 d^{-1} for the 240 km study coastline. The average salinity of the SGD was assumed to be 22.9, the average of the beach groundwater, while that of the surf zone was 34.6. Thus, based on salt balance, 3.1×10^7 m^3 d^{-1} is the freshwater component (Q_{fw}) of the SGD. The remainder (5.4×10^7 m^3 d^{-1}) must be balance by recirculated seawater, so that:

$$SGD = Q_{fw} + Q_{sw}$$

where Q_{sw} is the seawater flux into the permeable sands that is ultimately recycled in the SGD. And for mass balance, assuming steady state, the net offshore transport of water due to mixing must equal Q_{fw}.

Referring to the conceptual model presented in Fig. 3, we can estimate the metal flux associated with the various water fluxes discussed above. These fluxes, which would include those in the freshwater moving toward the ocean (F_1), in SGD (F_2), in seawater recirculated through permeable sediments (F_3), the flux to the inner shelf (F_4), and the flux (not shown in Fig. 5) necessary to sustain the observed mean concentrations of metals observed in inner shelf waters (F_5). These can be estimated using the following equations:

$$F_1 = [C]_{fw} \times Q_{fw} \tag{1}$$

$$F_2 = [C]_{bgw} \times SGD \tag{2}$$

$$F_3 = [C]_{sz} \times (SGD - Q_{fw}) \tag{3}$$

$$F_4 = \{[C]_{bgw} \times SGD\} - \{[C]_{sz} \times (SGD - Q_{fw})\} = F_2 - F_3 \tag{4}$$

$$F_5 = [C]_{shelf} \times (V_{surface\ layer}/\tau) \tag{5}$$

where $[C]_{fw}$, $[C]_{bgw}$ $[C]_{sz}$ and $[C]_{shelf}$ are the mean concentrations of a given nutrient in permanent wells, beach groundwater, surf zone, and inner shelf, respectively; $V_{surface\ layer}$ is the volume of the inner shelf surface layer (240 km × 22 km × 10 m) used by Windom *et al.* (2006) to calculate a residence time, τ, of 8.3 days using radium isotopic tracers. Because we only have radium tracer data from our December 2004 campaign, from which we can estimate SGD, the flux estimates presented below are based only on the results from samples collected during that campaign. Table 2 gives the results of these calculations along with estimated metal fluxes from the nearby Parana River based on average metal concentrations in world rivers (Martin & Whitfield, 1981; Martin & Windom, 1991) and an average discharge of 14 000 m^3/s). From these data we make the following conclusions:

The estimated fluxes of Fe, Mn and Co in SGD, F_2, cannot be sustained by the F_1 flux to the freshwater–seawater mixing zone represented by beach groundwater. The estimated sea water flux back to the groundwater, F_3, is small and leads to the implication that the F_4 flux to the shelf is more than sufficient to maintain observed metal concentrations. We must therefore conclude that a large part of the SGD flux of these metals is removed relatively rapidly in the near shore to reconcile these flux estimates. The accumulation of iron associated with the fresh-sea water boundary in subterranean mixing zone has been demonstrated by Charette *et al.* (2005) and Windom *et al.* (2006) have shown that the concentrations of iron in surface waters of this region decrease exponentially going offshore. Nonetheless, the data plotted in Fig. 2, demonstrate that the groundwater source provides and important third end-member input, along with surface freshwater and ocean end members, necessary to explain observed surface shelf water Fe, Mn and Co concentrations. The SGD flux for these three metals is also about twice that estimated for the Parana.

Flux estimates for U suggest that this metal is being removed in the near shore (i.e. more is entering the groundwater system in recirculated sea water than is being removed in SGD). And while this estimated removal rate is on the order of one percent

Table 2 Estimated fluxes, as identified in Fig. 5, compared to those for the Parana River. Underlined flux estimates likely having the greatest uncertainty.

Fluxes Moles/day	Fe 10	Mn $\times 10^4$	Co	U	V	Mo	Cu	Cd	Zn
F1	66	6.3	740	4.0	280	110	53.0	17.0	590
F2	205	37	380	250	810	3850	190	4.5	660
F3	5.1	0.1	50	640	2140	3790	400	10.0	410
F4	200	37	3300	–390	–1330	60.0	–210	–5.5	250
F5	36.4	1.5	20000	86000	16600	58600	14000	267000	4000
Parana	87	18	2000	1200	24000	7000	28000	120	11000

of the flux necessary to maintain the observed mean shelf concentration, the results shown in Fig. 2 for the U-salinity relationship, suggests that this removal may be affecting shelf water concentrations more than the calculations in Table 2 reflect. A similar conclusion was made by Charette & Sholkovitz (2006) for a coastal system along the northeast Atlantic coast of the USA.

For V, Cd and Cu, results of flux calculations suggest that these metals are being removed in the near shore. Unlike U, which is conservative in the ocean, V, Cd and Cu are depleted in surface waters. During the Austral summer, the influence of the Brazil current on shelf waters was significant (Windom *et al.*, 2006) and the ocean water end member of surface waters in this region should be depleted in these three elements (Bruland, 1983) and the results shown in Fig. 2 (for V) and Fig. 4 (for Cu and Cd) indicate a low ocean end member and suggest that SGD acts as a source for these elements. A likely explanation is that these metals are relatively insoluble in suboxic groundwater (Cu and Cd as sulphides) and the resuspension of particles in the near-shore leads to their remobilization. Thus redox processes in the near shore would invalidate the flux calculations for these elements, suggesting that processes associated with SGD provided a sink for these elements when, in fact, this system acts as a source providing a third end member for the surface water concentrations shown in Fig. 2. It is interesting to note that the estimated Cd flux from the Parana is clearly insufficient to support the shelf (F5) flux, suggesting another source, whereas its Cu and Zn fluxes are of the order of magnitude.

For Zn, the flux calculations suggest that SGD is a source and the relation between surface water concentrations shown in Fig. 2 suggests this source provides a third mixing end member similar to the results shown for V, Cu and Cd.

The calculated flux to the shelf, F_4, suggests that that SGD may provide a positive input of Mo. But as indicated in Table 2, this calculation has a high uncertainty, while results shown in Fig. 2 suggest that the near shore acts as a sink, similar to the U results. Windom & Neincheski (2003) demonstrated that U and Mo, which are conservative in the ocean, were depleted in groundwater mixtures of freshwater and seawater and attributed this removal as being mediated by microbial processes.

CONCLUSIONS

The results reported here show that SGD may provide both sources and sinks for metals in coastal ocean regions characterized by permeable sediments such as barriers of coastal lagoons. They also suggest that the subterranean transport pathway may be important in explaining surface water mixing relationships of metals. Although flux calculations based on water balance are informative regarding the magnitude of fluxes, they do not take into account processes occurring at interfaces. Groundwater transport, seawater cycling through permeable coastal/shelf sediments and SGD along the coast of southern Brazil is a complex process deserving greater attention both here and in other coastal areas dominated by permeable sands.

Acknowledgements This work was support in part by the National Science Foundation (OCE-0233465, HLW & OCE-0233657, WSM) and CNPq (Brazil– Grants 490126/2003-0 and 301219/2003-6).

REFERENCES

Bruland, K. W. (1983) Trace elements in seawater. In: *Chemical Oceanography*, vol. 8, 157–220. Academic Press, London, UK.

Charette, M. A. & Sholkovitz, E. R. (2006) Trace element cycling in a subterranean estuary: Part 2. geochemistry of the pore water. *Geochim. Cosmochim. Acta* **70**(4), 811–826.

Charette, M. A., Sholkovitz, E. R & Hansel, C. M. (2005) Trace element cycling in a subterranean estuary: Part 1. geochemistry of permeable sediments. *Geochim. Cosmochim. Acta* **69**(4), 2095–2109.

Dillenburg, S. R., Roy, P. S., Cowell, P. J. & Tomazelli, L. J. (2002) Influence of antecedent topography on coastal evolution as tested by the Shoreface Translation-Barrier Model (STM). *J. Coast. Res.* **16**, 71–81.

Martin, J.-M. & Whitfield, M. (1981) River input of elements to the ocean. In: *Trace Metals in Sea Water* (ed. by C. S. Wong, E. Boyle, K. W. Bruland, J. D. Burton & E. D. Goldberg), 265–296. Plenum Press, New York, USA.

Martin, J.-M. & Windom, H. L. (1991) Present and future role of ocean margins in regulating marine biogeochemical cycles of trace elements. In: *Ocean Margin Processes in Global Change* (ed. by R. F. C. Mantoura, J.-M. Martin & R. Wollast), 45–67. John Wiley & Sons, Chichester, UK.

Moore, W. S. (2003) Sources and fluxes of submarine groundwater discharge delineated by radium isotopes. *Biogeochemistry* **66**, 75–93.

Piola, A. R., Matano, R. P., Palma, E. D., Möller, O. O., Jr & Campos, E. J. D. (2005) The influence of the Plata River discharge on the western South Atlantic shelf. *Geophys. Res. Lett.* **32**, LXXXXX, doi:10,1029/2004GL021638.

Windom, H. & Niencheski, L. F. (2003) Biogeochemistry in a freshwater–seawater mixing zone in permeable sediments along the coast of Southern Brazil. *Mar. Chem.* **83**, 121–130.

Möller, O. O., Lorenzzentti, J. A., Stech, J. L. & Mata, M. M. (1996) The Patos-Lagoon summertime circulation and dynamics. *Cont. Shelf Res.* **16**, 335–351.

Windom, H. L., Niencheski, L. F. & Smith, R. G., Jr (1999) Biogeochemistry of nutrients and trace metals in the estuarine region of the Patos Lagoon (Brazil). *Estuar. Coast. Shelf Sci.* **48**, 113–123.

Windom, H., Moore, W. S., Niencheski, L. F. & Jahnke, R. (2006) Submarine groundwater discharge: a large, previously unrecognized source of dissolved iron to the South Atlantic Ocean. *Mar. Chem.* **102**, 252–266.

Chemical and isotopic characteristics of stagnant water isolated in a coastal area

YASUNORI MAHARA[1], EIJI NAKATA[2], TAKAHIRO OOYAMA[2], KIMIO MIYAKAWA[2], YOSHIHISA ICHIHARA[3] & HIROYUKI MATSUMOTO[3]

1 *Research Reactor Institute, Kyoto University, Kumatori, Osaka 590-0494, Japan*
mahara@hl.kyoto-u.ac.jp

2 *Civil Engineering Research Lab., CRIEPI, Abiko, Chiba 270-1194, Japan*

3 *Kushiro Coal Mine Co. Ltd, Kushiro, Hokkaido 085-0811, Japan*

Abstract Groundwater was investigated at the Kushiro Coal Mine after the mining area was extended by up to 8.5 km off the Pacific shore and 700 m below m.s.l. Three different types of water were found in the mine. The first is freshwater, which flows from the land and is present in the shallow mining area down to 150 m depth. The second is saline water with chloride ranging 5 000–22 000 mg/L, which was found in water drops from the tunnel ceilings throughout the entire undersea mining area and has the chemical properties of altered present-day seawater. The third is estimated to be fossil seawater; it is very saline water from boreholes, and has homogeneous chemical and isotopic properties. Based on the ratio of $^{36}Cl/Cl$, the altered seawater had been in the Cretaceous formations and isolated from groundwater–seawater mixing for more than 2 million years, despite being located in a coastal area.

Key words groundwater–seawater interaction; fossil seawater; groundwater residence time; radiogenic chlorine-36; secular equilibrium ratio

INTRODUCTION

The Kushiro Coal Mine (formerly called the Taiheiyou Coal Mine) has been mining from the land, offshore, under the sea for approximately 86 years since 1920. The mining area under the Pacific Ocean has been extended to a maximum of 8.5 km offshore and 700 m depth below mean sea level (b.m.s.l). According to mining records, mining tunnels were excavated along the Harutori coal-bearing layers formed in the Palaeogene period. Also, the inside tunnel has stayed very dry, and there is little seepage of water from the tunnel walls and floor except for water-drops from the ceilings. But as the underlying Cretaceous beds, which are rich in natural gas and pressurized very saline water, is below the coal formation, some boreholes were drilled from the tunnel floor towards the formation to prevent gas explosions and flooding. Water flow from these boreholes stops within a few months or years after drilling, and dries up. In this study, we investigated the chemical and isotopic characteristics in flowing saline water compared with other water collected in the mining area and the present seawater overlying the undersea coal mine. Furthermore, we estimated the isolated duration from the growth of the radiogenic $^{36}Cl/Cl$ ratio under the *in situ* weak neutron activation.

MATERIALS AND METHODS

Geological setting of study area and chloride ion concentration contour map

The geological setting around the Kushiro Coal Mine has the upper Cretaceous Nemuro formation (Mesozoic era) composed of sandstone and mudstone, as the basement rock. Palaeogene, Neogene and Quaternary formations widely overlay the basement rock from the bottom towards the top, sequentially (Sato & Sato, 1980). The mining tunnels were excavated along the coal-bearing formation, located just above the Cretaceous (Fig. 1).

The chloride-ion concentration varies horizontally from low (0–1000 mg/L) to high (>17 000 mg/L), increasing in the offshore direction (Fig. 2). However, vertically it increases up to 22 000 mg/L in the Palaeogene formation and decreases to 13 300 mg/L in the Cretaceous, increasing with depth below sea level (Fig. 1).

Groundwater sampling method

We collected nine seepage water samples from tunnel walls and floors, 45 water-drop samples from the ceiling of the tunnel, and 10 flowing water samples from boreholes drilled toward the Cretaceous. The flowing groundwater from the boreholes were characterized by chloride-ion concentrations of 13 000–15 000 mg/L, shown by the grey-interior circles in Fig. 2. We paid great attention to using the closed sampling system in order to prevent evaporation from the small water-drop samples. We only collected groundwater samples for measurement of $^{36}Cl/Cl$ ratio from relatively large flowing or seepage volumes to avoid air-borne ^{36}Cl contamination.

Analytical methods

Tritium The concentration of tritium was measured by counting beta rays after electrolytic enrichment. One litre of groundwater was reduced to about 40 mL by electrolysis using Ni/Fe electrodes. Forty millilitres of distillate and 60 mL of scintillation cocktail were mixed in a 100-mL Teflon vial. This mixture was analysed over 1000 min using a low-background liquid-scintillation counter. The detection limit for this method is 0.3 TU.

Cations and anions The concentrations of dissolved cations (Na^+, K^+, Ca^{2+} and Mg^{2+}) and anions (Cl^-, SO_4^{2-} and HCO_3^-) were measured using ICP and liquid chromatography, respectively. The analytical procedures followed the standard methods for the examination of water (Clesceri *et al.*, 1989).

Stable isotopes The ratio of deuterium to hydrogen (D/H) was measured by mass spectrometry after reducing water to hydrogen gas using metallic uranium heated to 600°C. The $^{18}O/^{16}O$ ratio was measured by mass spectrometry after reaching isotopic exchange equilibrium between water and gaseous CO_2. Both D/H and $^{18}O/^{16}O$ ratios were expressed as deviations per thousand (‰) from standard mean ocean water (SMOW), following the conventional method defined by Craig (1961). The measurement uncertainties were ±1‰ for D/H and ±0.1‰ for $^{18}O/^{16}O$.

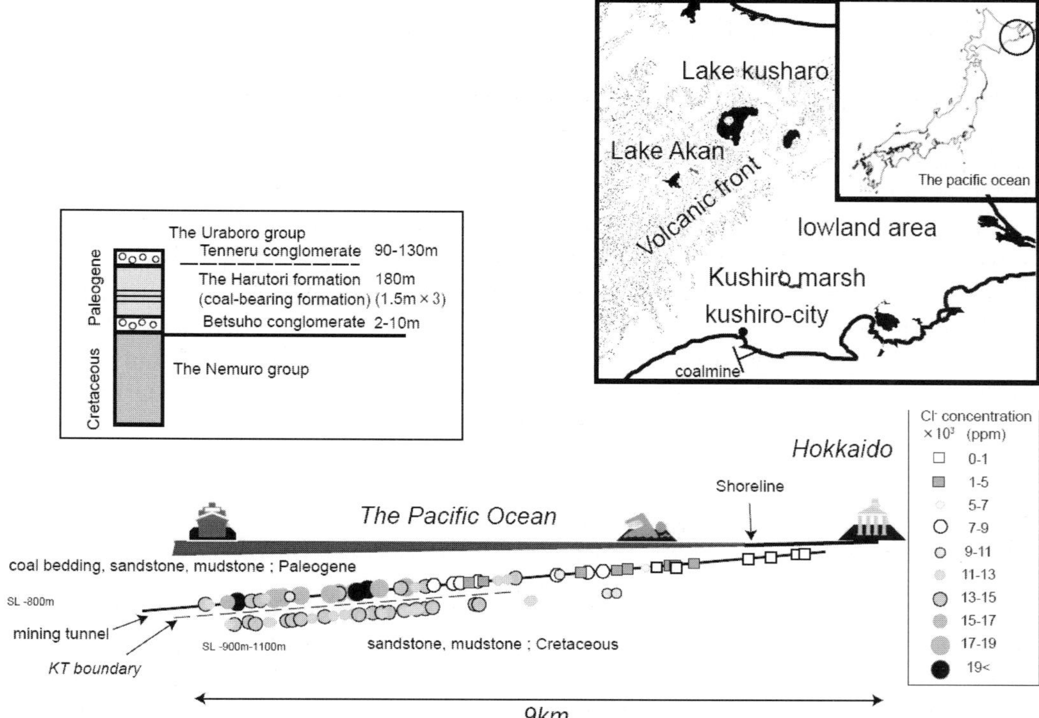

Fig. 1 Location of the study area in Hokkaido in Japan and geological setting of study area (vertical section of the Kushiro Coal Mine) and vertical distribution of chloride ion concentration in the mining area.

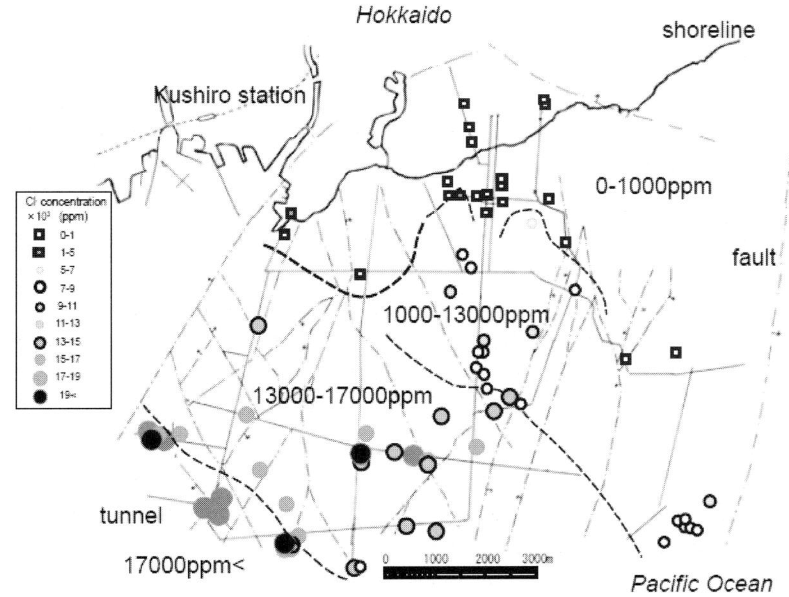

Fig. 2 The horizontal distribution of chloride ion concentration in the mining area.

$^{36}Cl/Cl$ ratio Chloride was precipitated as AgCl in a clean room. The precipitate was purified by repeated re-dissolution in NH_4OH and re-precipitation (after the removal of sulfur as $BaSO_4$). All AgCl precipitation samples for acceleration mass spectrometry (AMS) measurement were made following the standard procedures used at the Australian National University (ANU) (Creswell, 2001). The $^{36}Cl/Cl$ ratio for the groundwater samples was measured using AMS facilities at ANU and ETH, Zurich.

U and Th in the rock A sample of 0.5 g of pulverized rock powder was placed in a Teflon beaker and completely dissolved by heating on a hot plate after adding concentrated HNO_3 and HF. The dissolved material was heated again until $HClO_4$ fuming after adding concentrated HNO_3 and $HClO_4$. The residual dried material was re-dissolved with a small amount of weak HCl, and diluted further to 100 ml using super-pure distilled water. The U and Th concentrations were measured using inductively coupled plasma-mass spectrometry (ICP-MS).

RESULTS AND DISCUSSION

Vertical distribution of chloride ion concentration

Figure 3 shows the vertical relationship between changes in chloride ion concentration and sampling depth. We can estimate the location of the interface of freshwater and seawater from the changing chloride ion concentration. The location of the boundary between the freshwater flow region and seawater can be deduced from a drastic change in the chloride concentration at around 450 m b.m.s.l. The chloride concentration increases to 22 000 mg/L, which is greater than that of the present seawater, down to a depth of 600–700 m b.m.s.l. Although different mechanisms of concentrating chloride are possible, e.g. strong evaporation in the very dry mining tunnel, long-term water rock interaction (Frape *et al.*, 1984), re-dissolution of residual sea salt by pore water,

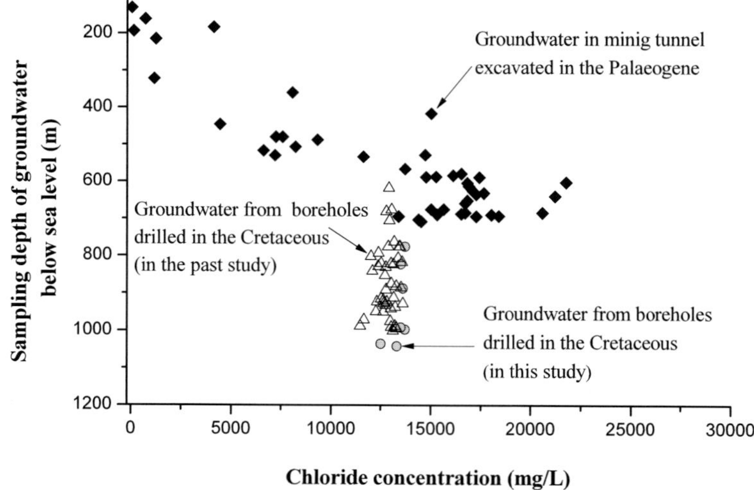

Fig. 3 Vertical chloride ion concentration with groundwater sampling depth.

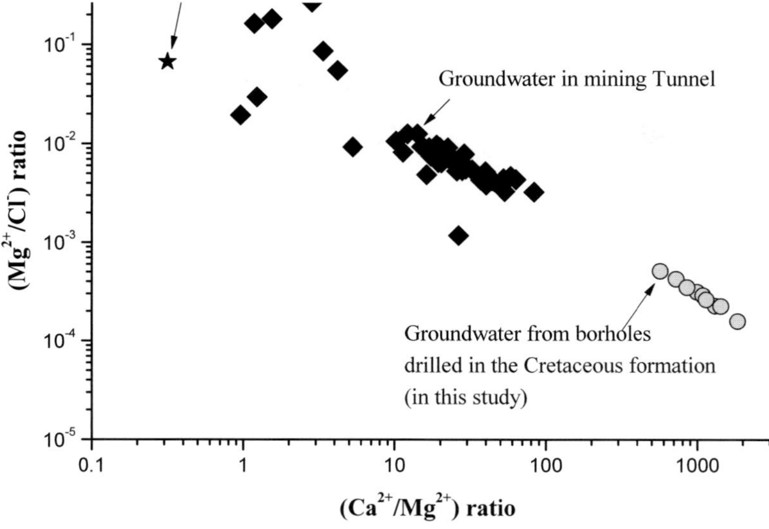

Fig. 4 Correlation between ratio of Mg^{2+} and Cl^- and ratio of Ca^{2+} and Mg^{2+} in the modern seawater and groundwater collected in the Kushiro Coal Mine.

or squeezing of pore water from the formation under a high geo-pressure, more detailed investigations and additional studies are required to adequately explain the formation of the excess chloride concentration.

In contrast, chloride concentration in the Cretaceous formation suddenly drops to 12 000–14 000 mg/L at depths of 800–1000 m b.m.s.l. The chloride concentration is very stable and nearly constant (an average of 13 300 mg/L) in groundwater flowing from boreholes. Its chemical properties are drastically different from the freshwater, shallow brackish water, and very saline water in the Palaeogene formation. The cation exchange between Mg^{2+} and Ca^{2+} is especially marked (Fig. 4), owing to water–rock interaction.

A certain boundary appears to exist between the Cretaceous and Palaeogene. The groundwater appears to not exchange or mix between the two different geological formations, as evidenced by the sudden change in chloride concentration, change in other chemical properties, and the fact that water flow from boreholes stops within a few years after drilling without recharging, and dries up.

Origin of waters

All the stable isotopes (δD and $\delta^{18}O$) data are displayed in Fig. 5. Most data from fresh and brackish groundwater, collected at shallower than 450 m b.m.s.l., are aligned with the global meteoric water line (GMWL). This indicates that the shallow groundwater originates from rainwater.

Saline water having chloride concentrations in excess of 5000 mg/L is lined up on a correlation straight line with a slope of 4.7. The magnitude of this slope is expected in evaporation, ion filtration (Coplen & Hanshaw, 1973) or mixing with seawater.

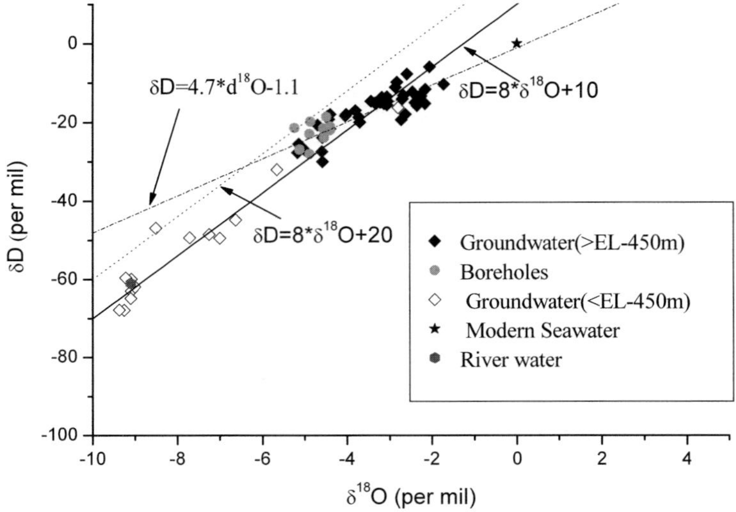

Fig. 5 Correlation between δD and $\delta^{18}O$ for estimation of origin of groundwater (the global meteoric water line: $\delta D = 8 \times \delta^{18}O + 10$, the meteoric water line in northeast Japan: $\delta D = 8 \times \delta^{18}O + 20$, the estimated mixing line in groundwater collected from 400 to 700 m b.m.s.l.: $\delta D = 4.7 \times \delta^{18}O - 1.1$.

Thus, we cannot conclude which mechanism was responsible solely from the data collected in this study.

All the data from flowing groundwater collected from boreholes drilled in the Cretaceous are plotted in a narrow zone on the aforementioned straight line. This suggests groundwater in the Cretaceous rock has a homogeneous original water quality, and is not created by mixing among different end-members. Consequently, this groundwater has one origin, when we consider the mixing of other groundwaters found in the surrounding region.

Residence time

We estimated groundwater residence time based on ^{36}Cl concentrations in groundwater. Usually, we estimate residence time using cosmogenic ^{36}Cl atoms. But, if the origin of groundwater collected in the Cretaceous is seawater or very saline water, we can ignore input of cosmogenic ^{36}Cl atoms due to an overabundance of stable chlorine atoms.

Furthermore, groundwater has been stored sufficiently deep in the Cretaceous rock to be free from the affects of cosmic ray interactions (Lehmann, *et al.*, 1993). We can expect a growing number of radiogenic ^{36}Cl atoms produced by neutrons, released in (α,n) reactions, caused by α-particles scattered from uranium and thorium radionuclides contained in the deep rock. The number of ^{36}Cl atoms in groundwater may be estimated by the following equation (Feige *et al.*, 1968; Andrews *et al.*, 1986; Lehmann & Loosli, 1991):

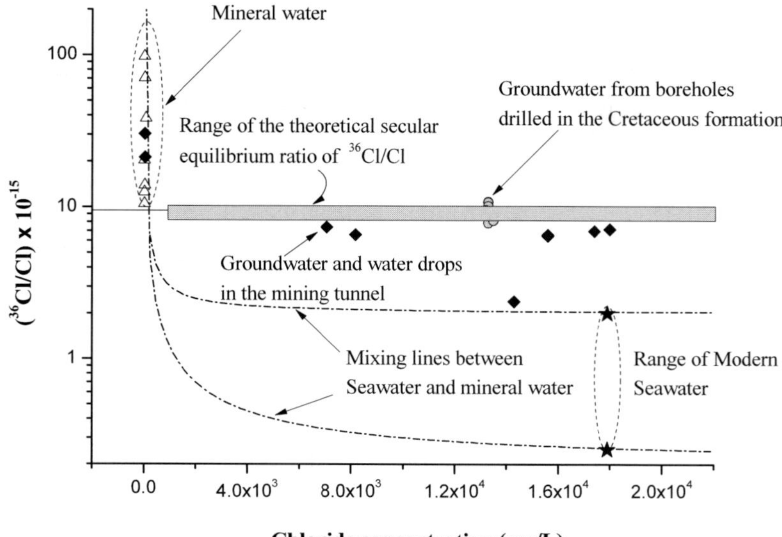

Fig. 6 The correlation between the chloride ion concentration and the ^{36}Cl/Cl ratio in groundwater from boreholes drilled in the Cretaceous, groundwater in the coal mining tunnel excavated in the Palaeogene formation, and the commercial mineral waters, showing the mixing water line between modern seawater and commercial mineral waters.

$$^{36}Cl = 4.55 \times 10^{-10} \times \Phi \times (1 - e^{-\tau t}) \times Cl \quad (1)$$

where, Φ is the neutron flux determined by the content of U, Th and other light elements, and the density of the surrounding rock, τ is the radioactive decay constant (7.3×10^{-14} s^{-1}) for ^{36}Cl, and *Cl* is the chloride ion concentration in groundwater.

Furthermore, as equation (1) reaches secular equilibrium conditions after 2 million years, we can rewrite this equation to:

$$\frac{^{36}Cl}{Cl} = 4.55 \times 10^{-10} \times \Phi \quad (2)$$

We calculated the neutron flux to be $(2.09\pm0.16) \times 10^{-15}$ n/cm^2 s^{-1} following Lehmann & Loosli (1991), and using the U content of 1.97±0.15 mg/g and the Th content of 6.17±0.85 mg/g measured in rock samples drilled in the coal mine. Using equation (2) we estimated the theoretical secular equilibrium ratio of ^{36}Cl/Cl to be $(9.5\pm0.73) \times 10^{-15}$. We can roughly estimate the residence time by comparing the measured ratio with the theoretical value.

The correlation between the ^{36}Cl/Cl ratio and the chloride ion concentration is illustrated in Fig. 6 for three different water types in the mine. This figure indicates that most of groundwater collected in the coal mine has not been produced by mixing of present seawater and, e.g. commercial fresh-mineral water, which represents young groundwater (Mahara *et al.*, 2004). Groundwater in the Cretaceous is located further from the mixing zone in Fig. 6 than other groundwater collected in the Palaeogene formation. The measured ^{36}Cl/Cl ratios range from 8×10^{-15} to 1.08×10^{-14}, the average value being $(9.46\pm1.36) \times 10^{-15}$. The average measured ^{36}Cl/Cl ratio is in line with the estimated theoretical secular equilibrium ratio of $(9.5\pm0.73) \times 10^{-15}$.

We may conclude that groundwater in the Cretaceous rocks has a residence time of more than 2 million years. In other words, the groundwater has been isolated in the geological formation without exchanging and mixing with other groundwater circulating in the upper formations.

CONCLUSIONS

In this study, we have revealed that groundwater in the Cretaceous formations, which have no contact with the overlying present seawater, has remained in the undersea basement rocks of the study area for a geological time scale. Groundwater quality has altered by ion exchange owing to long-term water–rock interactions. The groundwater has been isolated for over 2 million years, as evidenced by *in situ* growth of ^{36}Cl activated by neutrons created naturally by α-decay of uranium and thorium contained in the deep rock.

REFERENCES

Andrews, N. J., Fontes, J.-Ch., Michelot, J.-L. & Elmore, D. (1986) *In situ* neutron flux, ^{36}Cl production and groundwater evolution in crystalline rocks at Stripa, Sweden. *Earth Planet. Sci. Lett.* **77**, 49–58.

Clesceri, L. S., Greenberg, A. E. & Trussell, R. R. (1989) *Standard Methods for the Examination of Water and Wastewater*, 17th edn. American Public Health Association, Washington, DC, USA.

Coplen, T. B. & Hanshaw, B. B. (1973) Ultrafiltration by a compacted clay membrane. I. Oxygen and hydrogen isotopic fractionation. *Geochim. Cosmochim. Acta* **37**, 2295–2310.

Craig, H. (1961) Isotopic variation in natural waters. *Science* **133**, 1702–1703.

Creswell, R. G. (2001) Manual for groundwater sample preparation for ^{36}Cl analysis (version: March 1999) (private communication).

Feige, Y., Oltman, B. G. & Kastner, J. (1968) Production rates of neutrons in soils due to natural radioactivity. *J. Geophys. Res.* **73**, 3135–3142.

Frape, K. S., Fritz, P. & McNutt, H. R. (1984) Water–rock interaction and chemistry of groundwaters from the Canadian Shield. *Geochim. Cosmochim. Acta* **48**, 1617–1627.

Lehmann, B. E. & Loosli, H. H. (1991) Isotopes formed by underground production. In: *Applied Isotope Hydrology – A Case Study in Northern Switzerland* (ed. by F. J. Pearson, W. Balderer, H. H. Loosil, B. E. Lehmann, A. Matter, Tj. Peters, H. Schmassmann & A. Gautschi), 239–265. Elsevier Sci. Publ. Co. Inc., New York, USA.

Lehmann, B. E., Davis, S. N. & Fabryka-Martin, J. T. (1993) Atmospheric and subsurface sources of stable and radioactive nuclides used for groundwater dating. *Water Resour. Res*. **29**, 2027–2040.

Mahara, Y., Ito, Y., Nakamura, T. & Kudo, A. (2004) Comparison of ^{36}Cl measurements at three laboratories around the world. *Nucl. Instr. and Meth. in Phys. Res. B* **223-224**, 479–482.

Sato, M. & Sato, S. (1980) Third report of the study on groundwater discharge at the coal face in the Taiheiyo Coal Mine – characteristics of the water bearing layers and behavior of and the stored gases in the layers. *J. Mining Inst. of Japan* **96** (1108), 391–396 (in Japanese).

Existence of stagnant fresh groundwater and diffusion-limited chloride migration in a sub-sea formation at Yatsushiro Bay, Japan

TOMOCHIKA TOKUNAGA[1], YUKI KIMURA[2] & JUN SHIMADA[3]

1 *Department of Environment Systems, University of Tokyo, Kashiwa 277-8563, Japan*
tokunaga@k.u-tokyo.ac.jp

2 *Department of Geosystem Engineering, University of Tokyo, Tokyo 113-8656, Japan*

3 *Department of Systems in Natural Environment, Kumamoto University, Kumamoto 860-8555, Japan*

Abstract We attempted to evaluate long-term behaviour of saline groundwater by analysing chloride concentration and chlorine isotopic ratios of porewaters obtained from the sub-sea formation at Yatsushiro Bay, southwest Japan. Chloride concentrations and stable chlorine isotopic ratios were measured on 13 porewater samples. Porewaters with chloride concentrations higher than 16 400 mg/L are found at depths shallower than 1.5 metres below sea floor (m b.s.f.). Chloride concentrations decrease downwards gradually, and become lower than 250 mg/L below 7.7 m b.s.f. The stable chlorine isotopic ratios show a minimum value of –1.27 ‰ at 5.5 m b.s.f., and those from other depths show minor fluctuation, from –0.45‰ to 0.11‰. From these results, diffusion is considered to be the dominant process for the transport of chloride at the location studied.

Key words seawater intrusion; diffusion; stable chlorine isotope; groundwater

INTRODUCTION

Understanding the mechanism of solute transport has become an important issue for studies on water and mass movement in the coastal area. Previous research results on the porewater chemistry of coastal sub-sea formations (e.g. Hathaway *et al.*, 1979; Groen *et al.*, 2000) have shown the existence of fresh to brackish porewaters. It has been suggested that the low permeability clayey formations that were deposited by transgression after the last glacial maximum, significantly delay the intrusion of sea-water into lower formations, so contributing to this occurrence (Groen *et al.*, 2000). Kooi *et al.* (2000) conducted numerical experiments and showed that the low permeability material controls the patterns of seawater intrusion into the sub-sea formation.

In this study, we tried to evaluate the physical mechanism of seawater intrusion into the sub-sea formation where marine clay is deposited, through drilling, porewater sampling, and analysing chlorine isotopic ratios. Chlorine isotopic ratios are used as a measure for evaluating the mechanism because differences in diffusion coefficients of chlorine isotopes caused by the difference in molecular mass create isotope fractionation (Senftle & Bracken, 1955; Desaulniers *et al.*, 1986; Kaufmann *et al.*, 1988; Eggenkamp *et al.*, 1994) but fractionations are not caused by hydrodynamic dispersion processes (Fig. 1). Thus, fractionation of the chlorine isotopic ratio caused by diffusion process can be measurable and be useful as direct evidence of diffusion.

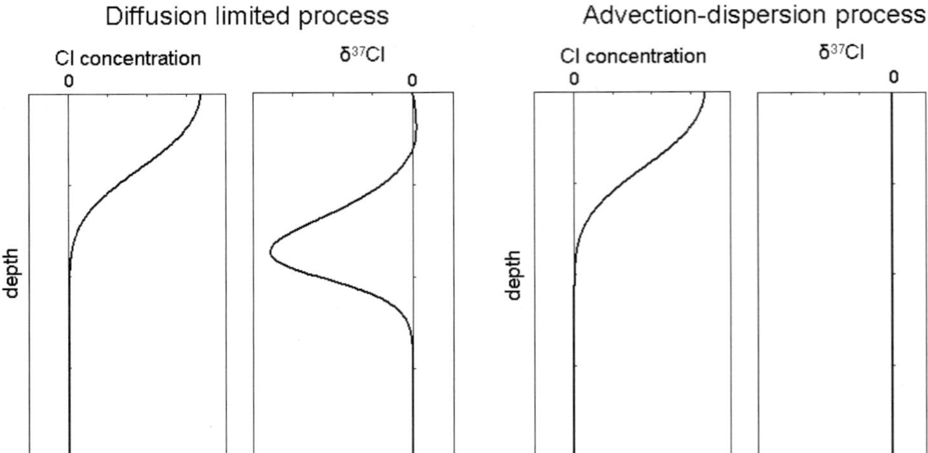

Fig. 1 Conceptual diagrams showing different profiles formed by the diffusion-limited process (left) and the advection–dispersion process (right). Note that chloride concentrations show very similar profile shapes while stable chlorine isotopic ratios show distinct differences. This diagram was constructed by a sedimentation–diffusion coupling model after Kimura (2006).

DATA OBTAINED

We drilled a 50-m borehole and took continuous sub-sea core samples where the average water depth is about 3 m, offshore of Eino-o, Kumamoto Prefecture, Japan (Fig. 2). At the study site, tuff breccia of late Pliocene to early Pleistocene is present from 3.8 m below sea floor (m b.s.f.) to the bottom of the borehole (50 m). Marine clay deposited after the last glacial maximum covers the tuff breccia. Porewaters were extracted from core samples both using a squeezing method for the rock cores and a centrifuge method for the marine clay. Then, we measured chloride concentrations and stable chlorine isotropic ratios on 13 porewater samples to construct these profiles.

DISCUSSION AND CONCLUSIONS

Figure 3 shows the results obtained. Porewaters with chloride concentrations higher than 16 400 mg/L are found at depths shallower than 1.5 m b.s.f. Chloride concentrations decrease downwards gradually, and to lower than 250 mg/L below 7.7 m b.s.f., showing an apparent diffusion profile. This result clearly shows the existence of fresh groundwater in the sub-sea formation at the location studied. The stable chlorine isotopic ratios show a minimum value of –1.27‰ at 5.5 m b.s.f., and those from other depths show minor fluctuation, from –0.45‰ to 0.11‰. Also, the profile of the stable chlorine isotopic ratios shows a very similar shape to that shown in Fig. 1. From these results, diffusion is considered to be the dominant process for the transport of chloride. Thus, fresh groundwater below 7.7 m b.s.f. is considered to be stagnant fresh groundwater, the existence of which has been controlled by the deposition of the low

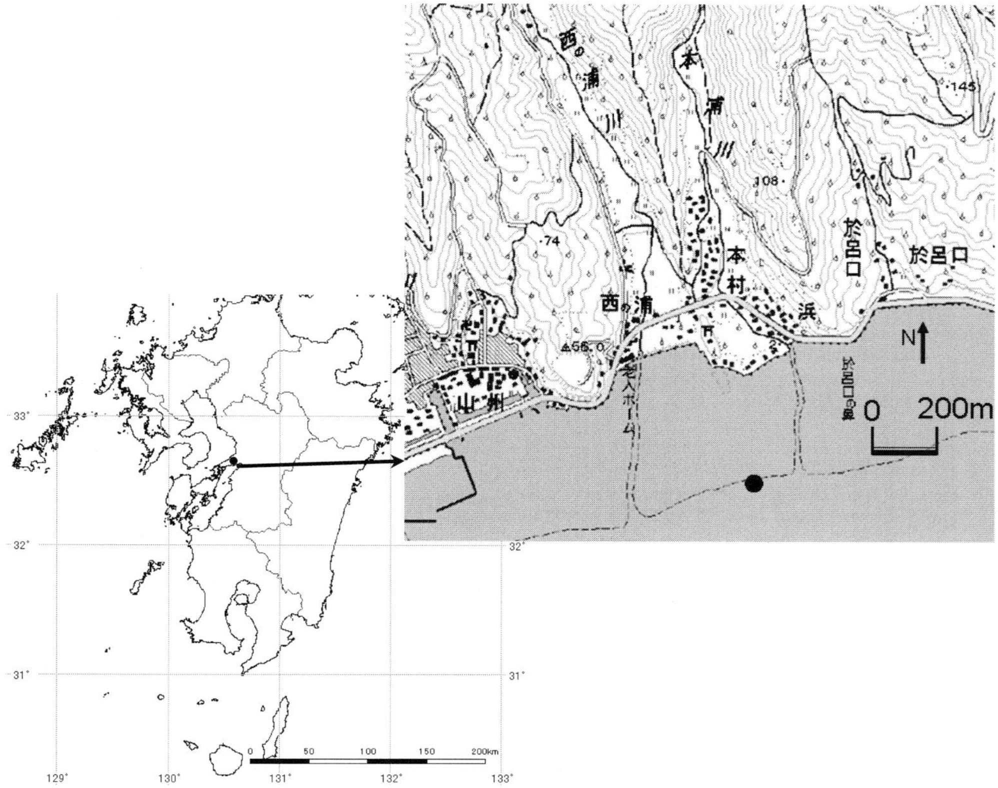

Fig. 2 Location map of the borehole studied. Topographic map (right) is a part of 1:25 000 scale "Matsuai" by the Geographical Survey Institute, Japan. The dot indicates the location.

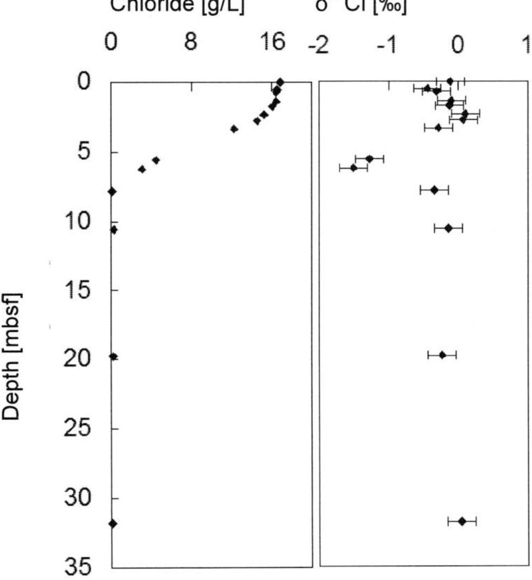

Fig. 3 Chloride concentration and stable chlorine isotopic ratios.

permeability clay formation. Further studies on modelling the process, and age determination of the fresh groundwater in the sub-sea formation are in progress, and we believe that quantitative understanding of the intrusion process is achievable.

REFERENCES

Desaulniers, D. E., Kaufmann, R. S., Cherry, J. A. & Bentley, H. W. (1986) ^{37}Cl–^{35}Cl variations in a diffusion-controlled groundwater system. *Geochim. Cosmochim. Acta* **50**, 1757–1764.

Eggenkamp, H. G. M., Middlburg, J. J. & Kreulen, R. (1994) Preferential diffusion of ^{35}Cl relative to ^{37}Cl in sediments of Kau Bay, Halmahera, Indonesia. *Chem. Geol.* **116**, 317–325.

Groen, J., Velstra, J. & Meesters, A. G. C. A. (2000) Salinization processes in paleowaters in coastal sediments of Suriname; evidence from $\delta^{37}Cl$ analysis and diffusion modeling. *J. Hydrol.* **234**, 1–20.

Hathaways, J. C., Wylie Poag, C., Valentine, P. C., Miller, R. E., Shultz, D. M., Manheim, F. T., Kohout, F. A., Bothner, M. H. & Sangrey, D. A. (1979) US Geological Survey core drilling on the Atlantic Shelf. *Science* **206**, 515–527.

Kaufmann, R. S., Long, A. & Campbell, D. J. (1988) Chlorine isotope distribution in formation waters, Texas and Louisiana. *Am. Assoc. Petrol. Geol. Bull.* **72**, 839–844.

Kimura, Y. (2006) Evaluation of diffusion processes through geological formation by stable chloride isotopes distribution MSc Thesis, University of Tokyo, Tokyo, Japan (in Japanese with English abstract).

Kooi, H., Groen, J. & Leijnse, A. (2000) Modes of seawater intrusion during transgressions. *Water Resour. Res.* **36**, 3581–3589.

Senftle, F. E. & Bracken, J. T. (1955) Theoretical effect of diffusion on isotopic abundance ratios in rocks and associated fluids. *Geochim. Cosmochim. Acta* **7**, 61–75.

Indicators and quality classification applied to groundwater management in coastal aquifers: case studies of Mar del Plata (Argentina) and Apulia (Italy)

E. M. BOCANEGRA[1], M. POLEMIO[2], H. E. MASSONE[1], V. DRAGONE[2], P. P. LIMONI[2] & M. FARENGA[1]

1 *CGCyC, UNMDP, CIC Funes 3350, 7600 Mar del Plata, Argentina*
ebocaneg@mdp.edu.ar

2 *CNR-IRPI, Via Amendola 122/i, I-70126 Bari, Italy*

Abstract The use of indicators is considered for environmental and ecological monitoring and in the general assessment of environmental sustainability at a local, national and international scale. They are used to briefly describe the interests and preoccupations of society with regard to environmental evolution, and to coherently aid in decision-making processes. Groundwater is affected by two types of degradation risks: quality and quantity degradation. In order to define the coastal environmental processes related to groundwater resources, some indicators addressed to decision makers and quality classification are proposed and discussed in relation to two different types of coastal aquifers located in different countries: the porous aquifer of Mar del Plata (Argentina) and the Salentine karstic aquifer (Italy). Beyond the strong hydrogeological differences between both areas, the analysis of indicators allows the identification of interesting similarities. The results show this approach could help in reaching a consensus to propose a methodology to deal with environmental quality assessment of water and establish groundwater exploitation criteria.

Key words coastal aquifer; management; monitoring; environmental sustainability; groundwater; indicators; degradation risk; seawater intrusion; pollution

INTRODUCTION

Two coastal aquifers are considered in two very different geographical, hydrogeological and socio-economic territories, the first being located in Argentina and the second in Italy. The use of indicators and quality classification criteria of groundwater are applied to test their usefulness in contributing to the sustainable management and reduction of degradation risk of coastal groundwater resources

The main and common environmental problems related to water resources are: high urban area expansion, seawater intrusion, pollution of groundwater, inadequate waste management, and recurrent or rare flash floods in urban and surrounding areas.

Mar del Plata, located on the Atlantic coast, is the main tourist centre in Argentina, and has a population of 600 000 inhabitants that increases threefold during the summer. Water for urban, agricultural and industrial uses is supplied solely by groundwater resources.

Due to a law on building (blocks of flats), especially since 1948, the urban growing process has led to an increase in population density, industrial activities (building, food, textile and fishing) and tourism in the central area.

Several studies have been carried out characterizing the process of seawater intrusion (Bocanegra *et al.*, 1993, 2002; Martínez *et al.*, 1996) and the impact of leachates at final waste disposal sites on groundwater (Massone *et al.*, 1993, 1994, 1998; Bocanegra *et al.*, 2001a ; Mascioli *et al.*, 2005).

The Apulia region is located in the southeastern portion of Italy. The whole Apulian groundwater has undergone a two-fold pollution, both originated by human action (Polemio, 2000, Polemio & Limoni, 2001): salinization has evolved progressively as it has affected increasingly larger portions of land; and biological and chemical-physical pollution has gained importance and is mainly concentrated around urbanized areas (Cotecchia, 1981; Cotecchia & Polemio. 1997).

Carbonate rock outcrops are widespread in three areas of the region; in these areas the natural protection of the aquifer by pollution is very low. The intrinsic vulnerability of main aquifers, which is spatially variable but significant everywhere, exposes groundwater almost directly to effects of potential pollution sources from anthropogenic activities at the land surface. The natural or intrinsic vulnerability is increased by using custom-bored wells, karstic pits and dolines to discharge underground wastewaters and runoff from urbanized surfaces.

HYDROGEOLOGICAL FEATURES OF MAR DEL PLATA AND SALENTO AQUIFER

Mar del Plata is located on the northeastern side of the Tandilia range, and it is the most important sea-side resort of Argentina (Fig. 1). The Tandilia range has a maximum altitude of about 40 m a.s.l. in the Mar del Plata area. In the study area, the range consists of lower Palaeozoic quartzites, grouped under the name of Balcarce Formation (Dalla Salda & Iñiguez, 1978). The quartzite bedrock is overlain by a sedimentary cover of Upper Tertiary and Quaternary silts and silty-to-sandy sediments. Miocene clayey-to-sandy sediments are found at a depth of 60 m in the grabens. The Quaternary deposits are called "pampean sediments" or "loess-like sediments" and, from a hydrogeological viewpoint, they constitute the most important sequence. They are a multi-layered phreatic aquifer with a thickness ranging from 70 to 100 m, and a hydraulic conductivity is 10 m/d. The transmissivity is about 600–800 m^2/d in the urban area and between 1000 and 1400 m^2/d in the rural area. The storage coefficient, estimated from pumping tests, is 0.001, and the porosity is 0.15.

The Apulia region is characterized mainly by the hydrogeological units of the Gargano and Tavoliere and the hydrogeological structures of the Murgia and the Salento. All of these areas are carbonate in nature, except for the Tavoliere, and constitute the largest coastal karstic aquifers of Italy, made up of Mesozoic rocks (Fig. 2). A detrital organogenic series (Tertiary and Quaternary) fills some troughs or partially overlaps the carbonate rocks, creating secondary aquifers in places (Cotecchia *et al.*, 2004).

The study area corresponds to a morphological-structural unit—the Salento—a lowland area (maximum height about 180 m a.s.l.), bounded by the Adriatic Sea and the Ionian Sea. The Mesozoic carbonate rocks are more widespread and covered by outcropping Quaternary soils and rocks where these outcrops are not very continuous (Fig. 1). The Salento peninsula (Fig. 2) corresponds roughly with the province of

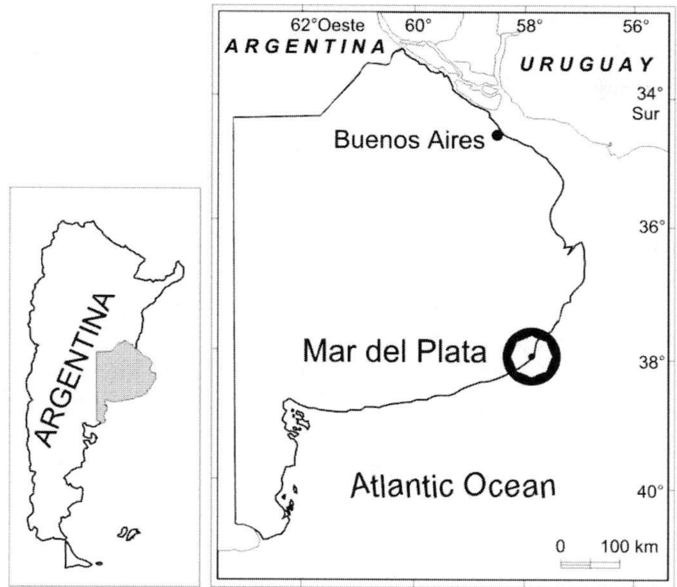

Fig. 1 Mar del Plata location map.

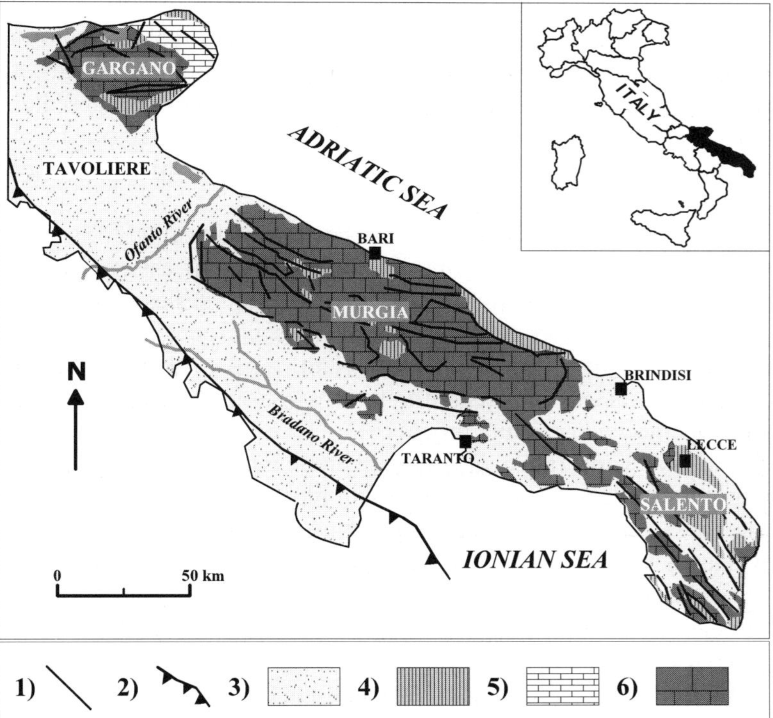

Fig. 2 Salento location map and geological scheme of Apulian hydrogeological units and structures. 1: Fault; 2: front of the Apennines; 3: recent clastic cover (Pliocene–Pleistocene); 4: Bioclastic carbonate rocks (Paleogene) and calcarenites (Miocene); 5: Carbonate platform rocks (Upper Jurassic-Cretaceous); 6: Scarp and basin chert-carbonate rocks (Upper Jurassic-Cretaceous).

Lecce, which is constituted by 97 municipalities, the population of which is equal to 787 825 inhabitants (Table 1) (Istat, 2001).

Several surveys have been undertaken to detect the chemical-physical-bacteriological characteristics of Apulian groundwater; the most important surveys have considered the coastal springs (Cotecchia & Tulipano, 1989) and wells of the Apulian groundwater monitoring system (Cotecchia & Polemio 1998, 1999). As part of their research, the IRPI has carried out their own surveys in the wells and springs mentioned above, until the end of 2003.

Groundwater heights of the Salento are less than 4–5 m a.s.l. The Salento groundwater divide is located along the SE–NW axis, parallel to the Ionian and Adriatic coastlines, with piezometric gradient generally equal to 0.2–0.3%.

A deep limestone aquifer can be distinguished in the whole Salento area where main shallow aquifers can be locally observed too, mainly composed of calcarenite and secondary sands and conglomerates. The groundwater flow in the limestone aquifer, considered hereinafter, is generally unconfined.

METHODOLOGICAL FRAMEWORK

In order to define the temporal–spatial evolution of coastal environmental processes related to groundwater resources, some indicators for decision makers were designed to assess the environmental pressure caused by human activity, the aquifer conditions or state, and the anthropic response to correct undesirable situations. This indicator selection will allow a temporal follow-up of the evolution of groundwater and integrate hydrogeological and social aspects, with respect to what was agreed on in Chapter 18, Section 2 of Agenda 21 (UNCED, 1992), in the mission statement of the Dublin Conference (ICWE, 1992) and Objective 7 of the Millenium Development Objectives (UNDP, 2003).

The design and use of indicators has been recently called for to be used in environmental and ecological monitoring and in the general assessment of environmental sustainability at local, national and international scales (Berger & Iams, 1996). Environmental indicators are variables with a social meaning attached to that of their own scientific entity, and whose aim is to briefly describe the interests and preoccupations of society with regard to the environmental evolution and coherently contribute to the decision-making process.

There is a wide range of environmental indicators (OECD, 1993; Cendrero & Fischer, 1997, Cendrero, *et al.* 2001), and most of them are useful to decision makers and are designed to assess: (a) environmental pressure caused by human activity; (b) evaluated environmental conditions or state; and (c) anthropogenic response to correct undesirable situations.

Groundwater quality represents a special case for indicators due to the speed of changes, the natural and anthropogenic origin of contamination sources, and the relevance of involved ecosystems, especially in the case of coastal aquifers.

In order to define the coastal environmental processes related to groundwater resources, some indicators for decision makers and a quality classification method are discussed in the case of the two selected coastal aquifers. For the first issue, a set of indicators for pressure, state and response were selected taking account their relevance

Table 1 Pressure indicators in 2000 for the two study areas.

Indicators	Mar del Plata	Salento
Population (10^3)	564.1	787.8
Urban area (km^2)	150	235.7
Population density (inhab/km^2)	3384	3342
Urban population (%)	90	90
Annual exploited volume (10^6 m^3)	97.6	91.0
Summer exploited volume (%)	37	n.a.
Daily water supply (L/inhab)	539	345
Population lacking water supply (%)	20.8	0.0
Population lacking sewage (%)	27.1	39.3
Irrigated areas (km^2)	141	291
Official urban damping (km^2)	1.2	n.a.*
Annual waste weight (10^3 ton)	255.5	374.3

n.a.: data not available.
* 1721m^3 annual authorized volume.

Table 2 State indicators in 2000 for the two study areas.

Indicators	Mar del Plata	Salento
Annual recharge (mm)*	198	300*
Wells <0 m a.s.l. (%)	14	0
Minimum piezometric level due to discharge (m a.s.l.)	–2.9	n.a.
Wells with Cl > 200 mg/L (%)	3	41
Max. chloride conc. (mg/L)	366	1610
Wells with NO3 > 45 mg/L (%)	18	1
Bacteriologically polluted wells (%)	> 83	13

* mean annual value, in the period 1930–2005.
n.a.: data not available.

Table 3 Response indicators in 2000 for the two study areas.

Indicators	Mar del Plata	Salento
Total wells *	310	20 369
Working wells	255	≈ 20 369
Abandoned wells	55	n.a.
Jurisdictional organization	Municipal	Public company and Municipal
Annual budget (US$/inhabitant)	30	n.a.
Cost of water systems (US$/m3)	0.10	1.00(**)
Staffing (employer/1000 inhabitant)	1.2	n.a.
Environmental regulations	Provincial Laws Local regulation on cesspools Register of well constructors Mechanisms of control Master Plan 2006–2016	European, national and regional regulations. Groundwater discharge managed on the basis of the regional Water quality recovery plan of 1981–1983 and the Protection Plan of 2007

* Private and domestic wells are included in the case of Salento.
** mean value.
n.a.: data not available.

and data availability (Tables 1–3). The year 2000 was used as a date of comparison. For the second one, the selected quality classification method is defined considering a shared and quite standard approach, as described by an Italian law (D.L.152/99) which applies to some European Directives (1991/271 and 1991/676). This approach could be easily modified to consider the application of European Water Framework Directive 2000/60, and the proposed European Directive on the protection of groundwater against pollution. The approach distinguishes two groups of parameters—basic and additional (Tables 4 and 5). They have been selected based on those suggested by law, and by considering the nature of pollution sources which are typical of Apulian and Mar del Plata groundwater. The so called chemical classes, from 1 to 4 in decreasing quality order, can be described as follows: Class 1, null or negligible anthropogenic impact, high groundwater quality; Class 2, low anthropogenic impact, long-term sustainable impact, good groundwater quality; Class 3, valuable anthropogenic impact, good groundwater quality subject to degradation risk; Class 4, marked anthropogenic impact, low groundwater quality; Class 0, null or negligible anthropogenic impact but with at least a parameter able to attain Class 4.

The mean concentration of each basic or additional parameter is used for the quality classification in each surveying period.

The quality classes are defined according to classes defined by Table 4 for each basic parameter, but also based on the threshold value of Table 5. The concentration of one or more pollutants higher than the thresholds of Table 5 determines the classify-cation in Class 4 otherwise the water is classified considering the parameters of Table 4. The final classification of the water sample is equal to the worst class selected by the whole set of basic and additional parameters. If the concentration of a parameter is high enough to attain Class 4 but its origin is natural, it is assigned Class 0.

Table 4 Quality classification on the basis of basic chemical parameters, Salento (D.L.152/99).

Parameter/Class	1	2	3	4
SEC (μS/cm at 20C°)	≤400		≤2500	>2500
Cl	≤25		≤250	>250
Mn (μg/l)	≤20		≤50	>50
Fe (μg/l)	≤50		≤200	>200
NO_3 (mg/l)	≤5	≤25	≤50	>50
SO_4 (mg/l)	≤25		≤250	>250
NH_4 (mg/l)	≤0.05		≤0.5	>0.5

SEC or specific electrical conductivity.

Table 5 Threshold of additional chemical parameters, Salento (D.L.152/99).

Parameter	μg/L
Arsenic	≤10
Cadmium	≤5
Chrome	≤50
Nitrite	≤500
Lead	≤10

The presence of significant concentrations of additional parameters lower than the threshold, except in the case of natural presence, is, however, a negative signal and unacceptable in terms of groundwater quality. In this case, actions to prevent further quality degradation, removing the pollution causes and reducing the pollution concentrations should be adopted.

RESULTS AND DISCUSSION FOR THE STUDY AREAS

The analysis and comparison of the pressure indicators (Table 1) for both cases shows very important similarities to population density, urban population and annual exploited volume; there are approximate similarities considering total population, population lacking sewage, annual waste weight and urban surface. Finally, marked differences were detected in daily water supply, population lacking water supply and irrigated areas.

The analysis of the state indicators (Table 2) for both cases allows us to identify some interesting issues in spite of the strong hydrogeological differences. It is notable how nitrates, chlorides and the biological pollution determine the present conditions. Whereas, in the case of Mar del Plata, nitrates and the biological pollution are more significant than chlorides, Salento shows the opposite effect. Also, there is a piezometric drawdown in Mar del Plata that is negligible in the case of Salento, if, in the latter case, the piezometric and groundwater quality data are available only for wells of the monitoring network, wells not used to discharge groundwater. For this reason the minimum piezometric level due to discharge is not available in the case of Salento. The percentage of wells hit by bacteriological pollution has been determined considering the bacteriological data from water samples taken from each of the wells and comparing these data with Italian drinking water criteria.

Response indicators (Table 3) show some remarkable differences considering the ratio of working wells to groundwater supply, the cost of running-water systems and the environmental regulations that have been applied. The present jurisdictional framework is quite similar in both cases.

In the quality assessment of Mar del Plata groundwater, 399 analyses of samples collected from 1995 to 2005 have been considered (Table 6). There is no quality class variation in chlorides between 1995 and 2005, 97% of the wells remaining in Class 3 and 3% in Class 4 during this period. On the other hand, nitrates (Fig. 3) show that the class is increasing in wells located in the suburban areas with high population density.

Table 6 Statistics of classification results for Mar del Plata.

	Main parameters						Others parameters
	SEC	Cl	Fe	NO3	SO4	NH4	NO2
PD (%)	100	100	55	100	100	94	100
Class1 (%)	0	0	48	6	46	96	
Class 2 (%)				53			
Class 3 (%)	99	97	29	17	54	4	
Class 4 (%)	1	3	23	24	0	0	1

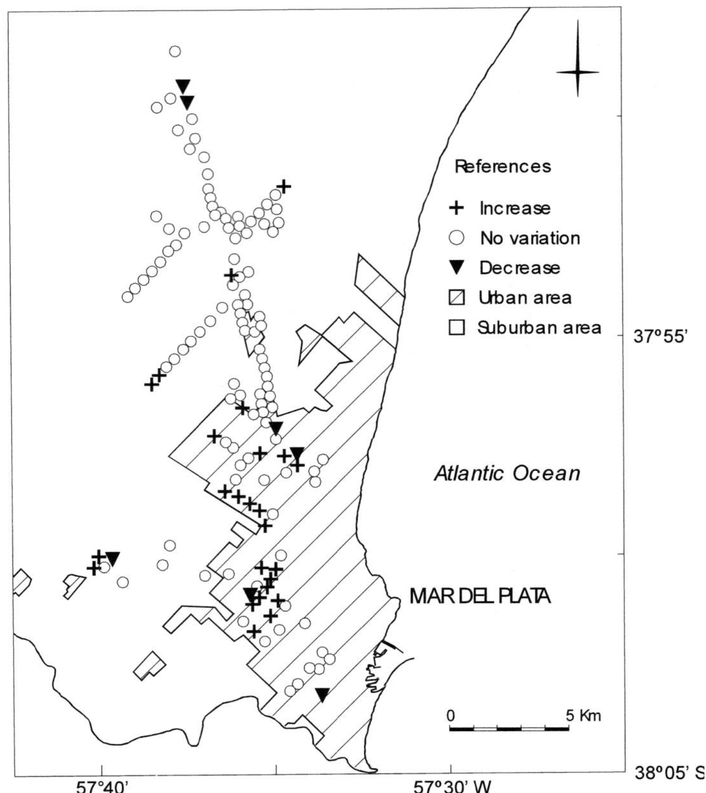

Fig. 3 Quality class variation of nitrate in 2005 compared to 1995 for Mar del Plata.

There is no class variation in rural areas and suburban areas with low population density, and finally, the quality class has decreased in few isolated wells.

The analysis of the data set from the monitoring network surveys has found the best quality groundwater is in some portions of the recharge area of Salento. These areas are located, as in the case of the 1984 situation (Apulia Regional Administration, 1984), inland, where the anthropogenic activities are not relevant. The extent of these areas appears to be decreasing since 1984.

The latest IRPI survey has been designed to characterize the quality modification of groundwater along the flow path from recharge areas to natural discharge areas along the coast line. For this purpose some wells and springs have been selected from that zone to represent the groundwater quality there.

The temperature, the specific electrical conductivity, pH, Eh, and the concentrations of ammonia, nitrite, nitrate, chloride, manganese, iron, mercury, lead, arsenic, cadmium zinc, arsenic and chromium of the groundwater have all been measured on site and in the laboratory by chemical analyses. The laboratory determination of the concentration of total coliforms, faecal coliforms and faecal streptococci permits quantifying the bacteriological quality of the groundwater. The high chloride concentration can be observed also far inland from the coast. The widespread presence of the saltwater has been caused by high rates of well discharge that have lead to seawater intrusion.

Pollution linked with nitrates in the study area falls within the limits established by the law. The sampled water presents a high level of bacteriological pollution determined mainly by total coliforms and secondly by faecal coliforms.

Wells located in core portions of recharge areas present the highest groundwater quality, of Class 1 or 2. In other portions of the aquifer, the predominant class is 4. In those portions belonging to the most polluted class, pollution is determined by specific electric conductivity and chloride, two correlated parameters, where the chloride content depends on the effect of seawater intrusion in these hydrogeological conditions. The parameters determining which wells belong to Class 4 are: specific electric conductivity, chloride, sulphate, ammonia, nitrate and lead.

Groundwater quality worsens along the flow paths from recharge areas to coastal discharge areas, surface water bodies, the hydrographic network, narrow lagoons, and the sea. This spatial trend is not continuous and homogeneous, due to local hydrogeological and, mainly, anthropogenic factors, since many urban areas are located around these areas. A greater spatial variability of quality is observed in Salento due to the higher aquifer vulnerability and density of villages, towns and areas of anthropogenic activities. As can be observed by the class variation of 2000 with respect to 1995, a huge quality worsening is observed in terms of chloride (Fig. 4(a)) and nitrate concentrations (Fig. 4(b)), especially in the northern part of the area.

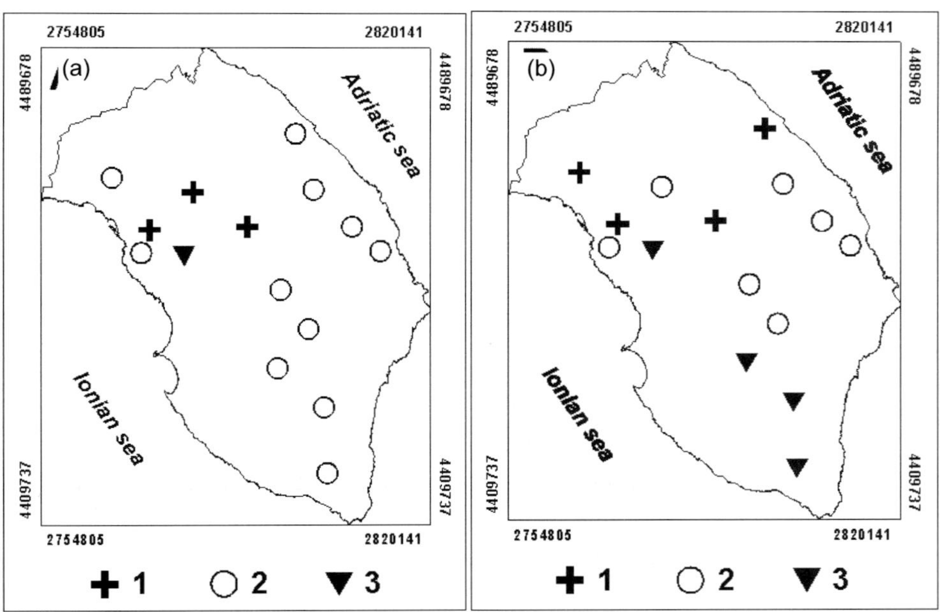

Fig. 4 Quality class variation of (a) chloride and (b) nitrate in 2000 compared to 1995 for Salento [1: class increase; 2: class unchanged; 3: class decrease].

CONCLUSIONS

The use of indicators of pressure, state and response to analyse groundwater conditions allows us identify several similarities between the Mar del Plata and Salento areas, in spite of the marked hydrogeological differences. The temporal analysis of the joint evolution of these indicators appears as the most effective tool for decision making; nevertheless, the combination of these indicators into one single quality index that will lead to the better adjustment of management policies is still under discussion. Some state indicators should constitute requisites of special value, since they show whether water is drinkable or not.

In the case of Mar del Plata, it can be appreciated that, despite the critical political and economic context, there have been satisfactory answers to the problems and demands originated around the issue of the groundwater resource. The application of quality class variation is new in Argentina and appears to be an efficient monitoring tool. The seriousness of the effects of natural and artificial phenomena which are highly relevant for the protection of the groundwater resources of Salento has been highlighted. Notwithstanding the complexity of the phenomena that threaten groundwater quality, the proposed quality classification clearly highlights that quality degradation hazards have already became a reality. High quality groundwater can now be found in inland and narrow Salento zones. Groundwater flows from these zones, which are hydrogeologically a portion of recharge areas, to the coastal areas under the huge effect of a progressive anthropogenic pollution.

Groundwater is loaded with pollutants that reach coastal springs without appreciable attenuation, due to the karstic nature of the aquifers. In the case of submarine springs, contamination is reduced by an appreciable dilution. Where spring outflow becomes inflow for lagoons or valuable humid habitats, the ecological equilibrium is subject to serious risks due to continuous degradation of groundwater quality.

REFERENCES

Apulia Regional Administration (1984) Piano Regionale Risanamento delle Acque. *Boll. Uff. Regione Puglia* **XV**, 57.

Berger, A & Iams, W. (1996) *Geoindicators*. Balkema, Rotterdam, The Netherlands.

Bocanegra, E. M., Martínez, D. E., Massone, H. E. & Cionchi, J. L. (1993) Exploitation effect and salt water intrusion in the Mar del Plata aquifer, Argentina. Study and modeling of salt water intrusion into aquifer. In: *Proc. of XII Salt Water Intrusión Meeting* (ed. by E. Custodio & A. Galofre), 177–191. Centro Internacional Métodos Numéricos-Universidad Politécnica de Cataluña.

Bocanegra, E. M., Massone, H. E., Martinez, D. E., Civit, E. M. & Farenga, M. (2001a) Groundwater contamination: risk management and assessment for landfills in Mar del Plata, Argentina. *Environ. Geol.* **40**(6), 732–741.

Bocanegra, E. M., Villamil, M. & Massone, H.E. (2001b) Aquifer vulnerability and bacteriological contamination hazard on Mar del Plata (Argentina) suburban area. In: *New Approaches Characterizing Groundwater Flow* (ed. by K. Seiler & S. Wohnlich) (Proc. XXXI IAH Congress), **1**, 457–461. Balkema, Rotterdam, The Netherlands.

Bocanegra, E. M., Martínez, D. E., Massone, H. E. & Benavente, M. A. (2002) Quantitative studies in coastal hydrogeology in Mar del Plata, Argentina. In: *Aguas Subterráneas y Desarrollo Humano. Págs* (ed. by E. Bocanegra *et al.*) (Proc. of XXXII IAH & VI ALHSUD Congress "Groundwater and Human Development", Mar del Plata, Argentina), 389–396. ISBN 987-544-063-9.

Cendrero, A. & Fischer, D. W. (1997). A procedure for assessing the environmental quality of coastal areas for planning and management. *J. Coastal Res.* **13**(3), 732–744.

Cendrero A., Francés, E., Panizza, M., Fabbri, A., Recatalá, L., Fermán, J. L., Fischer, D. W., Quintana, C., Latrubesse, E., Tecchi, R., Cantú, M. P., Hurtado, M. A. & Cecioni, A. (2001) A conceptual background and methodological approach for assessing and monitoring environmental quality. III Reunion de Geología Ambiental y I del Area del Mercosur. Mar del Plata. Published on CD.

Colucci, V., Limoni, P. P. & Serravezza, C. (1998) La rete di controllo idrometrografico e qualitativo delle falde idriche pugliesi. *Acque Sotterranee* **59**, 55–61.

Cotecchia, V. (1981) Methodologies adopted and results achieved in the investigations of seawater intrusion into the aquifer of Apulia (Southern Italy). In: *Salt Water Intrusion Meeting* (Hannover, 15–18 October 1981), 1–68.

Cotecchia, V. & Polemio, M. (1997) L'inquinamento e il sovrasfruttamento delle risorse idriche sotterranee pugliesi. In: VI Workshop del Progetto Strategico *Clima, Ambiente e Territorio nel Mezzogiorno* (Taormina, Italy, December 1995), **I**, 447–484.

Cotecchia, V. & Polemio, M. (1998) The hydrogeological survey of Apulian groundwater (Southern Italy): salinization, pollution and over-abstraction. In: *Proc. Int. Conf. on Hydrology in a Changing Environment* (British Hydrological Society, Exeter, UK, 6–10 July 1998), **II**, 129–136. John Wiley & Sons, Chichester, UK.

Cotecchia, V. & Polemio, M. (1999) Apulian groundwater (Southern Italy) salt pollution monitoring network. *Flemish J. Natural Science*, Ghent, Belgium, 197–204.

Cotecchia, V. Grassi, D. & Polemio, M. (2004) Carbonate aquifers in Apulia and seawater intrusion. In: *Some Engineering Geology Case Histories in Italy* (32nd IGC), 1–16.

Cotecchia, V. & Tulipano, L. (1989) Le emergenze a mare individuate anche con tecniche di telerilevamento, come vettori di carichi inquinanti dagli acquiferi carbonatici e carsici pugliesi all'ambiente costiero. *I Workshop Progetto Strategico Clima Ambiente e Territorio nel Mezzogiorno* (Taormina, Italy), 133–144.

Dalla Salda, L. & Iñiguez, A. (1978). La Tinta: Precámbrico y Paleozoico de Buenos Aires. Procc. VII Congreso Geológico Argentino, Actas:1. 539-550.

ICWE (International Conference on Water and the Environment) (1992) *Development Issues for the 21st Century*. The Dublin Statement and Report of the Conference, Dublin, Ireland, 26–31 January 1992. World Meteorological Organization, Geneva, Switzerland.

Istat (Istituto Nazionale di Statistica) (2001) Popolazione e movimento anagrafico dei comuni: anno 2000. (Annuario n. 13).

Martínez, D. E., Bocanegra, E. M. & Cionchi, J. L. (1996) Modelación hidrogeoquímica de procesos de mezcla. Su aplicación a casos de estudio en el acuífero de Mar del Plata. *Serie Correlación Geológica* **11**, 69–80.

Martínez, D. E., Massone, H. E. & Bocanegra, E. M. (2005) Groundwater salinization in the harbour area graben in Mar del Plata, Argentina. Hydrogeochemical perspective. *IGME, Serie Hidrogeología y Aguas Subterráneas 15*, 585–595. ISBN 84-7840-588-7.

Mascioli S., Martinez D. & Bocanegra E. (2005). Determinación del coeficiente de partición de Zn en sedimentos loessicos y su utilización en la simulación de transporte reactivo. In: *IV Congreso Argentino de Hidrogeología* (Río Cuarto, Córdoba, 25 al 28 de octubre de 2005), Actas I: 221–230. ISBN 950-665-346-1.

Massone, H., Del Rio, J.L., Fajardo, D., Cionchi, J., Martinez, D. & Bocanegra, E. (1993) Los Residuos Sólidos Domiciliarios del Partido de General Pueyrredon (Provincia de Buenos Aires) desde una Perspectiva Geológico-Ambiental. Parte I: Aplicación de la Cartografía Geocientífica a la Selección de Sitios de Disposición Final. *XII Congreso Geológico Argentino, Mendoza*, Actas VI, 303–310.

Massone, H., Martinez, D., Cionchi, J. L. & Bocanegra, E. M. (1994) Procesos de contaminación del acuífero de Mar del Plata, Argentina: diagnóstico y pautas de prevención y control. In: *II Congreso Latinoamericano de Hidrología Subterránea* (Santiago, Chile), vol. 1, 81–95.

Massone, H., Martínez, D. E., Cionchi, J. L. & Bocanegra, E. M. (1998) Suburban areas in developing countries and its relation with groundwater pollution. Mar del Plata (Argentina) as a case study. *Environ. Manage.* **22**(2), 245–254.

OECD (Organisation for Economic Co-operation and Development) (1993) Environmental indicators for environmental performance reviews. OECD, París.

Polemio, M. (2000) Degradation risk owing to contamination and overdraft for Apulian groundwater resources (Southern Italy). In: *Water Resources Management in a Vulnerable Environment for Sustainable Development* (ed. by K. Andah), 185–194. UNESCO - International Hydrological Programme, Grifo Publishers, Perugia, Italy.

Polemio, M. & Limoni, P.P. (2001) L'evoluzione dell'inquinamento salino delle acque sotterranee della Murgia e del Salento. *Memorie della Società Geologica Italiana* **56**, 327–331.

Ruíz Huidobro, O. & Tofalo, O. R. (1975) La intrusión de agua de mar en acuíferos litorales. Su control en Mar del Plata (República Argentina). *VI Cong. Geol. Arg.*, Actas, 515–523. Buenos Aires.

Sala, J. M., Hernández, M., González, N., Kruse, E. & A. Rojo (1980) Investigación geohidrológica aplicada en el área de Mar del Plata. Convenio Obras Sanitarias de la Nación - Universidad Nacional de La Plata. V Tomos, (unpublished).

UNCED (United Nations Conference on Environment and Development) (1992) *Report of the United Nations Conference on Environment and Development*, Rio 1992.

UNDP (2003) *UN Millennium Declaration*. Millennium Development Goals.

Isotopic characterization of saline intrusion into the aquifers of a coastal zone: case study of the southern Venice Lagoon, Italy

JULIE C. GATTACCECA[1], CHRISTINE VALLET-COULOMB[1], ADRIANO MAYER[2], OLIVIER RADAKOVITCH[1], ENRICO CONCHETTO[3], CORINNE SONZOGNI[1], CHRISTELLE CLAUDE[1] & BRUNO HAMELIN[1]

1 *CEREGE, Université Paul Cézanne, Europôle Méditerranéen de l'Arbois, BP80, F-13545 Aix en Provence, France*
jgatta@cerege.fr

2 *IDPA, CNR, Milano, Italy*

3 *AATO Laguna di Venezia, Mestre (Ve), Italy*

Abstract This study deals with the geochemical characterization of salinization in the semi-confined aquifer of the southern part of the Venice Lagoon, Italy. Twelve boreholes reaching the aquifer were sampled for stable isotopes ($\delta^{18}O$ and δD). Electrical conductivity (EC) displays a large range of variation (0.7–40 mS/cm). The more saline groundwaters are located at up to 2 km from the lagoon and Adriatic Sea shorelines. In the $\delta^{18}O$ *vs* δD diagram, the more saline groundwaters plot along a well-defined mixing line, passing through a continental and a seawater end-member. The brackish and fresh groundwaters do not have a clear spatial distribution. Heterogeneous $\delta^{18}O$ and EC compositions of brackish and fresh groundwater reflect complex exchanges between deep groundwater, surface and/or rain waters. Characterization of the continental end-member(s) in the mixing is not straightforward, revealing a complex hydrodynamic behaviour in this aquifer.

Key words salinization; coastal groundwaters; stable isotopes; Venice Lagoon

INTRODUCTION

Coastal zones, globally, are highly impacted by the consequences of increasing population. Numerous studies have been recently undertaken to evaluate various impacts of the anthropogenic influence on these fragile ecosystems, and to assess policies for management. Coastal aquifers are especially vulnerable to enhanced pumping for water supply, as this leads to saline water intrusions into the freshwater aquifers and in some cases to land subsidence in coastal flatlands (Andreasen & Fleck, 1997; Barbecot *et al.*, 2000; Capaccioni *et al.*, 2005). In addition, particular attention to saline intrusion is also required given the prospect of sea level rise induced by the expected global warming. The inland areas surrounding the Venice Lagoon represent a dramatic example of these negative effects. In this paper, we focus on the salinization process in the shallow, partially unconfined aquifer in the southern part of the Venice Lagoon. Although substantial extraction of groundwater has not occurred for many years (it has been forbidden by law since the late 1970s), land subsidence has continued in this area driven by natural mechanisms, such as the compaction of surface sediments and oxidation of organic carbon in soils (peat land) (Gambolati, 2003;

Brambati *et al.*, 2004). Saltwater intrusion has been previously documented in this zone using geophysical methods (Carbognin *et al.*, 2003; Di Sipio *et al.*, 2006). Our study aims in particular to characterize the saline intrusion and the exchanges between groundwater and surface waters in this area by using chemical and isotopic tracers. We present here initial results obtained from groundwater samples taken from the partially unconfined aquifer.

STUDY AREA

The study area is situated in the lower Venetian plain, in the coastal flatland of the southern part of the Venice Lagoon (Fig. 1). It extends 20 km east to west and 10 km north to south, delimited by the Venice Lagoon itself on the northeast, the Adriatic Sea on the southeast and the Adige River on the south. One important characteristic of this sector is that the ground level lies mostly below sea level (down to 4 m b.s.l.). The present-day surface water system is artificially regulated by a complex network of pumping plants and canals, which aim to introduce freshwater for irrigation, and to drain the phreatic aquifer. The canal network takes water upstream from the Brenta, Adige, and Bacchiglione rivers, the three main rivers of the area which rise in the Alps and the pre-Alps (Piedmont). In the area, the so-called "Venetian Aquifer system" is mostly composed of sand and silt of the coastal (Holocene) and continental (late Pleistocene) deposits, which cover the Quaternary basement at 230 m b.s.l. The aquifer contains intermittent clay lenses and peat layers (Brambati *et al.*, 2003; Bassan *et al.*, 1994). Despite the overall heterogeneity, six successive confined aquifers have been described between 50 m and 300 m b.s.l. (Carbognin & Tosi, 2003; Dal Prà *et al.*, 2000). The upper 50 m are the phreatic and semi-confined aquifer, and will be referred as A0.

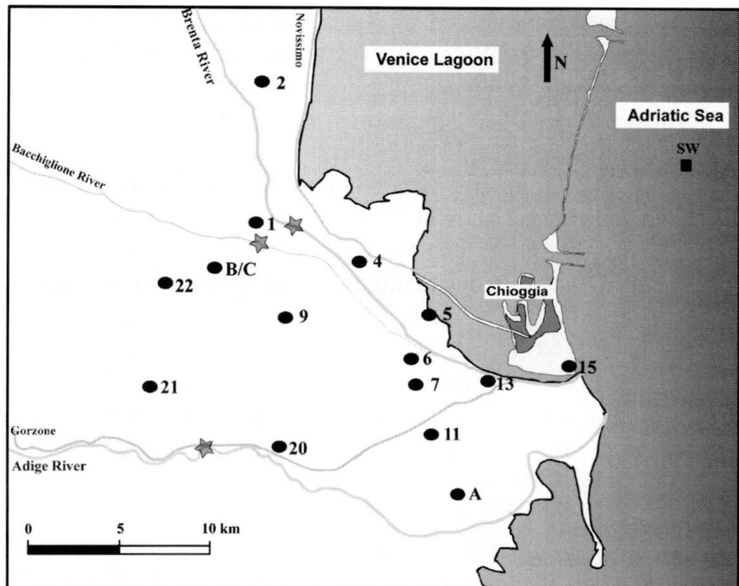

Fig. 1 Study area and water samples location. Circles stand for ISES boreholes; square point stands for seawater; stars stand for river waters.

Samples and analytical techniques

Sample collection was carried out during four field trips in September 2004, February, April and October 2005. Figure 1 displays the location of the boreholes used for this study, which are part of a monitoring network set up in 1999 for a project on salinization and subsidence (ISES). Eleven shallow boreholes screened down to 20 m b.s.l. and one screened between 32 and 53 m b.s.l. (C) were used to sample the groundwater in A0. Samples are identified with ISES boreholes numbers. Samples of groundwater from the first confined aquifers (A1) were taken from two artesian deeper boreholes screened at 76–94 m b.s.l. (A) and at 61–85 m b.s.l. (B). Vertical profiles of electric conductivity (EC) were obtained *in situ* before and after purging water in most boreholes (Fig. 2). In all cases, an immersion pump was set at 1 m above the borehole's bottom and water samples were collected after a stable EC of the pumped water was reached. Surface waters were also collected in the main rivers of the area (Adige, Brenta and Bacchiglione rivers) and in the coastal seawater. Physical parameters (temperature and EC) were measured in the field. Stable isotope analysis ($\delta^{18}O$ and $\delta^{2}H$) are performed at CEREGE on a Delta Plus MS, by measuring a gas (CO_2 and H_2 successively) equilibrated with water samples. Precisions obtained are 0.05‰ and 1‰ for $\delta^{18}O$ and $\delta^{2}H$, respectively.

RESULTS AND DISCUSSION

EC Data

The electrical conductivity of Adriatic seawater is 58 mS/cm. Brenta, Adige and Bacchiglione rivers were sampled in April and October of 2005. The mean EC for these three rivers is 0.44 mS/cm, 0.33 mS/cm, and 0.74 mS/cm respectively. The electrical conductivity in the groundwater samples of A0 is highly variable (0.7 to 40 mS/cm). Vertical profiles of EC, performed in 10 boreholes tapping A0 are shown on Fig. 2. Fresh groundwaters (I1, I2, I22) display EC values that vary between 0.7 to 1.3 mS/cm. The EC vertical profiles in boreholes ISES2 and ISES22 show a constant EC below 2 mS/cm without any vertical stratification. These groundwaters are found in the northwestern area, where the ground level lies slightly above sea level. They are located upstream of the general groundwater circulation. In this area, brackish to saline groundwaters (I4, I5, I7, I11, I13, I15 I20, I21) display highly variable salinity, with EC ranging from 3.5 to 40 mS/cm. In contrast to fresh groundwaters, EC vertical profiles show a vertical stratification with a sudden increase of EC, locally between 5 and 13 m depth. The upper parts of all profiles converge towards low salinity values, between 2 and 8 mS/cm. The more saline groundwaters (I5, I6, I13 and I15) are found in the boreholes closest to the seawater and the lagoon, within 2 km of the shorelines. Farther inland, neither the EC nor the transition level in brackish waters show a clear spatial distribution. This is likely to be due to the spatial heterogeneity of the aquifer conductive layers. By contrast, in the underlying confined aquifer, the groundwater keeps a constant salinity level, with the EC ranging between 3.4 and 3.7 mS/cm.

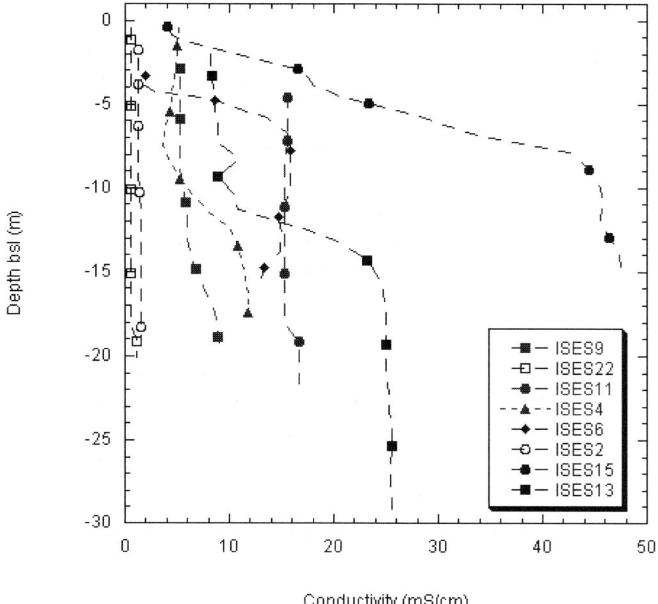

Fig. 2 Vertical EC profiles in ISES boreholes. Black, grey and white dots represent saline, brackish and fresh groundwaters, respectively.

Stable isotopes

The isotopic data are reported in the conventional δ (‰) notation as a deviation from the V-SMOW in the SMOW-SLAP scale. Precipitation was sampled monthly between October 2004 and September 2005, and had a weighted average composition of $\delta^{18}O$ = –6.10‰ and δD = 39.1‰, close to the long-term average composition measured in Trieste ($\delta^{18}O$ = –6.60‰ and δD = 42.6‰, Longinelli & Selmo, 2003). A meteoric water line has been defined for northern Italy from long-term average of several sampling stations (North Italy Meteoric Water Line, NIMWL, $\delta D = 7.7094 \times \delta^{18}O + 9.40343$; Longinelli & Selmo, 2003). The measured Adriatic seawater composition is $\delta^{18}O$ = 1.29 ‰ and δD = 8.4‰, in agreement with previous studies (Stenni *et al.*, 1995). The isotopic compositions of the three main rivers of the Venice region, Adige, Brenta and Bacchiglione rivers, are $\delta^{18}O$ = –11.36‰, –9.53‰, and –8.67‰, and δD = –81.2‰, –63.3‰ and –57.2‰, respectively. There is a significant difference between these three rivers, reflecting the different recharge altitudes: the Adige River originates from the Alps, while the Brenta and the Bacchiglione rivers have their sources at lower altitudes. The isotopic composition of groundwater from the A0 aquifer ranges from –10.4 ‰ to –1.84‰ for $\delta^{18}O$ and from –61.4‰ to –13.7‰ for δD. Deep groundwater shows a $\delta^{18}O$ and δD isotopic composition ranging from –10.22‰ to –10.66‰ and from –69.2‰ to –72.3‰ respectively. These values are intermediate between the Brenta and Adige samples and are similar to the most depleted A0 groundwater samples.

In δ space (Fig. 3), freshwater samples cluster along the NIMWL, with a composition intermediate between the Brenta River and local precipitation values.

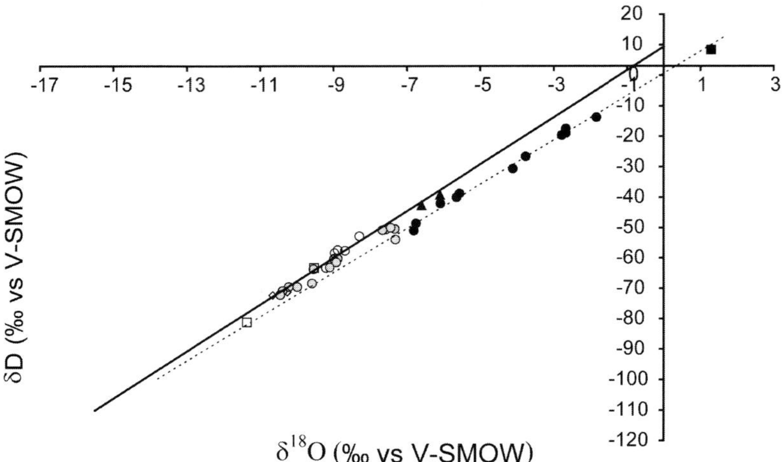

Fig. 3 δD *vs* $\delta^{18}O$ diagram. Black, grey and white circles stand for saline, brackish and fresh groundwaters from A0 respectively; black square stands for seawater; white squares stand for Adige, Brenta and Bacchiglione rivers, diamond shaped dots stand for deep groundwater, triangle dot stands for rainfall, (—) defines the NIMWL, (---) defines the regression line through the more saline groundwaters from A0.

Brackish samples present a more important range of variation (from $\delta^{18}O = -10.45$ to -7.33‰) and are located along, or slightly lower than, the NIMWL. The saline groundwaters are enriched in stable isotopes compared with the two previous groups, and deviate significantly from the NIMWL (I5, I6, I13, I15). They define a mixing line ($\delta D = 7.27 \times \delta^{18}O + 0.67$; $n = 11$; $r^2 = 0.992$) between the seawater composition and a continental water end-member, isotopically depleted. The isotopic composition of this continental end-member, as estimated by the intercept of the mixing line and the NIMWL, is very sensitive to the slope of the mixing line. By taking into account the saline samples only, its composition could correspond to that of the Adige River (Fig. 3). However, taking brackish groundwater samples into account for the linear regression, in addition to the more saline groundwaters, would lead to a lower slope, and to a higher isotopic composition for the continental end-member.

A linear trend also appears in the relationship between EC and $\delta^{18}O$ for the more saline groundwater samples ($EC = 5.3041 \times \delta^{18}O + 52.43$; $r^2 = 0.88$; Fig. 4). The mixing line crosses the seawater plot, suggesting that the high salinity originates mainly from the mixing with seawater. However, the linear correlation is not robust enough, and it is not possible to identify a unique continental end-member. Freshwater samples compositions reflect a mixing between waters originating from the main rivers, with a potential influence of local rainwater. The situation is more complex for the brackish samples: their position on both sides of the mixing line defined by the saline group may reflect the contribution of surface water, deep groundwater, or/and rainfall. A detailed geochemical study will be necessary to identify the respective contributions of these different waters.

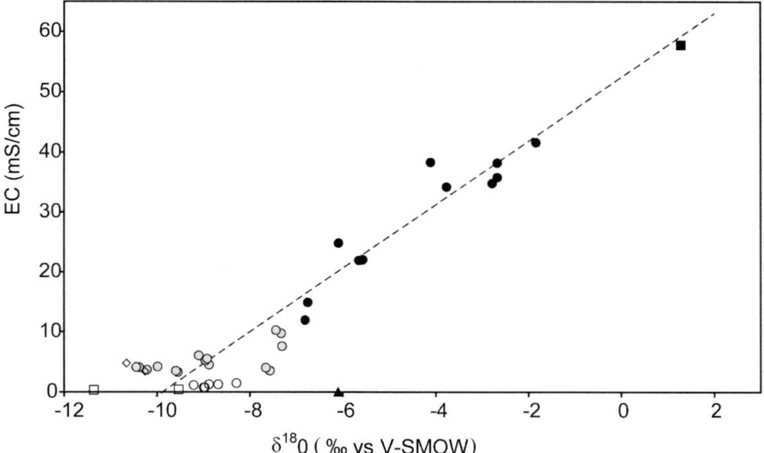

Fig. 4 δ^{18}O *vs* EC (same symbols as in Fig. 3) for the water samples representation. The dotted black line stands for the linear trend between δ^{18}O and EC for the saline groundwaters (EC = 5.3041 × δ^{18}O + 52.43; $r^2 = 0.88$).

CONCLUSION

A first investigation of the salinization processes has been carried out by using EC and isotopic tracers in the semi-confined aquifer of the southern Venice Lagoon system. Saline groundwaters are shown to result from a mixing between continental and marine water end-members. However, the dynamic process of saltwater intrusion in the semi-confined aquifer has to be clarified as the salinization may occur via intrusion of seawater, either directly or through the lagoon or the rivers system. Further data are also necessary to characterize the origin of the continental end-member(s) in the mixing process with seawater: surface water, lateral groundwater flow and/or deep groundwater. The presence of fresh to brackish groundwater associated with a water table below the sea level is not compatible with a natural steady-state system. It may be explained by a non-steady state saline intrusion in response to the subsidence, or by the influence of the artificial canal network which induces a permanent renewal of groundwater. Indeed, the isotopic data reveals an important influence of surface water in fresh groundwater samples. Brackish samples have shown an important heterogeneity which was not clearly related to their geographical position, nor to a simple mixing process. The origin of their salts has to be studied further.

Acknowledgements This study was supported by 2004–2006 Corila programme. The authors warmly thank the Morgan Rilievi SARL, the Provinzia di Venezia and the Consorzio Adige Bacchiglione for their permanent support and their precious help on the field. We would also like to thank Professor Tosi Luigi (ISMAR-CNR) for long discussions that helped us in our research.

REFERENCES

Andreasen, D. C. & Fleck, W. B. (1997) Use of bromide:chloride ratios to differentiate potential sources of chloride in a shallow, unconfined aquifer affected by brackish-water intrusion. *Hydrogeol. J.* **5**(2), 17–26.

Barbecot, F., Marlin, C., Gibert, E. & Dever, L. (2000) Hydrochemical and isotopic characterisation of the Bathonian and Bajocian coastal aquifer of the Caen area (northern France). *Appl. Geochem.* **15**, 791–805.

Bassan, V., Favero, V.,Vianello, G. & Vitturi, A. (1994) Studio geoambientale e geopedologico della Provincia di Venezia, parte meridionale. Provinzia di Venezia, Italy.

Brambati, A., Carbognin, L., Quaia, T., Teatini, P. & Tosi, L. (2003) The Lagoon of Venice: geological setting, evolution and land subsidence. *Episodes* **26**(3), 264–268.

Capaccioni, B., Mariano, D., Carmela, P. & Lia, D. (2005) Saline intrusion and refreshening in a multilayer coastal aquifer in the Catania Plain (Sicily, Southern Italy): dynamics of degradation processes according to hydrochemical characteristics of groundwaters. *J. Hydrol.* **307**, 1–16.

Carbognin, L. & Tosi, L. (2003) Il progetto ISES per l'analisi dei processi di intrusione salina e subsidenza nei territori meridionali delle province di Padova e Venezia. Istituto per lo Studio della Dinamica delle Grandi Masse, CNR, Italy.

Dal Prà, A., Gobbo, L.,Vitturi, A. & Zangheri, P. (2000) Indagine idrogeologica del territorio provinciale di Venezia. Provinzia di Venezia, Italy.

Di Sipio, E., Galgaro, A. & Zuppi, G. M. (2006) New geophysical knowledge of groundwater systems in Venice estuarine environment. *Estuar. Coast. Shelf Sci.* **66**, 6–12.

Gambolati, G. (2003) Subsidence due to peat oxidation and its impact on drainage infrastructures in the farmland catchment south of the Venice Lagoon. *Materials and Geoenvironment* **50**(1), 125–128.

Longinelli, A. & Selmo, E. (2003) Isotopic composition of precipitation in Italy: a first overall map. *J. Hydrol.* **270**, 75–88.

Stenni B., Nichetto, P., Bregant, D., Scarazzato, P. & Longinelli, A. (1995) The $\delta^{18}O$ signal of the northward flow of Mediterranean waters in the Adriatic Sea. *Ocean Acta* **18**(3), 319–328.

Modélisation de l'intrusion marine dans l'aquifère côtière du Gabès (sud tunisien)

BADIAA CHULLI[1], ABDALAH TAHERI TIZRO[2] & NASIME JABNOUN[3]

1 *Centre de Recherches et Technologies des Eaux, Labo Hydrogéologie, Technopole Borj Sadria, Tunisie*
bchoulli@yahoo.fr

2 *Dept of Water Engineering, College of Agriculture, Razi University, Kermanshah, Iran*

3 *Faculté des Sciences de Tunis, El Manar II, Tunisie*

Résumé L'aquifère côtière du Gabès sud fait partir du système aquifère multicouche de Jeffara nord. Elle est formée par des sédiment alluvionnaire et détritiques d'age Quaternaire et par des dépôt argilo-sabeuses d'age Mio-Pliocène. La nappe phréatique de Gabès sud est la plus anciennement exploitée dans le gouvernorat de Gabès. Bien qu'elle présente de bonnes caractéristiques hydrogéologiques et hydrochimiques, cette nappe est de plus en plus sollicitée. La plus grandes densités des puits de surface sont localisées dans la région de Kettana et de Mareth. La mauvaise répartition de l'exploitation dans l'espace provoque un déséquilibre hydrodynamique et hydrochimique et aussi le risque d'une intrusion des eaux marines si en se trouve dans un stade d'exploitation avancé. Dans le cadre de la présente article nous abordant les principales caractéristiques, lithologiques, hydrodynamiques et géochimiques de l'aquifère du Gabès sud dans le but d'élaborer une exquise de comportement de ces paramètre, dans l'espace et dans le temps au sein de cette ensemble stratigraphique.

Mot clefs systeme aquifère de Jeffara; hydrodynamisme; hydrochimie, Tunisie

Seawater intrusion modelling for the Gabes coastal aquifer system (southern Tunisia)

Abstract The south Gabes coastal aquifer is a multilayer aquifer system located in northern Jeffara in southern Tunisia. Unconsolidated alluvial deposits, mainly Mio-Pliocene and shale of Quaternary age, underlain by a thick sequence of Mesozoic and Cenozoic formations, characterize the area. Over-abstractions of groundwater in the Kettatna and Mareth areas have resulted in severe disequilibrium of its water balance. Many deep and shallow wells have been constructed by farmers in order to cover the water demands for irrigation supply. The increase in groundwater abstraction, combined with seawater intrusion, have been accompanied by a fall in water table depth. Along the coastal area, increased groundwater salinity is reported. In this study, an attempt has been made to decipher the lithological formations, hydrodynamic properties and geochemical characteristics of the south Gabes aquifer in order to establish the causes. All existing geological, hydrological, and hydrogeological data were evaluated and reworked.

Key words Jeffara aquifer system; saline intrusion; hydrodynamic; hydrochemical; Tunisia

INTRODUCTION

La région d'étude couvre la totalité de la plaine côtière de la Jeffara nord (Fig. 1). Elle est limitée au nord-est par la mer méditerranéenne et au sud-ouest par les Jebels

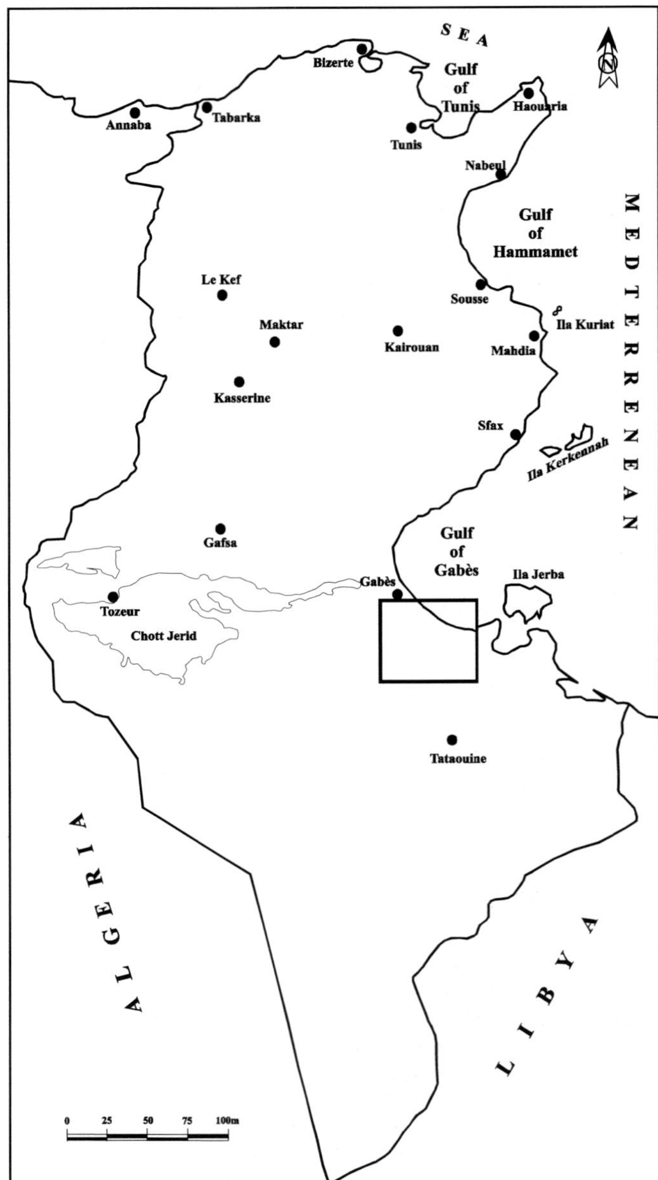

Fig. 1 Localisation géographique de domaine d'étude.

de Matmata. Les reliefs de Jebels Matmata jouent le rôle d'écran et favorisent le ruissellement des eaux pluviales vers la mer.

Cette région est caractériser par un climat aride a semi aride, des pluviométries qui varie entre 183 mm et 220 mm par an, une température moyenne annuelle de 19.7°C et un taux d'évaporation annuelle de 2128 mm.

La plaine côtière de Gabès sud est traversée par quatre oueds importants qui sont Oued Zigaou, Oued Zerkine Oued El Ferd et Oued Sourrag.

CARACTERISTIQUE HYDROGEOLOGIQUE

Les eaux de la nappe phréatique de Gabès sud se logent Ayadi (1986):

- Soit dans les formations allivionnaires et détritiques résultant d'une ancienne activité hydrographique, des oueds descendant des Matmatas. Ces formations sont d'age Quaternaire.
- Soit dans le formations argilo-sableus du Mio-Pliocène.

Dans ces formations alluviales les variation latérales de faciès sont toujours fréquentes, à cause de l'apport en matériaux des oueds des Matmatas.

L'alimentation de la nappe phréatique de Gabès sud ce fait à partire des eaux des crues, infiltrées au niveau des lits des principaux oueds descendant des Matmatas et par infiltration directe des eaux météoriques tombant sur la plaine côtière et aussi l'infiltration des eaux d'irrigation au niveau des oasis.

La décharge de la nappe ce fait par les sources, les Sebkhas, la mer et les puits de surfaces qui constituent les exutoires de la nappe.

HISTOIRE DE L'EXPLOITATION DE LA NAPPE (1979–1995)

Tableau 1 Récapitulation de l'histoire de l'exploitation de la nappe de Gabès sud de 1979 à 1995.

Année	Puits équipés	Puits non équipés	Puits abandonnés	Total puits	Exploitation (Mm^3/an)
1977	104	416	0	520	3.37
1979	217	319	147	707	4.80
1985	492	262	224	978	5.83
1987	560	307	230	1097	6.60
1990	598	405	272	1275	7.94
1995	994	48	409	1451	10.40

ETUDE PIEZOMETRIQUE

Etat de la piézomètrie en 1979 La carte piézométrique dressée par Mammou en 1979 (Fig. 2) montre que le sens d'écoulement de la nappe suit la topographie du terrain et se fait à partir des reliefs vers la côte. Les axes principaux d'écoulement suivent sensiblement les lits des principaux oueds descendant de la chaîne des Matmats. Les lits de ces oueds représentent ainsi des zones préférentielles d'alimentation de la nappe. La pente hydraulique de la nappe est de l'ordre de 5% ; elle se réduit vers l'amont de la plaine pour atteindre 1.4%.

Etat de la piézomètrie en 1995 La carte piézomètrique (Fig. 3) montre que le sens d'écoulement des eaux galonnent la topographie de la nappe dans un sens orienté SW–NE. Le gradient hydrolique présente des valeurs comprises entre 2.8 et 4% dans les zones amonts et 8% prés de la côte où les courbes isopièzes deviennent plus serrées. Ceci peut être du à la variation lithologique des formations aquifères qui deviennent plus détritiques. L'écoulement de la nappe est influencé par l'exploitation intense au niveau des régions à grande concentration de point d'eau, localisées autour des lits des oueds qui offrent des conditions pédologiques meilleures pour la culture.

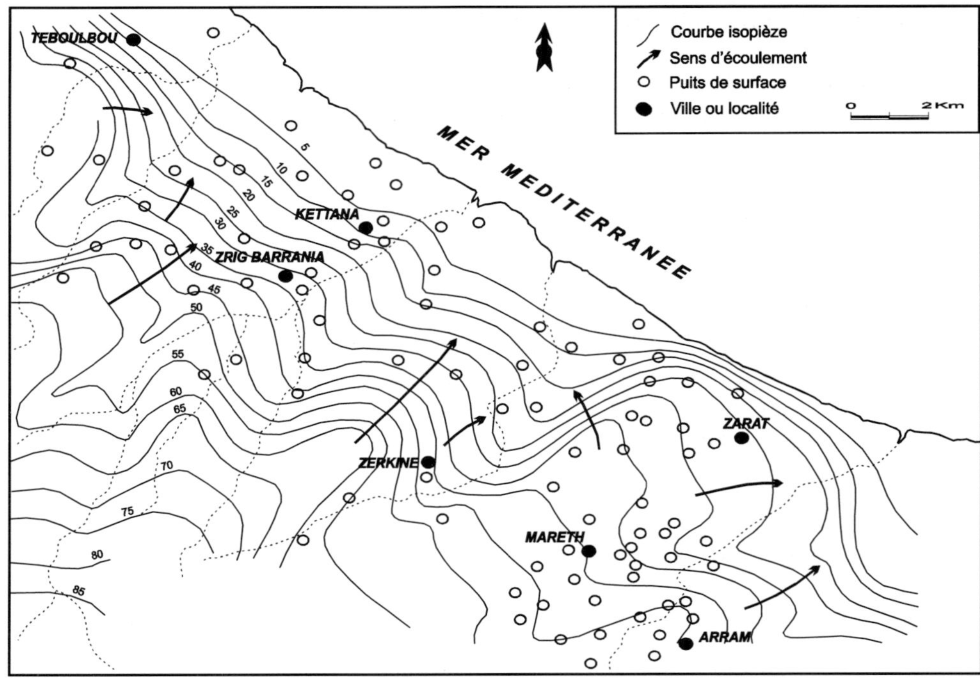

Fig. 2 Carte piézométrique de la nappe phréatique de Gabès sud Mammou: 1979.

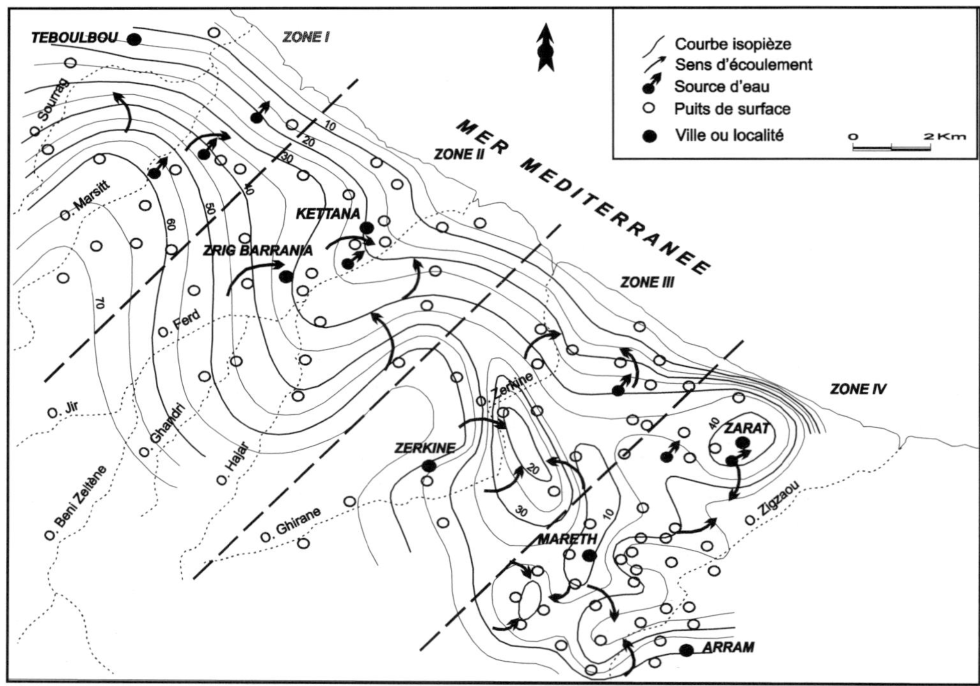

Fig. 3 Carte piézométrique de la nappe phréatique de Gabès sud Mammou: 1995.

EVALUATION DES RESSOURCES DE LA NAPPE

Afin de pouvoir cerner les ressources de la nappe, quatre zones représentant des sous-bassins des principaux oueds sont identifiées (Fig. 3).

Tableau 2 Représentation des nombres de puits et de leurs débit d'exploitation pour chaque zone.

Zones	Sous-bassins	Nombre de puits	Débit d'exploitation (L/s)
Zone I	Oued Marsitt et Sourrag	134	30.70
Zone II	Oued Ferd	240	54.45
Zone III	Oued Zerkine	181	41.25
Zone IV	Oued Zigzaou	896	203.60

Les ressources potentielles réelles

On va applique la Loi de Darcy:

$$Q = T\,L\,I \tag{1}$$

Tableau 3 Récapitulation des ressources potentielles réelles avec le calcul de débit amont et aval dans chaque zone.

	Zone I: Amont	Aval	Zone II: Amont	Aval	Zone III: Amont	Aval	Zone IV: Amont	Aval
T (m²/s)	2 10⁻³	0.5 10⁻³	2 10⁻³	0.5 10⁻³	2 10⁻³	05 10⁻³	1.44 10⁻³	0.35 10⁻³
L (m)	9 10³	7.65 10³	14.4 10⁻³	8.37 10⁻³	10.8 10⁻³	6.75 10⁻³	21.6 10⁻³	14.85 10⁻³
I (%)	4 10⁻³	8.10⁻³	4 10⁻³	6 10⁻³	4.5 10⁻³	7.3 10⁻³	3.23 10⁻³	3.64 10⁻³
Q(m³/s)	72	30,6	115.20	25.11	97.20	24.64	100.46	20
R_{pr} (L/s)	41.4		90.09		72.56		80.46	

T: Transmissivité horizontale; L: Largeur du front de la nappe (m); I: Gradient hydraulique;
Q: Débit traversant le front de la nappe; R_{pr}: ressources potentielles réelles.

Les ressources potentielles réelles (R_{pr}) est égale à la différence entre un débit amont (Q_m) et un débit aval (Q_v):

$$R_{pr} = Q_m - Q_v \tag{2}$$

La somme des débits amont de tout les sous bassins égale (Q_m = 385 L/s).
La somme des débits aval de tout les sous bassins égale (Q_v = 100 L/s).
On applique l'équation (2) R_{pr} = 285 L/s.

Dans un cas idéal on doit trouver $Q_{ex} = R_{pr}$. C'est à dire que le volume perdu entre l'amont et l'aval doit égaliser le volume exhaure au niveau des puits. Or dans notre cas le volume calculé par la méthode de Darcy et celui d'exploitation sont différents ($Q_{ex} > R_{pr}$). Cette différence qui s'élève de 45 L/s (330–285 = 45) Pourrait être expliquée par:

- l'infiltration par endroits du surplus des eaux d'irrigation au niveau des oasis,
- le déversement des eaux de la nappe profonde du à la drainance verticale facilitée par l'existence des failles.

Les ressources renouvelables:

$$R_r = V_{id} + V_{ir} \tag{3}$$

Le Tableau 4 suivant résume les résultats de calculs des ressources renouvelables.

Avec:

$$V_{id} = P\, S\, I_1 \quad (4)$$

$$V_{ir} = V_r\, I_2 \quad (5)$$

P est la pluviométrie moyenne du bassin versant considéré (m); S est surface du bassin versant (m^2); V_r est volume ruisselé (m^3/an); I_1, I_2 sont coefficients d'infiltration, $I_1 = 2\%$ et $I_2 = 45\%$.

Cette valeur de ressources renouvelables donne un débit $Q_r = 350$ L/s.

Tableau 4 Récapitulation des résultats de calculs des ressources renouvelables.

Bassin	V_{id} (m^3/m)	V_{ir} (m^3/an)	R_r (m^3/an)
O. Zigzaou	812 820	992 925	1 805 745
O. Zerkine	568 800	585 855	1 154 655
O. Ferd	1 866 800	2 309 175	4 175 975
O. Sourrag	1 329 620	1 125 585	2 455 205
O. Segui Mareth	672 600	778 410	1 451 010
Total	5 250 640	5 791 950	11 042 590

R_r: ressources renouvelables (m^3/an); V_{id}:volume d'infiltration directe (m^3/an); V_{ir}: volume d'infiltration après ruissellement (m^3/an).

Bilan hydrique de la nappe

Le bilan global de la nappe montre que les sorties excèdent les entrées; il s'agit donc d'un bilan déficitaire. La zone IV est la seule touchée par cette surexploitation. Dans cette région l'exploitation représente 222.60% des ressources dynamiques. Cette situation nous pousse à prendre des mesures rigoureuses pour interdire toute nouvelle création de puits d'eau dans cette région.

Tableau 5 Résume de bilan hydrique de la nappe.

Zones	Entrées (L/s)	Sorties (L/s)	Differences (L/s)
Zone I	72.00	61.30	10.70
Zone II	115.20	79.56	35.64
Zone III	97.20	65.89	31.31
Zone IV	100.46	223.60	–123.14
Total	384.86	432.35	–47.49

ETUDE DE LA SALINITE DE LA NAPPE

Etat de la salinité en 1979 Sur cette carte (Fig. 4) on remarque que les salinités de la nappe varies entre 2 et 10 g/L (Mamou, 1979). Les zones à fortes salinités correspondent à:

- des lieux d'évaporation (sebkhas);
- des lieux de contamination marine comme l'embouchure sur l'Oued Ferd, Oued Sourr (Mamou, 1995); et
- des zones de surexploitation (cas de Mareth).

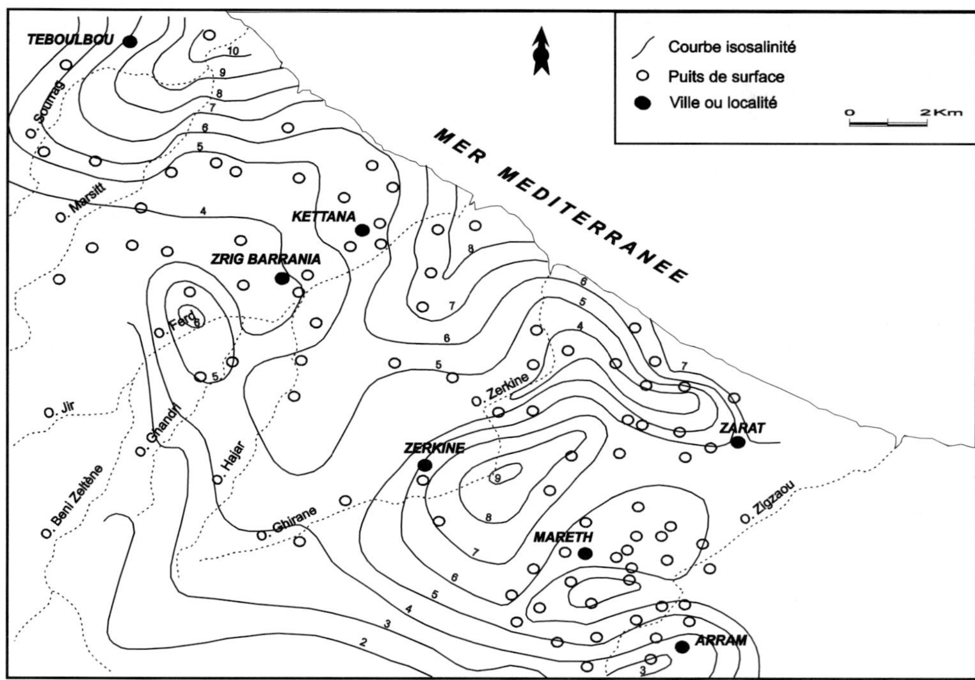

Fig. 4 Carte des isosalinités de la nappe phréatique de Gabès sud: 1979.

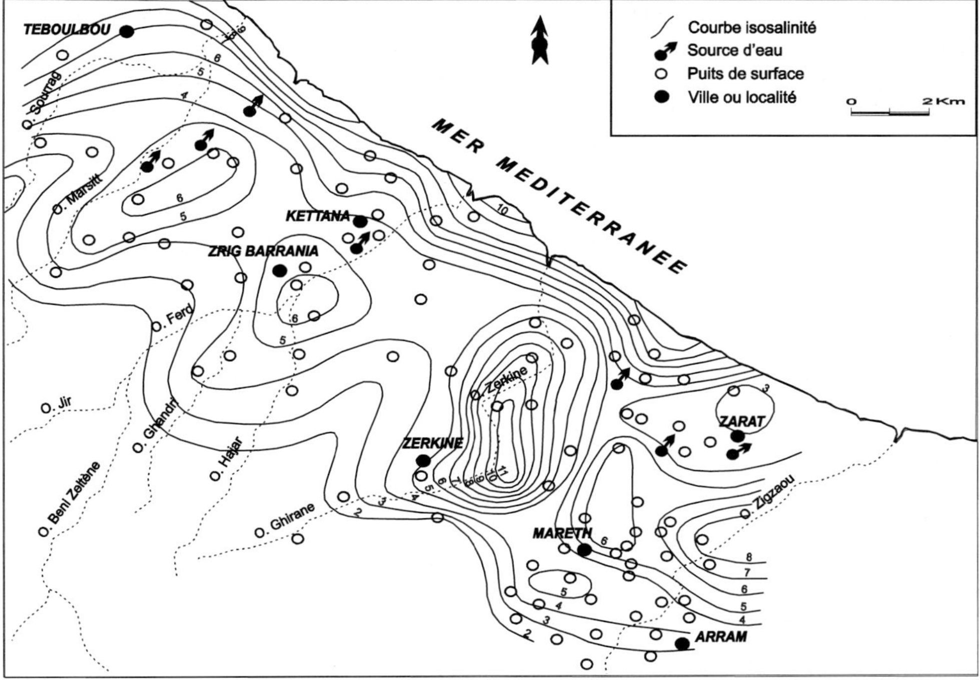

Fig. 5 Carte des isosalinités de la nappe phréatique de Gabès sud: 1995.

Etat de la salinité en 1995 Sur le littoral entre Teboulba et Zarat la salinité (Fig. 5) varie entre 5 et 10 g/L (Mamou, 1995). Ces valeurs sont expliquées par le phénomène de contamination par les eaux marines et de dissolution de sels par suite de l'écoulement des eaux souterraines.

La partie amont, considérée comme sone d'alimentation de la nappe se caractérise par des valeurs de salinité plus faible (2 à 4 g/L). Pour ces deux parties de la nappe l'allure des courbes isosalinités suit parfaitement celle dessinée par les isopièzes; on en conclu donc que la salinité évolue dans le même sens que l'écoulement de la nappe.

La forte salinité 11 g/L correspond soit à une sone d'évaporation, comme la Sebkha de Zerkine, soit à une zone à grande concentration de points d'eau, comme Ketana et Mareth. La surexploitation combinée à l'évaporation des eaux ne font qu'augmenter la salinité des eaux.

Les faibles valeurs de salinité enregistrées dans la région de Zarat et la zone se trouve entre Oued Zerkine et Oued Ferd ne font que confirmer le phénomène de drainage verticale mis en évidence par la piézomètre de la nappe. Ces deux zones sont donc des zones d'alimentation à partir de la nappe profonde.

CONCLUSIONS

L'aquifère phréatique de Gabès sud est logé dans les formations alluvionnaires et détritiques du Quaternaire et dans les formations argilo-sableuses du Mio-Pliocène.

L'écoulement global de la nappe se fait des reliefs vers la cote dans un sens SW–NE. Cet écoulement de la nappe est perturbé par la surexploitation régionale, par l'existence de zones de décharges (Sebkhas) et de zones d'alimentation à partir de la nappe profonde par alimentation verticale (Chulli, 1998).

Les ressources dynamiques sont estimées à 385 L/s, les ressources potentielles réelles sont de l'ordre de 285 L/s.

A cause de la surexploitation, le bilan global de la nappe est déficitaire et le déficit s'élève à 1.43 Mm^3/an.

Le suivie de l'évolution de la salinité des eaux à montré qu'elle évolue dans le même sens que l'écoulement de la nappe et que la qualité chimique des eaux s'est dégradée entre 1985 et 1995 dans les régions côtières en aval de Kettana (Chulli, 2002).

REFERENCES

Ayadi, M. (1986) Etude de la nappe phréatique de Mareth, Rap. Int., D.G.R.E.

Chulli, Z. B. (1998) Hydrodynamisme et géothermie de l'aquifère triasique du sud tunisien. *Africa Geoscience Review* **5**(3), 297–311.

Chulli, Z. B. (2002) Structuration profonde et géothermie de la marge orientale de la Tunisie. *Africa Geoscience Review* **9**(1), 39–46.

Mamou, A. (1979) Note sur la situation d'exploitation des nappes du gouvernorat de Gabès, Rap. Int. D.G.R.E.

Mamou, A. (1995) Caractéristiques, évaluation et gestion des ressources en eau du sud tunisien. Thèse Doct.ès-Sciences naturelle, Université de Paris-Sud, Centre d'Orsay.

4 MODELLING APPROACHES

Driving while under the influence: pumping-driven circulation under the influence of regional groundwater flow

M. BAYANI CARDENAS[1] & JOHN L. WILSON[2]

1 *Geological Sciences, University of Texas at Austin, Austin, Texas 78712, USA*
cardenas@mail.utexas.edu

2 *Earth and Environmental Sciences, New Mexico Inst. of Mining and Technology, Socorro, New Mexico 87801, USA*

Abstract Regional submarine or ambient groundwater discharge (AGD) along sea coasts interacts with current-driven interfacial pumping of seawater through shallow sediments. We use numerical simulations to investigate this interaction for a turbulent current flowing over dune topography. AGD reduces the extent of the current-topography driven interfacial exchange zone (IEZ) and may prevent its development when AGD overpowers interfacial exchange. Under upwelling-AGD conditions (upward flux of deep groundwater into the water-column), the IEZ is centred on the stoss face of bedforms and AGD occurs near the crest. Under downwelling-AGD conditions (downward flux of seawater deep into the aquifer), the IEZ forms around the crest and water infiltrating along the stoss face into the sediments does not return to the sediment–water interface. The IEZ depth, flux and residence time are functionally related to current Reynolds number (Re). For example, the IEZ water residence time, which follows a power-law distribution, is larger at lower current Re.

Key words interfacial exchange; bedform; groundwater discharge; turbulent flow

INTRODUCTION

Ocean and estuarine water are connected to and interact with coastal groundwater. Regional (0.1–10 km) interactions and connections occur when groundwater discharges as diffuse flow from coastal aquifers (diffuse submarine groundwater discharge) or, when the flow is reversed, through intrusion of seawater into aquifers. Depending on direction, this net deep groundwater flow can be referred to as "groundwater upwelling" into the ocean or estuary, when the flow is upward, or as "groundwater downwelling" when the flow is from the ocean or estuary to the groundwater. Hydrogeologists, who normally work with terrestrial freshwater lakes and streams, refer to these situations as "gaining" or "losing", respectively, where the lake or stream gains water by upward groundwater flow, and loses water to the aquifer by downward flow. This paper will use that terminology, but generalize by referring to both situations as examples of regional ambient groundwater discharge (AGD).

Seawater and groundwater interact locally (0.1–10 m) within permeable bottom sediments through pumping and circulation of seawater from the ocean or estuarine water column, downward through sediments, and back up again into the water column. This small-scale circulation is driven by a combination of waves, tides, and currents (Burnett *et al.,* 2003). This paper focuses on current-driven exchange, caused by the interaction of the water-column current with bottom topography, and which induces eddies and bottom-pressure variations along the sediment–water interface (SWI). Local-scale downwelling occurs in zones of high bottom-pressure where seawater enters the sediments, while local-scale upwelling of water from the sediments occurs in low-pressure zones. When the sediments are permeable enough, the current-driven bottom-pressure variations drive significant advective transport of seawater across the SWI and through the sediments.

The rates and spatial distribution of local current-topography-driven exchange and regional-scale AGD influence many important biogeochemical and ecological processes taking place along the SWI (Boudreau *et al.*, 2001), with measurable impacts from local to global scales for both estuaries (Webster *et al.*, 1996) and oceans (Riedl *et al.*, 1972). The influence is strongest when advection dominates diffusion. When this happens, insight into understanding and observing the biogeochemistry and ecology requires an appreciation of the hydrodynamics (Burnett *et al.*, 2003; Huettel *at al.*, 2003). This paper investigates the hydrodynamics of current-bedform driven exchange under the influence of regional ambient groundwater discharge (AGD) when the current is turbulent. When the current is laminar, refer to Cardenas & Wilson (2006, 2007b).

METHODOLOGY

We numerically model the flow above and below SWIs in a two-dimensional cross-section and at steady state (Fig. 1). The procedure and system formulation follows that of Cardenas & Wilson (2007a) where the turbulent water column is governed by the Reynolds-averaged Navier-Stokes (RANS) equations coupled with the k–ω closure scheme (Wilcox, 1991), and the interstitial flow within the sediments is assumed to be Darcian and described by the groundwater flow equation. Briefly, we solve the RANS equations first (using the commercial CFD-ACE+ code) while considering the bottom of the water column (the SWI) to be a no-slip wall. We then impose the RANS-based bottom-pressure solution along the wall as a Dirichlet boundary to the top of the groundwater flow model that represents the sediments (solved via the commercial Comsol Multiphysics code). The lateral boundaries for both domains (water column and sediments) are spatially periodic but with a prescribed pressure drop resulting in mean flow from left to right. This mean flow in the sediments is typically referred to as "underflow". Details (e.g. water depth, water column mean velocity field, etc.) and validation of the methodology are presented in Cardenas & Wilson (2007a), in which we considered current-bedform driven exchange but ignored AGD, and are not repeated here. The main difference between our previous work and this study is that we impose a prescribed flux at the lower boundary of the sediments to represent AGD (compare our Fig. 1 to Fig. 1(c) in Cardenas & Wilson (2007a)). At this boundary, we consider a prescribed flux both into and out of the domain representing, respectively, "gaining" and "losing" water columns. Fluid properties are those of freshwater at 20°C and isothermal conditions are assumed (we ignore effects of salinity and temperature on fluid density and viscosity). The sediment, assumed to be sand, is assigned a homogeneous, isotropic permeability (k) of 10^{-10} m^2 in all simulations. We consider the presence of a triangular bedform along the SWI with length L of 1.0 m and height H of 0.05 m, whose crest is invariantly located at $0.9L$, typical of subaqueous sand dunes under a unidirectional current.

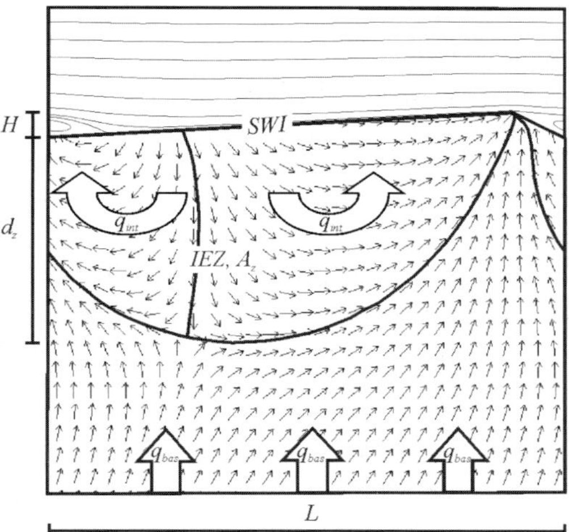

Fig. 1 Schematic representation of the periodic flow domain for the water column and sediments, depicting geometric parameters: height (H) and length (L), and fluxes: basal (q_{bas}) and interfacial (q_{int}). The depth of the interfacial exchange zone (IEZ), d_z, is the vertical distance from the trough to the deepest portion of the streamline enveloping the region wherein water flows from and then back into the water column. Streamlines are presented in the water column while small arrows in the sediments correspond to flow directions but not magnitude. SWI: sediment–water interface.

Sensitivity analysis is performed via multiple CFD simulations by varying the parameter values. We vary the water-column current, measured by a Reynolds numbers (*Re*), and the prescribed ambient groundwater flux density at the base of the groundwater flow domain, q_{bas}, which we refer to as "basal flux" to differentiate it from other flux terms (Fig. 1). The total basal flux is then given by the product $q_{bas}L$. The Reynolds number is defined as $Re = U_{ave}H/v$, where U_{ave} is the average horizontal water-column velocity along a vertical-section taken above the crest of the bedform, and bedform height H is the characteristic length scale. The current U_{ave} and Reynolds number are varied by imposing different horizontal pressure drops across the domain.

RESULTS AND DISCUSSION

We limit our discussion to the flow through the permeable sediments. The water-column current and resulting bottom-pressure distribution are described in Cardenas & Wilson (2007a). The area within the permeable sediments that is physically influenced by current-bedform induced exchange across the SWI is the interfacial exchange zone (IEZ). This is the zone created by the local downwelling of seawater, which then circulates though the sediments and upwells back to the water-column. It is characterized by streamlines that originate and end at the SWI (Fig. 1) and represents the volume of sediments advectively swept by exchanged seawater. Freshwater ecologists and hydrologists commonly refer to this as the "hyporheic zone". The depth of the IEZ, d_z, is taken as the distance between the deepest portion of the streamline which envelopes all streamlines originating from and returning to the SWI, and the trough of the bedform (Fig. 1). The IEZ flux through the SWI is computed as follows: (i) first, a volumetric flux through the SWI is computed by integration of the magnitude (absolute value) of the normal flux along the bedform surface; (ii) from this we subtract the basal flux ($q_{bas}L$); and (iii) the resulting quantity is divided by two because the integration does not discriminate between current-bedform-driven flux going in and out of the bed, which are in balance. Dividing the IEZ flux by L yields an IEZ flux density, q_{int}, based on bedform length. The total IEZ flux is given by the product $q_{int}L$ and takes place only for that portion of the SWI subjected to current-driven flushing (bounded by the dividing streamline discussed above and illustrated in Fig. 1). The basal and interfacial flux densities, q, which are schematically represented in Fig. 1, are nondimensionalized by $q^* = q/K$, where $K = (kg/\nu)$ is the hydraulic conductivity of the sediments, ν is kinematic viscosity, and g is the acceleration due to gravity.

Flow fields for scenarios with different Re and fixed basal flux, q_{bas}^*, are presented in Fig. 2. The first important observation is that IEZs can form in the presence of AGD. Secondly, the flow fields within and outside the IEZ are a result of the competitive interaction between current-bedform driven flow and ambient groundwater discharge. The results for both gaining and losing conditions are elaborated below.

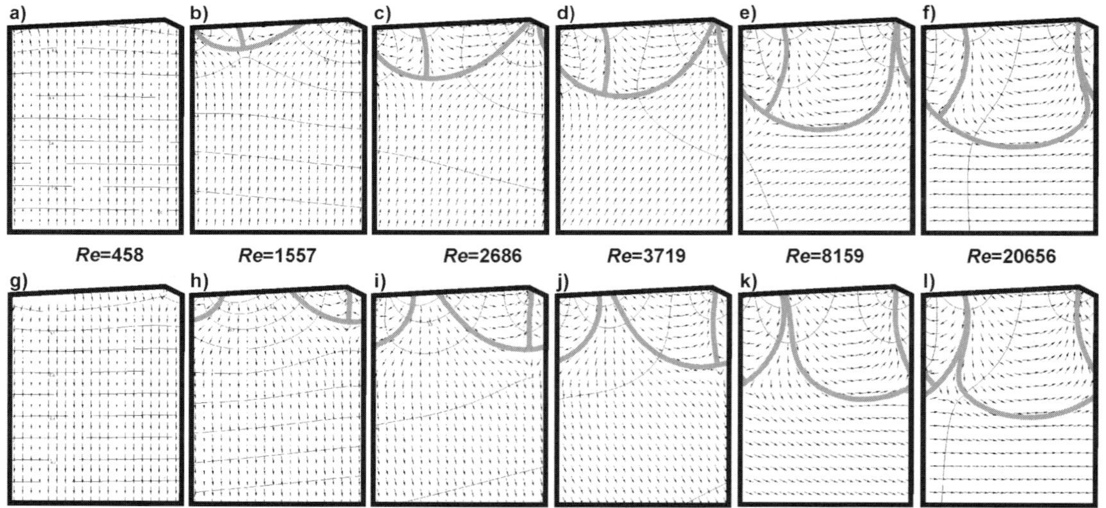

Fig. 2 Top row: pressure fields (contours with intervals of 0.1) and flow directions (arrows) in the sediments for gaining conditions ($q_{bas}^* = 9.8 \times 10^{-2}$) at different Reynolds number (Re). Bottom row: the same for losing conditions ($q_{bas}^* = -9.8 \times 10^{-2}$). Pressure is normalized as: $p^* = (p - p_{min})/(p_{max} - p_{min})$ where p^* is normalized pressure, p is pressure and p_{max} and p_{min} the maximum and minimum pressures for each simulation, respectively. The grey streamlines delineate the interfacial exchange zones. H/L = 0.05 and L = 1.0 m for all cases where H is the bedform height and L is its length.

IEZ spatial configuration and fluxes: gaining conditions

For "gaining" conditions with upward AGD, the IEZ, when it is present, is centred around the bottom-pressure maximum which is located along the stoss or upstream-facing surface of the dune (Fig. 2, top row). This maximum pressure point is located approximately where the water-column eddy re-attaches (see figures and discussion in Cardenas & Wilson, 2007a). There are two flow cells in the sediments, similar to cases without AGD (Cardenas & Wilson, 2007a). For lower *Re*, the IEZ terminates at points along the SWI that are between the location of the bottom-pressure maximum and the pressure minimum located at the crest (e.g. Fig. 2(b)). These termination points move closer to the crest as the *Re* increases, with the AGD becoming more focused towards the crest. For high *Re* these points almost coalesce, but a thin streamtube still connects the AGD to the pressure minimum near the crest; eventually the IEZ becomes similar to that without AGD (compare to Fig. 7 in Cardenas & Wilson (2007a)). Figure 2(a)-(b) also suggests the presence of a threshold or critical Reynolds number, Re_{crit}, below which there is no IEZ and the sediments are completely filled with discharging deep groundwater. Below Re_{crit}, the dominant vertical pressure gradients result in essentially one-dimensional flow upwards (Fig. 2(a)). Just above Re_{crit}, the IEZ is very sensitive to *Re* (Fig. 2(b)). For much larger Reynolds numbers, current-bedform driven flow dominates and the IEZ approaches its asymptotic depth (Fig. 2(f)), similar to the base case without AGD.

Following Cardenas & Wilson (2007a), a curve-fitting algorithm was used to find a functional expression for the simulated interfacial depth, $d_z(Re)$. Simulations and fits are plotted on the left side of Fig. 3. The Morgan-Mercer-Flodin (MMF) growth model (Morgan *et al.*, 1975), $d_z/L = (ab + cRe^d)/(b + Re^d)$, consistently provides good fits ($R^2 > 0.99$) to the simulation results for all values of q_{bas}. The x-intercept of the fitted MMF models, a, represents Re_{crit}. As shown in these plots and in Fig. 2, the IEZ depth, d_z, approaches an asymptotic, maximum depth at high *Re*. That maximum depth is always less than ~0.75L and is less for higher AGD. The flux density through the IEZ, q_{int}*, is plotted on the right side of Fig. 3. It continually increases with *Re*, following a power model, for all values of AGD. This resembles the model fit for the no-AGD relationship (Cardenas & Wilson, 2007a).

IEZ spatial configuration and fluxes: losing conditions

When AGD is downwards along the bottom boundary, the water column is losing water to deep groundwater. Looking only at IEZ depth, d_z, as a measure of the IEZ, there is almost no discernable difference between losing (Fig. 3, bottom row) and gaining (Fig. 3, top row) conditions across varying q_{bas}*. The simulations can again be fitted with an MMF model ($R^2 > 0.99$) with an asymptote. The dependence of IEZ flux density q_{int}* on *Re* for various q_{bas}* conditions is likewise similar between gaining and losing conditions (compare top and bottom of right column in Fig. 3).

Differences, however, are apparent when viewing the flow-field configuration, as shown in Fig. 2. Under losing conditions, the IEZ is centred around the bottom-pressure minimum at the crest (Fig. 2, bottom row), whereas it is centred around the pressure maximum for gaining conditions (Fig. 2, top row). The termination points of the IEZ along the SWI are areas where water is flowing down into the IEZ for the losing scenario; these are areas where water is flowing up into the water column for the gaining case. These termination points get closer to each other with increasing *Re* until, for the losing case, they almost coalesce at the location of the bottom-pressure maximum and the IEZ again looks similar to the non-AGD case. AGD from the

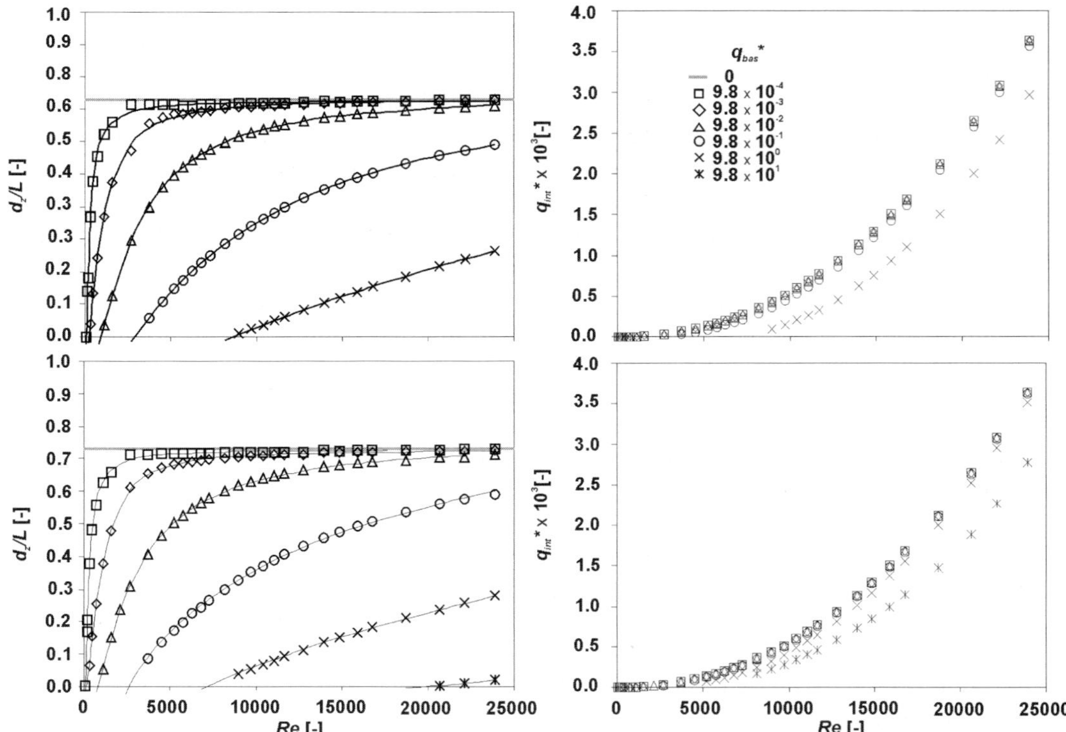

Fig. 3 Dimensionless interfacial exchange zone (IEZ) normalized depths and interfacial fluxes plotted as a function of Reynolds number (*Re*) for different q_{bas}* for gaining (top) and losing (bottom) cases. The horizontal grey lines in the left-hand graphs indicate the IEZ depth for the base case without ambient groundwater discharge (Cardenas & Wilson, 2007a). The solid curves are Morgan-Mercer-Flodin fits to the simulation results. The plots with the open triangles correspond to the simulations shown in Fig. 2. $H/L = 0.05$ and $L = 1.0$ m for all cases.

overlying water column is focused into a narrow streamtube that originates around the pressure maximum at the SWI (Fig. 2(l)).

There is also a threshold Re_{crit} for losing conditions. When the downward AGD is dominant, there is no return flow to the SWI (Fig. 2(g)) and all water downwelling from the water column flows towards deeper groundwater. As *Re* increases, the lateral pressure gradient along the SWI becomes sufficiently large such that some of the downwelling water near the bottom-pressure maximum is influenced by the pressure minimum and flows up towards it, returning to the SWI and creating a local IEZ, instead of following the ambient downward flow.

Exchange zone residence times for both cases

Mean residence times, t_r, of fluids flowing through the IEZ can be readily computed from the IEZ flux and porous area, $t_r = nA_z / q_{\text{int}} L$, where is the mean residence time, n is porosity, and IEZ area, A_z, is defined by the SWI and the bounding streamlines that surround all water originating from and returning to the SWI (Fig. 1). We studied dimensionless mean residence time, defined by $t_r^* = t_r K/nL$, as a function of *Re* and q_{bas}^*, for both gaining and losing conditions (not shown). We found that residence time is large at low *Re*s and decreases rapidly with *Re*. Even though the IEZs are smaller in extent at low *Re*, the fluxes are so small that it takes a long time for fluids to circulate through these small zones. AGD decreases the IEZ residence time, with a larger decrease for lower *Re*. At the lowest *Re* the decrease is exceptionally large. Residence time eventually approaches 0 as the *Re* approaches its critical value, when there is no current-bedform driven exchange and the IEZ shrinks to zero (Fig. 2(a),(g)). We also studied residence time distribution and found it to be fractal (not shown).

SUMMARY

Currents interact with bedform topography to drive advective-interfacial exchange across the sediment–water interface (SWI). The interfacial exchange zone (IEZ) is the area within the sediments characterized by local (0.1–10 m) downwelling and upwelling of seawater, with streamlines originating and ending at the SWI. Current-bedform driven advection of water through the sediments is further complicated by the presence of ambient groundwater discharge (AGD), the upwelling of deeper groundwater, such as in coastal areas influenced by regional (0.1–10 km) submarine groundwater discharge, or the deeper, regional downwelling of seawater intruding into aquifers. Hydrogeologists refer to these two situations relative to whether the water column is "gaining" or "losing" regional flow, respectively. This paper describes the hydrodynamics of IEZs under the joint influence of current-bedform driven exchange and AGD using high-fidelity multiphysics numerical models. We simulate turbulent flow above the SWI and Darcy-groundwater flow below the SWI.

The simulations show that IEZ can develop even in the presence of AGD, under both gaining and losing conditions, provided that the forcing due to current-bedform interaction is sufficiently strong. The competition between current-bedform interaction forcing and AGD forcing controls the IEZ shape, depth, fluxes, and residence times. The IEZ spatial extent (depth and area) increases with current Reynolds number (*Re*) but is diminished by ambient groundwater discharge (AGD) for both gaining and losing conditions. As current *Re* increases, the IEZ depth, area and shape are asymptotic to those for neutral conditions (where the water column is not gaining or losing). The relationship of IEZ depth and area to *Re* is described by simple (MMF) functional models no matter what the relative magnitude of AGD. Under gaining conditions, the IEZ is centred around the bottom-pressure maximum, which is located on the stoss face of dunes, near where the eddy in the water column re-attaches. The deep groundwater upwells near the bottom-pressure minimum which is located at the crest. Under losing conditions, the IEZ forms around the pressure minimum at the crest. Seawater entering the sediments near the bottom-pressure maximum, along the stoss face, downwells into the deep groundwater below the sediments and does not return to the SWI.

The IEZ flux and the mean residence time both depend on current *Re.* Flux increases with *Re* (following a power law), while residence time decreases. Ambient groundwater discharge (AGD) reduces IEZ flux and residence time, for both gaining and losing conditions.

Our high-fidelity multiphysics numerical modelling study shows that current-bedform driven exchange interacts and competes with ambient groundwater discharge to control the hydrodynamics below sediment–water interfaces. Understanding the hydrodynamics opens a door towards an integrated physical–biological–chemical perspective of biogeochemical and ecological processes along sediment–water interfaces, as well as interfaces with other porous bottom materials including coral and debris.

We have submitted for publication elsewhere a much more detailed account of interfacial exchange when AGD competes with the effects of turbulent current-bedform driven exchange (Cardenas & Wilson, 2007c), including the influence of varying bedform shape.

Acknowledgements This research was funded by an AGU Horton Research Grant and a New Mexico (NM) Water Resources Research Institute Student Grant. MBC was supported by the Frank E. Kottlowski Fellowship of the NM Bureau of Geology and Mineral Resources at the NM Institute of Mining and Technology. We acknowledge technical support by Comsol, Inc. and ESI-Group.

REFERENCES

Boudreau, B. P., Huettel, M., Forster, S., Jahnke, R. A., McLachlan, A., Middelburg, J. J., Nielsen, P., Sansone, F., Taghon, G., Van Raaphorst, W., Webster, I., Weslawski, J. M., Wiberg, P. & Sundby, B. (2001) Permeable marine sediments: overturning an old paradigm. *EOS* **82**, 133–136.

Burnett, W. C., Bokuniewicz, H., Huettel, M., Moore, W. S. & Taniguchi, M. (2003) Groundwater and pore water inputs to the coastal zone. *Biogeochemistry* **66**, 3–33.

Cardenas, M. B. & Wilson, J. L. (2006) The influence of ambient groundwater discharge on exchange zones induced by current-bedform interactions. *J. Hydrol.* **331**, 103–109. doi:10.1016/j.jhydrol.2006.05.012.

Cardenas, M. B. & Wilson, J. L. (2007a) Dunes, turbulent eddies, and interfacial exchange with permeable sediments. *Water Resour. Res.* (in press).

Cardenas, M. B. & Wilson, J. L. (2007b) Hydrodynamics of coupled flow above and below a sediment–water interface with triangular bed forms. *Adv. Water Resour.* **30**, 301–313, doi:10.1016/j.advwatres.2006.06.009.

Cardenas, M. B. & Wilson J. L. (2007c) Exchange across a sediment-water interface with ambient groundwater discharge. *J. Hydrol.* (in press).

Huettel, M., Roy, H., Precht, E. & Ehrenhauss, S. (2003) Hydrodynamical impact of biogeochemical processes in aquatic sediments. *Hydrobiologia* **494**, 231–236.

Morgan, P. H., Mercer, L. P. & Flodin, L. W. (1975) General model for nutritional responses of higher order mechanisms. *Proc. Nat. Acad. Sci. USA* **72**, 4327–4331.

Riedl, R. J., Huang, N. & Machan, R. (1972) The subtidal pump: a mechanism of interstitial water exchange by wave action. *Mar. Biol.* **13**, 210–221.

Webster, I. T., Norquay, S. J., Ross, F. C. & Wooding, R. A. (1996) Solute exchange by convection within estuarine sediments. *Est. Coast. Shelf Sci.* **42**, 171–183.

Wilcox, D. C. (1991) A half century historical review of the k-ω model. *AIAA Paper 91-0615*, Reston, Virginia, USA.

A New Focus on Groundwater–Seawater Interactions
(Proceedings of Symposium HS1001 at IUGG2007, Perugia, July 2007). IAHS Publ. 312, 2007.

Evaluation of the hydraulic gradient at an island for low-level nuclear waste disposal

PREM ATTANAYAKE & MICHAEL SHOLLEY
Bechtel Corporation, San Francisco, California 94119-3965, USA
pmattana@bechtel.com

Abstract The geographic and hydrological isolation of small islands makes them enticing candidates for the subsurface disposal of low-level radioactive waste. Placement of waste deep below the seabed is a scenario under which such repositories have been considered. One of the key hydrogeological factors influencing the suitability of such a repository is the groundwater gradient across the island and beneath the offshore area. The hydraulic gradient affects the direction and velocity of groundwater flow and, hence, the potential transport of radionuclides. In this study, the hydraulic gradient at a small island off the coast of China was evaluated for the performance assessment of a potential low-level nuclear waste repository. A preliminary assessment of the hydrogeology for the proposed offshore disposal chambers indicated that an unfavourable hydraulic gradient might occur under several scenarios.

Key words hydraulic gradient; saltwater–freshwater interface

INTRODUCTION

This paper is intended to provide a conceptual understanding of hydraulic gradients at a small island off the coast of China. For the proposed offshore disposal chambers (shown in Fig. 1), a sustained hydraulic gradient might occur under several circumstances. Possible temporary hydraulic gradients imposed during disposal chamber construction, operation and closure activities are unlikely to have an impact on the long-term performance of the facility, and are not evaluated in this document. This paper evaluates the possibility of sustained hydraulic gradients under the following circumstances of dynamic equilibrium:

- the potential for a confined aquifer to receive recharge at high elevations on the mainland and to extend beneath the sea to the island (located 20 km offshore), resulting in mainland groundwater flow reaching the island;
- the presence of a freshwater lens overlying saline groundwater at the island or extending from a nearby island;
- a hypothetical pumped well inducing flow of saline groundwater toward the island; and
- a low hydraulic gradient in the saline groundwater, with flow toward the freshwater-saline water mixing zone.

POSSIBLE REGIONAL HYDRAULIC GRADIENT

The formations underlying the coast of China in the vicinity of the island, and beneath the island, consist mostly of Jurassic-Cretaceous igneous rocks, mainly granitic

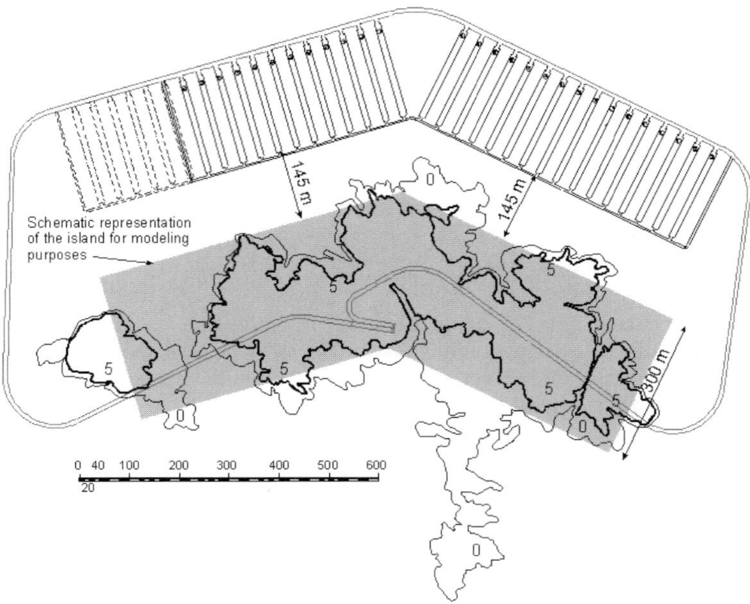

Fig. 1 Proposed offshore disposal chambers.

intrusives with some volcanics. In the study area, the rocks are predominantly granite and gabbro. Groundwater of the mountainous coastal area, approximately 20 km west of the disposal island, is primarily designated as "fissure water in magmatic rocks in mountains and hills" of low yield (Chinese Academy of Geological Sciences, 1988). In the mountainous coastal area the rocks "at shallow depths often abound with unconfined water in weathered fissures" (UNESCAP, 1983). However, the igneous rocks "at depth contain little water or are impermeable". In the coastal terrain, low-flow seasonal springs commonly appear within the weathered bedrock. Small quantities of groundwater also occur in alluvial and marine deposits of limited extent within local coastal plain areas, with low yields and containing brackish water near the coastline. There is no evidence of confined freshwater at the coast to provide the necessary head to drive freshwater offshore toward the island.

Figure 2 shows a cross-sectional conceptual model of mainland hydrogeology and expected conditions extending into the Taiwan Strait. Fine-grained seabed sediments overlie the igneous rocks beneath the Taiwan Strait between the mainland and the island and might provide a confining layer, a barrier to upward groundwater flow from the underlying fractured bedrock. However, the seabed sediments do not extend onshore to the "fissured" bedrock of the mainland or of the island. The absence of seabed sediments near the shore of the island is discussed further in evaluating the "local hydraulic gradient". As discussed previously, there is no evidence of confined water at the coast to sustain groundwater beneath the seabed sediments. Groundwater of the mainland primarily discharges to springs and streams. Any groundwater that might reach the coastline would discharge to the near-shore sea through a freshwater–saline water interface. Flow of mainland freshwater extending offshore 20 km to the island is considered unlikely.

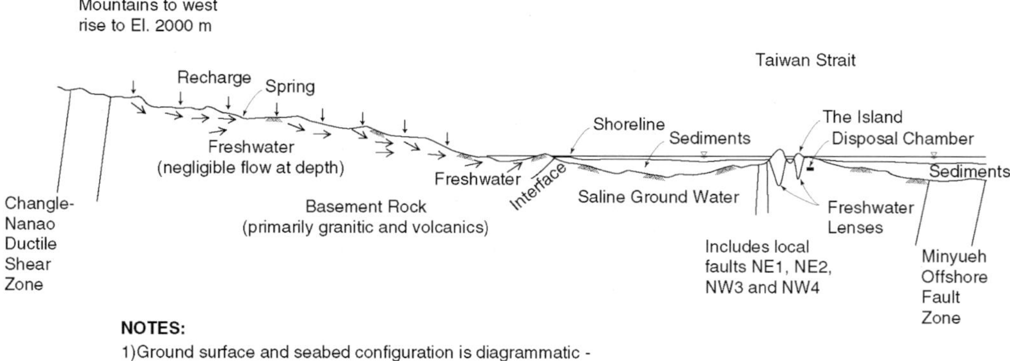

Fig. 2 Regional groundwater flow patterns.

EVALUATION OF LOCAL HYDRAULIC GRADIENT

The bedrock at the island consists of fractured granite intruded by dioritic dikes. The bedrock at a nearby second island consists primarily of fractured gabbro. Potentially confining seabed sediments appear to be absent offshore from the islands where seawater depth is less than 30 m. The disposal chambers underlie this area of shallow seawater. Further offshore from the islands, where seawater depth is >30 m, the sediment thickness ranges from 10 to 140 m.

Water-level measurements are available from shallow and deep open boreholes drilled during the preliminary design investigation. Figure 3 shows approximate average groundwater levels and interpreted water level contours. The highest water levels are shown as El. +9 to +14 m mean sea level (m.s.l.) near the highest hills on the island. Wells at low ground surface elevations show significant response to tidal fluctuations.

The island groundwater consists of a freshwater lens overlying saline groundwater. The interface between the freshwater and denser saline groundwater essentially forms a no-flow boundary for freshwater under conditions of dynamic equilibrium. The recharge rate from precipitation, the hydraulic conductivity, and the island size affect the thickness and offshore extent of the freshwater lens.

The location of the freshwater–saline water interface was estimated analytically with a method described by Cooper *et al.* (1964) and Todd (1980). The estimate shows that the freshwater in an equilibrium condition could extend to El. –194 m beneath the island, with freshwater discharge extending to 48 m offshore. Seabed sediments are absent at this distance, corresponding to a seawater depth of 10 m, so a near-shore confining layer is considered absent. These results indicate freshwater flow might extend within approximately 50 m of the disposal chambers, based on an offshore distance of about 100 m to the chambers.

A cross-sectional analytical Flownet model of the freshwater lens based on the interface location provides an evaluation of the flow pattern and hydraulic gradients within the freshwater lens. Figure 4 shows the Flownet results with groundwater flow lines and head drops. Hydraulic gradients estimated from Fig. 4 are typically 0.01 to 0.02 within the freshwater lens.

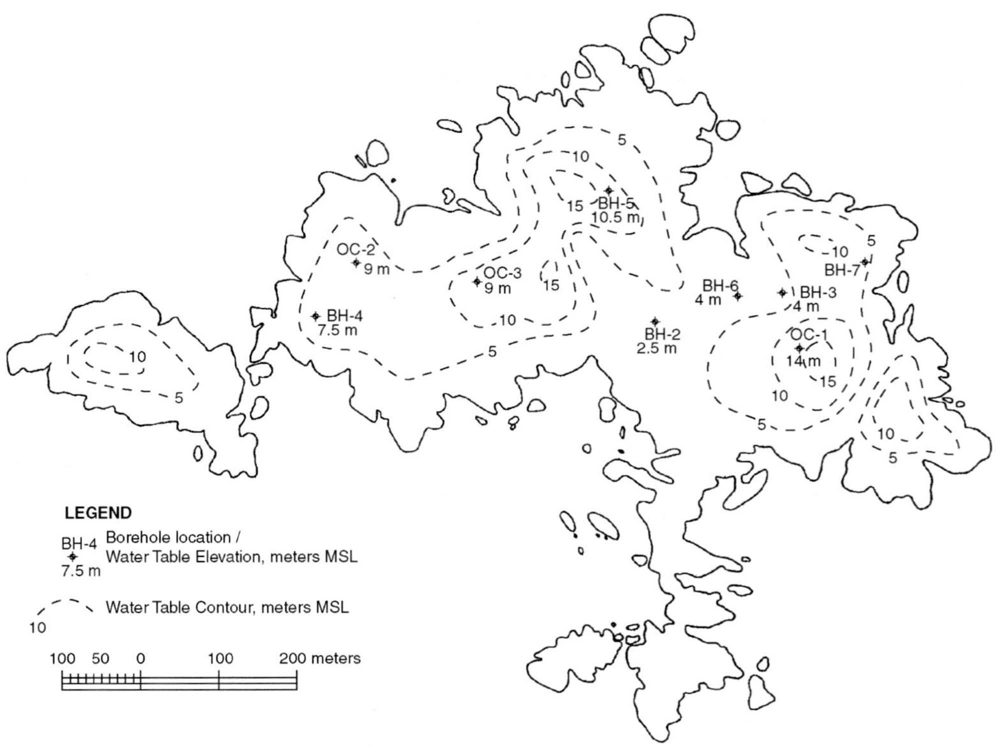

Fig. 3 Groundwater contours at the island.

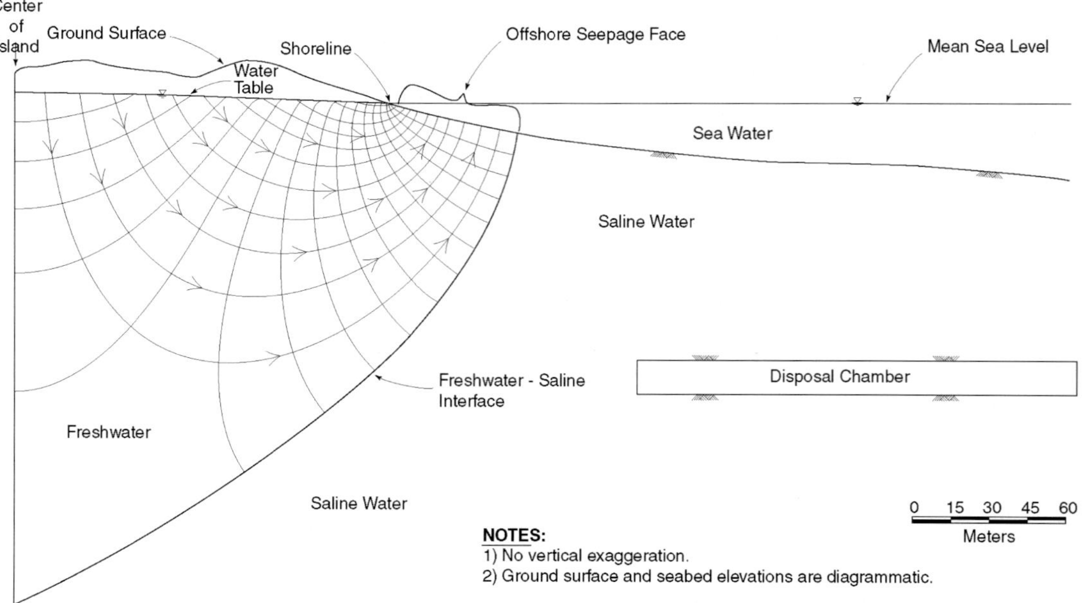

Fig. 4 Freshwater–saltwater interface.

The calculation and the analytical Flownet model assume an average recharge rate of 16 cm/year, an isotropic and homogeneous hydraulic conductivity of 3.2×10^{-5} cm, and an island width of 300 m. The island width is approximate because the actual shoreline is irregular. The hydraulic conductivity value is the mean of packer test results for depths of 10 to 178 m at BH-7. A water table elevation of approximately 4 m was back-calculated based on an assumed reasonable recharge rate of 16 cm/year. This recharge rate corresponds to 16% of an assumed precipitation rate of 100 cm/year for the island. Isohyetal contours indicate an annual precipitation of 100 cm/year at the coastline (China State Bureau of Geology, 1979).

EFFECTS OF MIXING ZONE

The actual interface between freshwater and saline groundwater will include a mixing zone, beyond which freshwater does not intermingle with saline groundwater. The mixing zone is shown as a zone of upward flow toward the seabed, with salinity greater than the freshwater and less than the seawater. The thickness of the mixing zone will be affected by seasonal and long-term variations in recharge rate, potential pumping of wells on the island, groundwater pressure fluctuations from daily tides and from storms, advective dispersion, and diffusion.

Figure 5 shows a Flownet sketch to demonstrate this conceptual model at the island. The sketch includes a freshwater lens, mixing zone, saline water beneath, and the disposal chamber location. Migration of the saline groundwater adjacent to the mixing zone would occur. Saline water entering the mixing zone is replaced by infiltration of seawater through the seabed beyond the seepage face. At the disposal chambers, the hydraulic gradient in saline water is expected to be much less than the freshwater hydraulic gradient. The hydraulic gradient within the saline water would decrease as the distance from the freshwater–saline water interface increases.

This conceptual model is consistent with field data and modelling studies by others. Figure 6 shows flowlines within freshwater and saline water based on field measurements of pressure head and salinity for a coastal aquifer in Florida (Cooper *et al.*, 1964). Saline groundwater was estimated to contribute one eighth of the total

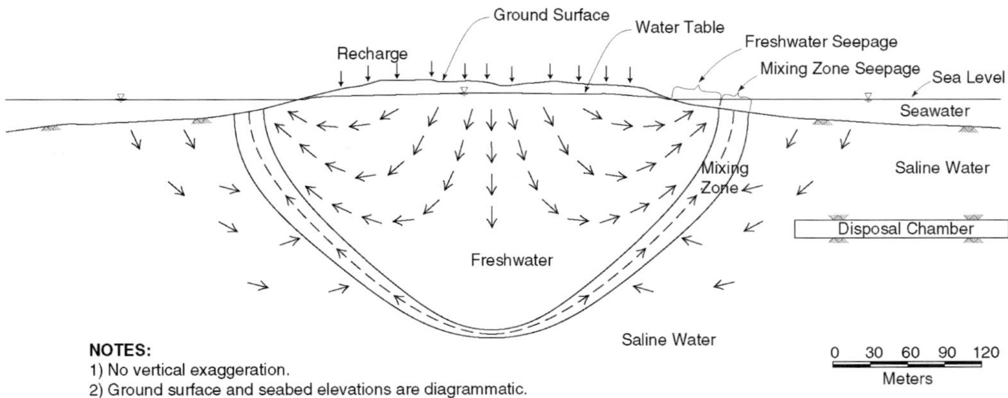

Fig 5 Hydrogeological conceptual model of the island.

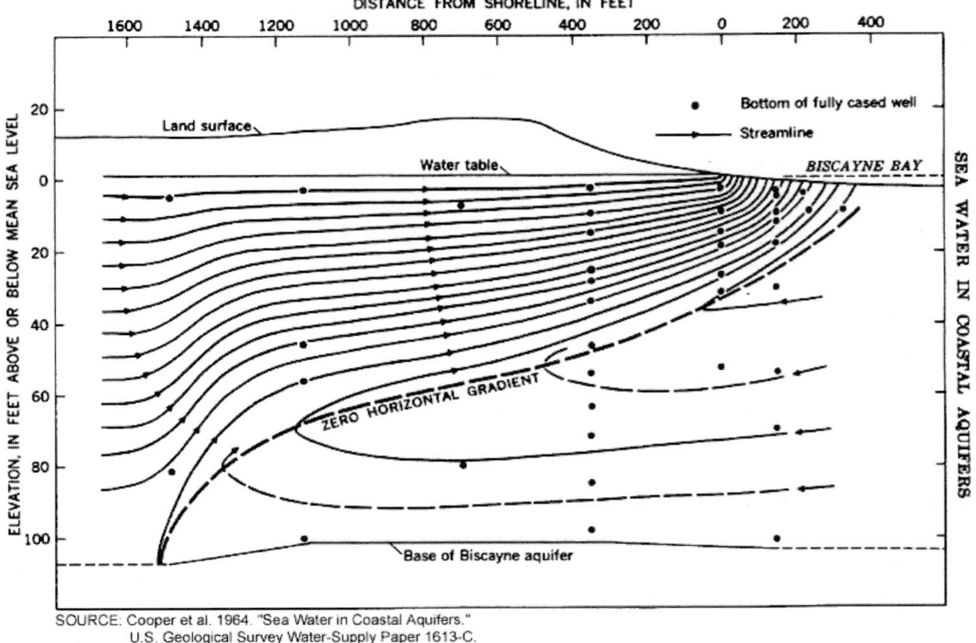

Fig. 6 Groundwater flow lines in a coastal aquifer in Florida, USA.

groundwater discharge at the seepage face. As noted by Cooper *et al.* (1964, p. v), "the circulation [of saline water] would warrant consideration in connection with proposals to inject radioactive waste into the salt water in coastal aquifers under the supposition that such water is essentially static".

EFFECTS OF WELL-INDUCED HYDRAULIC GRADIENT

The rate of potential pumping within the freshwater lens must be less than the island-wide recharge rate. Pumping at rates greater than the average recharge would deplete the freshwater lens. Pumping is also limited by "upconing" of saline groundwater into the well, making the water quality unsuitable for consumption. At a limited pumping rate (e.g. 10 L/min), the freshwater lens will reach a new dynamic equilibrium with reduced thickness and reduced seepage face.

Figure 7 shows a revised conceptual flow model to demonstrate the effect on the freshwater lens corresponding to a permanent pumping well. Compared to Fig. 5, the interface shifts upward and inward to the island, as upconing occurs and a new equilibrium interface location is reached.

A transient condition would be imposed at the start of pumping, and saline water would move toward the island to replace the shrinking extent of the freshwater lens. When dynamic equilibrium is reached and the interface is permanently shifted upward and landward, then the zone in which saline water migrates toward the mixing zone may also be shifted further from the disposal chamber location.

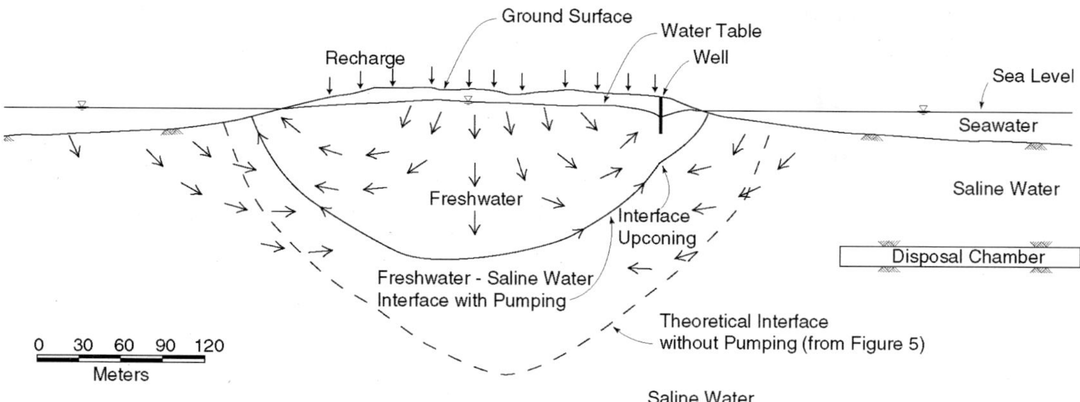

Fig. 7 Freshwater–saltwater interface with groundwater extraction.

EFFECT OF FRESHWATER LENS

The freshwater lens from a nearby second island is expected to have much less effect at the disposal chambers than the freshwater lens at the disposal island. The distance between the islands is about 1.7 km. The second island is slightly larger than the disposal island and therefore may have a slightly larger freshwater lens.

CONCLUSIONS

The evaluation of potential causes of hydraulic gradients at the disposal chambers indicates the following:

- a regional hydraulic gradient-effect unlikely;
- a local hydraulic gradient from freshwater lens-effect unlikely;
- a mixing zone-hydraulic gradient is induced within the saline ground water, but is probably low to negligible at the disposal chambers;
- pumping within freshwater lens is unlikely to affect the gradient after a new dynamic equilibrium is achieved.

REFERENCES

Chinese Academy of Geological Sciences (1988) *Hydrogeologic Map of China*. Scale 1:4 000 000. Institute of Hydrogeology and Engineering Geology. China Cartographic Publishing House, Beijing.

China State Bureau of Geology (1979) *Hydrogeologic Atlas of the People's Republic of China*. Institute of Hydrogeology and Engineering Geology. Chinese Geologic Cartographic Printing House.

Cooper, H. H., Jr, Kohout, F. A., Henry, H. R. & Glover, R. E. (1964) Sea water in coastal aquifers. *US Geol. Survey Water-Supply Paper 1613-C*.

Todd, D. K. (1980) *Groundwater Hydrology* (second edn). John Wiley & Sons, Inc., New York, USA.

UNESCAP (United Nations Economic and Social Commission for Asia and the Pacific) (1983) Hydrogeological Mapping in Asia and the Pacific Region. In: *Proceedings of the ESCAP-RMRDC Workshop*, Bandung.

Climatic variation, recharge and freshwater lens salinity of a coral atoll in the Pacific Ocean

M. VAN DER VELDE[1,5], M. VAKASIUOLA[2], S. R. GREEN[3], V. T. MANU[4], V. MINONESI[4], M. VANCLOOSTER[5] & B. E. CLOTHIER[3]

1 *Lamont Doherty Earth Observatory, Columbia University, PO Box 1000, 61 Route 9W, Palisades, New York 10964, USA*
marijn@ldeo.columbia.edu; marijnvandervelde@gmail.com

2 *Tonga Water Board, Nuku'alofa, Tongatapu, Kingdom of Tonga*

3 *Environment and Risk Management Group, HortResearch Institute, Private Bag 11-030, Palmerston North, New Zealand*

4 *Ministry of Agriculture, Forestry and Food, PO Box 14, Nuku'alofa, Kingdom of Tonga*

5 *Department of Environmental Sciences and Land Use Planning, Université Catholique de Louvain-la-Neuve (UCL), Croix du Sud 2 BP2, B-1348 Louvain-la-Neuve, Belgium*

Abstract The El Niño-Southern Oscillation (ENSO) exerts a moderate control on the temporal fluctuations of the salinity of water pumped from the subterranean water resources of the raised coral atoll of Tongatapu (175°12′W, 21°08′S; Kingdom of Tonga). The lens reacts buoyantly to recharge events and the saltwater transition zone moves vertically with the buoyant response. Here we show, using data obtained at the main well field of Tongatapu, some preliminary observations that can be made on the combination of the travel time of water infiltrating through the vadose zone, coupled with the hydraulic buoyant response time of the lens in the aquifer, and the errors associated with the identification of the transfer function parameters identified with a simple inverse procedure from well salinity data.

Key words El Niño-Southern Oscillation; coral atoll; prediction; salinity; freshwater lens; salt water intrusion; Tonga; wells; pumping; climate change

INTRODUCTION

Coastal aquifers around the world suffer from salt water intrusion caused by natural as well as human-induced processes (Oude Essink, 2001). Human activities that threaten coastal lowlands include mining for natural resources (water, sand, oil and gas) and land reclamation that leads to subsequent subsidence. Coastal aquifers are affected by mean sea level (MSL). The Intergovernmental Panel on Climate Change (IPCC) estimates a global averaged sea-level rise of about 0.6 m in its present estimate for the coming century under the "Year 2000 constant concentrations" scenario, with a likely range of 0.3–0.9 m (IPCC, 2007). However, other scenarios predict increases that range from 1.1 to 6.4 m. It is thus expected that the salinization of coastal aquifers will accelerate. Consequently, the availability of fresh groundwater resources in coastal areas will decrease, putting serious constraints on present populations and future coastal development.

Seawater intrusion is reported for areas all around the world, including for example, Mexico (Halvorson *et al.*, 2003), Greece (Lambrakis & Kallergis, 2001), and Cypress (Ergil, 2000). Salt water intrusion is also a severe problem in several islands located in the Pacific region. Salt water intrusion has been reported for Kiribati, Kingdom of Tonga, Salomon Islands, Fiji, Hawaii, amongst others (see SOPAC website: http://www.sopac.org). One of the main reasons for salt water intrusion, or local upconing, is (over)pumping at well sites.

Logically, integrated coastal management plans are essential for small islands and they need to explicitly recognize the important links between geological setting, oceanographic conditions, reef health and the coastal zone (Solomon & Forbes, 1999). Solomon & Forbes (1999) further note that, although oceanographic hazards are often associated with catastrophic events, such as tsunamis and cyclones, they are also caused by processes that operate more slowly and over longer time scales, such as geological subsidence or uplift and regional and global sea-level changes with superimposed tides.

Although minimally responsible, small islands will be among the first to be struck by the effects of climate change and increased climatic variability. In particular, the availability of freshwater resources of low-lying atolls is vulnerable to climate change (Meehl, 1996). It is predicted that an increase in the frequency of El Niño events will occur in a warming global climate (Timmermann *et al.*, 1999; Tsonis *et al.*, 2005), although predictions remain uncertain (Cane, 2005). This may then lead to an increase in the occurrence of dry spells (Meehl, 1998).

Recently, the seasonality in the motion of the freshwater–saltwater interface has been shown to potentially have an important impact on the chemical loading of coastal waters (Michael *et al.*, 2005). In the South Pacific, rainfall exhibits seasonal as well as interannual variability related to the El Niño-Southern Oscillation (ENSO), with a consequent impact on the subterranean water resources (van der Velde *et al.*, 2006). It may be expected that interannual variability will have a comparable effect to that of seasonality on the temporal chemical loading of nearby coastal regions in areas affected by ENSO. For further understanding of small islands' marine ecosystems, and optimal water resources utilization on small islands, it is essential that a thorough understanding of the influence of climate variability on the interaction and water exchange between groundwater, surface water, lagoon, and seawater is obtained.

We studied the subterranean water resources of the raised coral atoll of Tongatapu (175°12′W; 21°08′S; Kingdom of Tonga). We will show some data on the freshwater lenses occurring in the atoll to deduce some preliminary observations that can be made on the combination of the travel time of water infiltrating through the vadose zone, coupled with the hydraulic buoyant response time of the lens in the aquifer, and the errors associated with the identification of the transfer function parameters identified with a simple inverse procedure from well salinity data.

METHODS

The water table of Tongatapu's freshwater lens is extremely flat. It is influenced by tides, sea-level change, barometric pressure, recharge, pumping and drought. Seawater level is mainly influenced by tidal processes, but is also affected by wind that causes build up of water masses of about 0.2–0.3 m at one side of the island and in the lagoon (Falkland, 1992). Furness & Gingerich (1993) showed that the water level of a well in Tonga showed a semi-diurnal response to the tidal constituents. Furness & Helu (1993) found that the barometric effect is of larger magnitude than the tidal pattern. Hunt (1979) presented 39 measurements of water table height in wells equally distributed over the island. Average water level above MSL was 0.32 ± 0.11 cm, or about 50% of the steady-state level presented by Pfeifer & Stach (1972).

In Tongatapu, the permeable limestone aquifer is overlain by a clay soil derived from volcanic ash with a variable thickness generally declining from 5 m in the west to 0.5 m in the east of the island (Cowie *et al.*, 1991). The filtering of the water that percolates towards the freshwater lenses depends on the hydraulic conductivity and the water holding capacity of the vadose zone. High values for saturated hydraulic conductivity were measured using disc permeates throughout a 1.2 m soil profile (van der Velde *et al.*, 2005).

Freshwater lens geometry is affected by recharge, hydrogeological characteristics of the aquifer and pumping well locations. Results from a sensitivity analysis by Griggs & Peterson (1993), show that the depth of the 50% salinity contour was most sensitive to permeability, and that the transition-zone thickness was most sensitive to dispersivity. The transition-zone thickness increases with increasing dispersivity values, and is most sensitive to changes in the transverse dispersivity. For a given recharge rate, the depth of the 50% salinity contour will decrease with increasing hydraulic conductivity.

Several authors have estimated the recharge towards the freshwater lens on Tongatapu (Pfeifer & Stach, 1972; Lao, 1978; Hunt, 1979 Kafri, 1989; Hasan, 1989; Falkland, 1992). The percentages of recharge these studies have estimated and calculated ranged from 5 to 35% with a generally adapted average estimation of 30%. Furness & Helu (1993) summarized the available recharge estimates (and also reported on the work of Kafri, 1989, and Hasan, 1989). Furness & Gingerich (1993) could not estimate the annual recharge due to a lack of rainfall during the measurement period. However, it was resolved from the residual fluctuation derived from the subtraction of the smoothed well water-level signal with the smoothed tide gauging signal, that an individual recharge event of less than 13 mm may pass undetected using water-level data.

Based on a pump test the saturated hydraulic conductivity (K) of the limestone was calculated by Hunt (1979) to be 1.5 cm s^{-1} (corresponding to 1296 m d^{-1}). This relatively large value of K related to the extremely porous nature of the coral limestone (especially since K was derived for the horizontal flow direction). Furness & Gingerich (1993) reported "from modelling of the aquifer and examination of drill-hole cores and exposures of rock in quarries, caves and cliffs it is apparent that the porosity and specific yield of the aquifer is very high. The latter probably averages as much as 0.4 (40%)". Limestone aggregate samples were obtained by Harrison (1993) to determine aggregate properties in relation to quarrying of the limestone for building and civil engineering purposes. Due to the weakness of the aggregates, and the ease by which they abrade, the limestone aggregates failed to meet specifications for concrete and road stone commonly used. Harrison (1993) further reports that the limestone is highly porous with a low density and that it contained large amounts of absorbed water. The physical quality (with respect to density, porosity, strength and durability) of the aggregates was also reported to be extremely variable. The limestone was found to be very pure, with percentages of $CaCO_3$ over 98.5%.

Wellfield salinity data

The data we present here are valuable, and obtained through the diligent monitoring efforts of the Tongan Water Board (TWB) and the Ministry of Lands, Survey and Natural Resources, from the well field of Mataki'eua. Because of its importance to the water supply for Nuku'alofa, the salinity measurements at the Mataki'eua well field have been carried out (although irregularly) over a long period of time. The data we use here include: (1) salinity measurements from three of 21 monitored and pumped wells; (2) several profiles of salinity through the freshwater lens, as measured at monitoring bores located in, or in the vicinity of, the well field; and (3) some information on the volumes pumped at several wells, as well as some water table heights. We use this information to derive transfer functions through the whole soil-limestone aquifer continuum for the pumped wells.

On the whole island of Tongatapu about 240 wells are in use. The depth has been measured for 90 of these wells and these depths generally increase from northwest to southeast, confirming the general tilt of the island's limestone. The wells on the island include: village wells with one or more wells in each village, which pump to an elevated water tank; and private wells which are mostly hand dug and not equipped with a motorised pump. However, most pumping is done at the main well field of Tongatapu, the Mataki'eua well field (see Fig. 1). The wells in Mataki'Eua well field are spaced approx. 150 m apart. Pumping is done from approx. 1–2 m below the average water level. Pumping data from the well site are scarce. The average pumping rate is about 3 L s^{-1}. Groundwater extraction from the Mataki'eua well field was approximated by Falkland (1992) to be 5.3×10^3 m^3 d^{-1}. Data from the TWB shows that monthly water production was quite stable, apart from a slight increase in 1998. The water production for 1996, 1997 and 1998 equalled 5.8×10^3 m^3 d^{-1}. However, leakage is reported to be high (above 30%) for the Tongan reticulation system. Also, when the water production is compared with the metered water, we find that roughly 50% of the pumped water is unaccounted for (JICA, 1999).

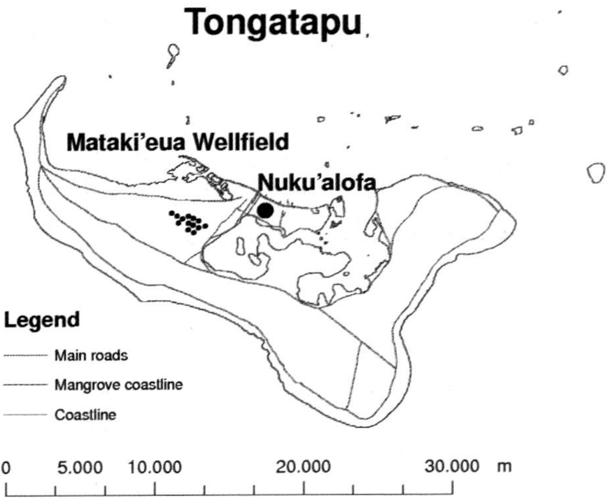

Fig. 1 The location of the Mataki'eua wellfield on Tongatapu.

Furness & Gingerich (1993) reported that the pumping at Mataki'eua has depressed the water table by about 0.25 m in the middle of the well field. According to Furness & Gingerich (1993) the water level in one individual well (105) falls about 0.1 m when the pump is operating. The electric conductivity of the water is inversely proportional to the height of the water table. The relation has a low correlation and conductivity is mostly affected by the time since the last heavy rainfall. In comparison, Furness & Gingerich (1993) noted that "it was surprising to see on Tongatapu falls of 0.3 m in the fresh water level without a drastic change in the conductivity of the water. Generally a rise of about 10% was noted in monitoring. It was concluded from this evidence that the movement of the transition zone from fresh to salty water is damped". We suggest that this delay mainly relates to the difference between vertical hydraulic conductivity and horizontal hydraulic conductivity (the latter being much higher in relation to the horizontal deposition of geologic layers). The relation between water table height and salinity was also investigated by Falkland (1992) with data taken during three days (6 August 1990, 11 January 1991 and 1 March 1991). As expected, increases in water table elevation were accompanied by expected decreases in salinity of the water at the water table. Falkland (1992) also reported on some of the uncertainty associated with the measurements. Most wells at the Mataki'eau well field show a decrease in salinity from 1990 to 1991 related to the high recharge during this period. However, it was reported by Falkland (1992) that "some wells (102, 106, 211 and 212) show an increase in salinity during this period".

Currently, about 30 wells operate at the well field. The salinity is estimated from electric conductivity (EC, $\mu S\ cm^{-1}$) measurements that were carried out sporadically since 1980 and approximately bi-monthly from January 1995 to June 2000. In this study we use measurements made since 1 January 1997 by the TWB. Twenty-one wells were monitored. The depth of the wells is influenced by the occurrence of relict coral patch reefs at the well field. The average salinity of each well shows the occurrence of a salt water wedge extending inland from the lagoon.

Monitoring bores and freshwater lens thickness

For a period of three years from 1997 to 2000 with an approximate 3-monthly time step the freshwater lens underneath the island was monitored for electric conductivity and temperature at different depths. Six points and later one additional point were monitored using 5–7 bore holes (MBs) at different depths to sample a profile of the freshwater lenses (see Fig. 2). One point (MB1) was located close to the lagoon. The other points are all located more inland. A polyethylene sampling tube with foot valve was used to obtain water samples from the monitoring tubes with an internal diameter of 38 mm. When approximately 5 L had been pumped the electric conductivity of the sample was measured, this was replicated three times, after which the temperature of the water was determined. A dipper was used to measure the depth to the water table. The depth to the base of the tube was determined and compared to the original depth measured after installation. The depth to the water level was also measured (Falkland, 1992).

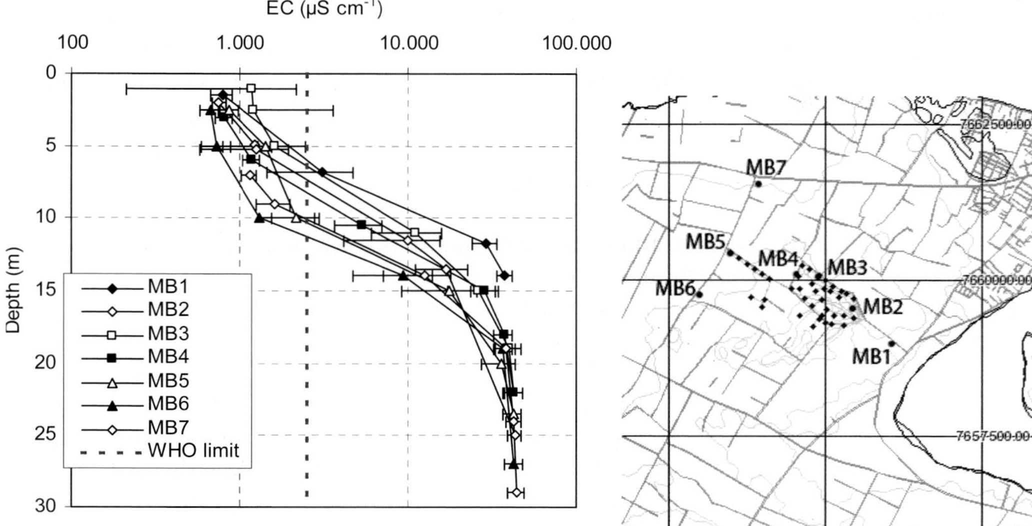

Fig. 2 Profiles of averaged conductivity measurements through the freshwater lens done at the monitoring bores (MB) located near or in the Mataki'eua well field, between May 1997 and July 2000 by the Tongan Waterboard. Error bars indicate standard deviation around the mean for the measurement period. The World Health Organization limit was defined at 2500 µS cm^{-1}. The right panel shows the location of the different MBs around the wellfield (cf. Fig. 1).

Temporal variations in salinity

From the measurements done in the MBs the depth of the freshwater lens can be inferred. The depth of the freshwater lens can be defined as the depth of potable water. defined by the World Health Organization (WHO) at 2000 µS cm^{-1}. From these measurements it can be shown that the thickness of the freshwater lens ranged between 5 and 12 m and that the lens becomes thinner towards the lagoon. The temporal variation in recharge of the lens relates directly to the temporal variations in rainfall. Mixing into the freshwater lens can be derived from the monitoring bores data (not shown). From these data, mixing into the lens can be observed deeper at 5–20 m below the water table, while at a certain depth (between 35–40 m) pure seawater is encountered. The TDS values from wells 102, 105 and 115, obtained by multiplying the EC data by a factor of 0.64, indicate a temporal pattern of the variation of TDS that relates to the amount of preceding rainfall (Fig. 3). It can be expected, however, that besides hydrogeological factors, pumping rates would also influence the fluctuations in salinity, especially if the pumps are not continuously operating.

Transfer functions

The salinity of the lens at a certain time is determined by the total mass of solute and the total mass of water in the pumped freshwater layer. If we consider that there exists an equilibrium state for the water lens corresponding to a continuous balance between averaged freshwater input and output and mean sea level, we then expect that every

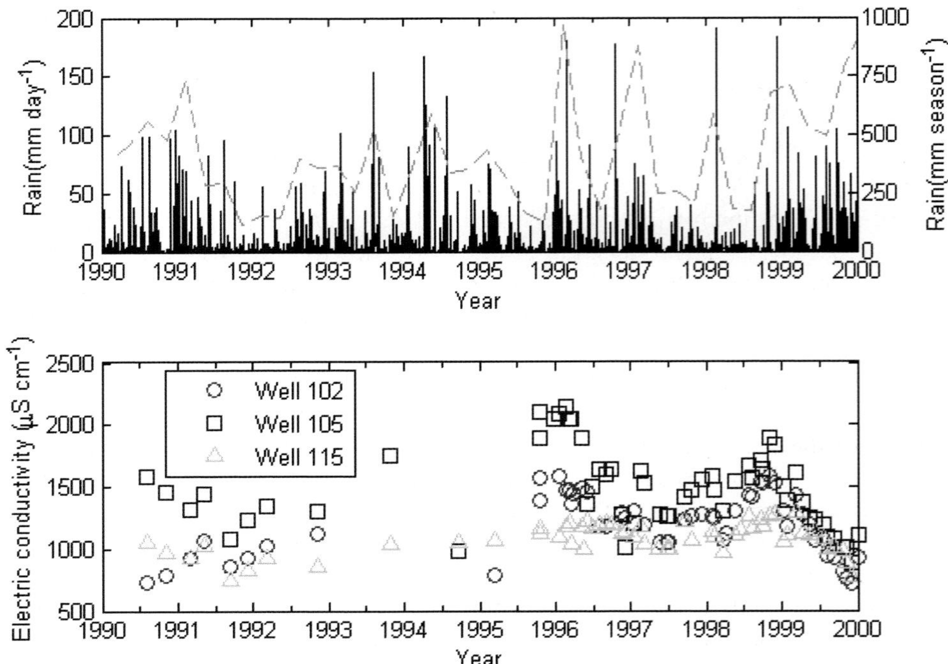

Fig. 3 Daily precipitation and three-monthly precipitation totals and total dissolved solids measured in 3 of 21 monitored wells on Tongatapu.

variation in the freshwater input from this equilibrium state will result in a change in the salt concentration. An increase in the rainfall rate would result in a decrease in salinity. Since the vadose zone between the surface and the aquifer smoothes out the variations of water flux, a transfer-function model can be used to describe the relation between the variation in rainfall and the variations in salinity. Transfer function theory was developed by Jury (1982) to simulate solute transport under natural field conditions where substantial variability exists in water transport properties. Jury's approach (1982) derives a distribution function based on the distribution of travel times of solutes from the soils' surface to a reference depth. Here we use the transfer function model in a slightly different setting then originally proposed. The transfer-function model describes the convolution of the variations in rain with the variations in the reciprocal of total dissolved solids (TDS) with a daily time step, so that

$$\mathrm{TDS}^{-1}(t) - \mu_{\mathrm{TDS}^{-1}} = \alpha \int_{-\infty}^{\infty} f(t) \cdot (P(t-\tau) - \mu_P)\mathrm{d}t + Z(t) \tag{1}$$

The integral in equation (1) expresses the convolution between the variations around the mean rainfall ($P(t) - \mu_P$) and the variations around the mean of the inverse of the total TDS ($\mathrm{TDS}^{-1}(t) - \mu_{\mathrm{TDS}}{}^{-1}$). Here t is time and τ the lag time between rainfall and dilution, α is a scaling factor, $f(t;\, \mu,\, \sigma)$ is the transfer function and $Z(t)$ is the residual error.

A lognormal probability density function was chosen for the transfer function $f(t)$ to reflect the transport and mixing of the percolating water as it moves through a porous medium exhibiting a log-normal distribution of pore-water velocities. Hence:

$$f(t) = \frac{1}{\sigma t\sqrt{2\pi}} \exp\left(\frac{-\ln(t) - \mu)^2}{2\sigma^2}\right) \tag{2}$$

where μ and σ respectively correspond to the mean and standard deviation of the log-normal distribution. The peak of the transfer function corresponds to the mode ($\exp[\mu - \sigma^2]$), the median is equal to ($\exp[\mu]$) and the mean to ($\exp[\mu + 0.5\sigma^2]$).

The use of a transfer function allows us to describe the whole system and the processes occurring in this system in a relatively simple way. Our system is a porous medium through which the infiltrated water moves and it includes the soil, the limestone, and the saturated limestone aquifer to a depth of 1–2 m below the water table. The processes in this system include the infiltration of rainwater in the soil, the downward movement of this water through the vadose zone, and the subsequent mixing of the drainage water into the freshwater lens to a depth of 1–2 m. The lens reacts buoyantly to recharge events, and the saltwater transition zone moves vertically with the buoyant response. Thus, the "response" or lag times we obtain with the transfer function encompass both the travel time through the vadose zone, and the hydraulic buoyancy response time.

The parameters α, μ, and σ in equations (2) and (3) were obtained by minimizing the sum of the sum of squared errors of simulations compared to the measurements available for each well. The objective function $\phi(\alpha, \mu, \sigma)$ was given by:

$$\varphi(\alpha, \mu, \sigma) = \sum_{i=1}^{N} (\mathrm{TDS}_{i,\mathrm{simulated}}^{-1} - \mathrm{TDS}_{i,\mathrm{observed}}^{-1})^2 \tag{3}$$

where N is the number of measurements in each well, $\mathrm{TDS}_{i,\mathrm{simulated}}^{-1}$ is the simulated variation of TDS^{-1} around the mean, and $\mathrm{TDS}_{i,\mathrm{observed}}^{-1}$ is the observed variation of TDS^{-1} around the mean. We used only the reliable data from the TWB obtained between January 1995 and June 2000 for 21 wells. The Nelder-Mead multidimensional unconstrained nonlinear minimization algorithm available in MatLab™ was used to minimize the objective function.

RESULTS

Measured and modelled variations in TDS are given for wells 102, 105 and 115 in Fig. 4. Although differences between the three wells are apparent, it appears that the transfer function approach used here gives reasonable results. The transfer function model is in good agreement with the variation of the measured TDS. The information contained in the rainfall signal thus seems sufficient to reflect the variation in salinity. The errors of the objective function will be discussed later. The variation in salinity differs between the wells, and the optimal parameters obtained by minimizing the objective function therefore yields different probability density functions (pdf).

The pdfs for the individual wells are graphed in Fig. 5. The thick line represents the pdf that is obtained when all well data are combined and modelled with one transfer function model (cf. van der Velde *et al.*, 2006). The occurrence of the peak of

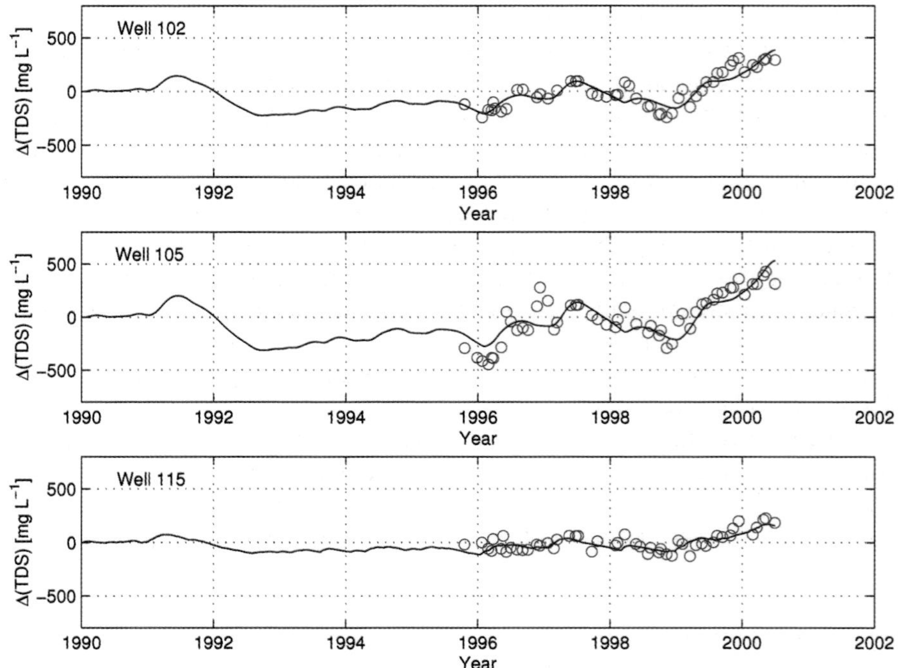

Fig. 4 Model and transfer function of wells 102, 105 and 115.

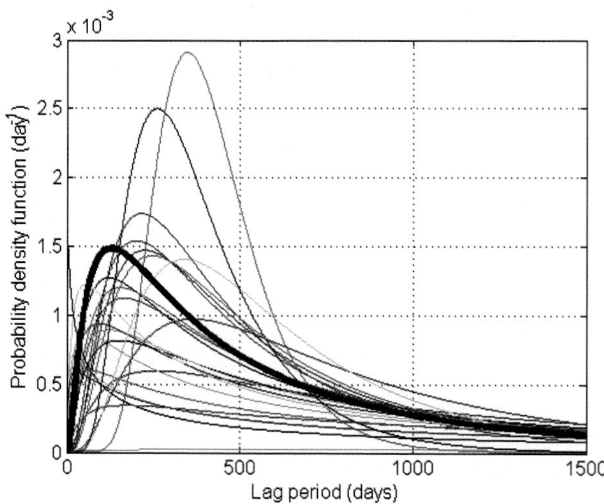

Fig. 5 The lognormal probability functions obtained for all the wells. The thick line is the probability density function obtained when all well data are combined with one transfer function model (cf. van der Velde *et al*., 2006).

the pdf (corresponding to the mode) ranges between 8 and 305 days. The median and mean lag periods respectively range between 305 and 4912, and 383 and 11572 days. However, most wells have a lag or response time below 2000 days, and a median response time below 1000 days. These lag times thus relate to a combination of the travel time through the vadose zone, as well as the hydraulic buoyancy response time.

These transfer times are relatively large. Hydrographs of well-level response plotted against daily rainfall by Jocson *et al.* (2002) showed that "the rate at which water is delivered to the lens is a function of rainfall intensity and the relative saturation of the vadose zone". This determines the portion of fast flow through preferred flow-paths that bypass the bedrock matrix, with respect to the water that percolates more slowly through the bedrock matrix. Jocson *et al.* (2002) reported significant buffering of recharge to the freshwater lens of Guam, similar processes will influence the recharge here. Nevertheless, Jones & Banner (2003) showed a threshold of rainfall of 190–200 mm month^{-1} before recharge occurs for three different limestone aquifers, and attributed this to similar climate and geology that produce soils with similar hydraulic properties.

Larger travel times may also be associated with parameters that were obtained from erroneous data, or from wells that are more severely influenced by other processes. Therefore it will be necessary to stringently evaluate the accuracy of the minimum error obtained for the objective function by the inverse procedure.

Objective function

The uniqueness of the parameters was evaluated by plotting the error surface for a range of parameter value combinations around the minimum error obtained by the inverse procedure. Data are plotted in Fig. 6 for well 105. Values are shown for pairs of parameters and the associated objective function error, as well as the minimum error that was obtained. The range of parameter values encompasses a range of

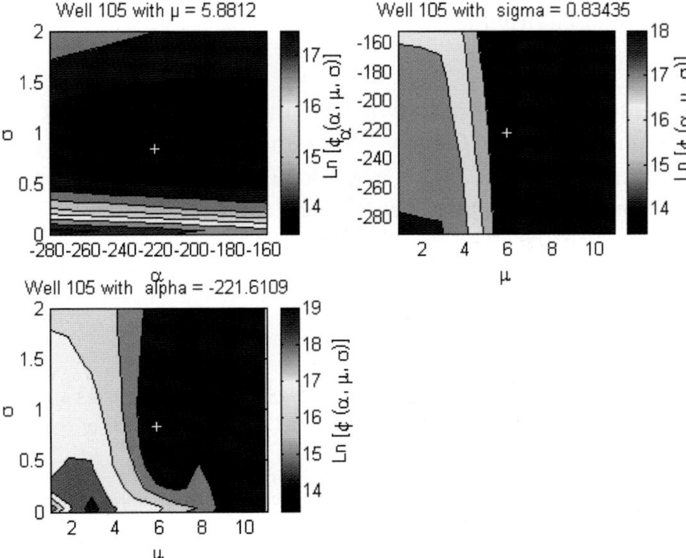

Fig. 6 Representation of the error surface of the objective function around the minimum error (white cross) obtained with the parameters for well 105.

probable parameter values. We find no artefacts in the objective function error surfaces. This shows that the minimum error was found by the inverse procedure. The value for μ is most clearly defined as indicated by the short range around the optimal value in both the μ vs α as well as the μ vs σ. μ relates strongly to the transfer times and corresponds directly to the median travel time (exp[μ]). A wider range around the optimal values for σ and α exists. Although for most wells the obtained model is in good agreement with the measurements (data not shown), differences between the error surfaces exist and problems of interpretation and fitting of the parameters give problems for some of the wells. We can choose an error threshold value of the objective function beyond which we would dismiss the obtained parameters, and thus the model for further analysis related to interpreting the influence on salinity variations of for example geographical location of the well with respect to topography and the coastal zone.

CONCLUSIONS

Two of the main issues that have to be dealt with by small islands that wish to sustain their water resources are intensification of agriculture (van der Velde *et al.*, 2007) and climate change and variability (van der Velde, 2006). Sustainable management of water resources must increasingly take into account human demands and pressures, and climatic influences upon these water resources. Sustainable water resource systems have been defined as "those designed and managed to fully contribute to the objectives of society, now and in the future, while maintaining their ecological, environmental and hydrological integrity" (UNESCO, 1999). An important aspect of this management is the understanding of climate variability and the influence on the freshwater resources of islands. This work and our previous work on the capability of the SOI to predict pumped water salinity may be beneficial for practical planning by water managers in similar environments.

Acknowledgements MvdV acknowledges the financial support of the European Commission under the INCO-DEV Programme (ICFP500A4PRO2) and NOAA's Postdoctoral Programme in Climate and Global Change, administered by UCAR. The monitoring efforts of the Tongan Waterboard and the Ministry of Lands, Survey and Natural Resources are greatly appreciated.

REFERENCES

Cane, M. A. (2005) The evolution of El Niño, past and future. *Earth Planet. Sci. Lett.* **230**(3/4), 227–240.

Cowie, J., Searle, P., Widdowson, J. & Orbell, G. (1991) *Soil of Tongatapu, Kingdom of Tonga.* DSIR Land Resources, Lower Hutt, New Zealand.

Ergil, M. (2000) The salinisation problem of the Guzelyurt aquifer, Cyprus. *Water Res.* **34**(4), 1201–1214.

Falkland, A. (1992) Tonga water supply masterplan project. Water Resources project. Prepared for PKK Consultants, Nuku'alofa, Kingdom of Tonga.

Furness, L. & Gingerich, S. (1993) Estimation of recharge to the freshwater lens of Tongatapu, Kingdom of Tonga. In: *Hydrology of Warm Humid Regions* (Proc. Yokohama Symp.) (ed. by J. S. Gladwell), 317–323. IAHS Publ. 216. IAHS Press, Wallingford, UK.

Furness, L. F. & Helu, S. P. (1993) The hydrogeology and water supply of the Kingdom of Tonga. Ministry of Lands, Survey and Natural Resources, Kingdom of Tonga.

Griggs, J. & Peterson, F. (1993) Groundwater flow dynamics and development at the atoll scale. *Ground Water* **31**(2), 209–220.

Halvorson, W., Castellanos, A. & Murrieta-Saldivar, J. (2003) Sustainable land use requires attention to ecological signals. *Environ. Manage.* **32**(5), 551–558.

Harrison, D. (1993) The limestone resources of Tongatapu and Vava'u, Kingdom of Tonga. British Geological Survey, Keyworth, Nottingham, Great Britain.

Hasan, R. (1989) Preliminary survey on rainwater harvesting. FAO, United Nations, Rome, Italy.

Hunt, B. (1979) An analysis of the groundwater resources of Tongatapu Island, Kingdom of Tonga. *J. Hydrol.* **40**, 185–196.

Japan International Cooperation Agency (1999) Basic design study report on the project for Nuku'alofa water supply system in the Kingdom of Tonga. TWB, Nuku'alofa, Kingdom of Tonga.

Jocson, J. M. U., Jenson, J. W. & Contractor, D. N. (2002) Recharge and aquifer response: northern Guam lens aquifer, Guam, Mariana Islands. *J. Hydrol.* **260**, 231–254.

Jones, I. C. & Banner, J. L. (2003) Hydrogeologic and climatic influences on spatial and interannual variation of recharge to a tropical karst island aquifer. *Water Res. Res.* **39**(9), 1253–1263.

Jury, W. A. (1982) Simulation of solute transport using a transfer function model. *Water Resour. Res.* **18** (2), 363–368

Kafri, U. (1989) Assessment of the groundwater potential in the island of Tongatapu, Kingdom of Tonga. Geological Survey of Israel and Ministry of Infrastructure, Kingdom of Tonga.

Lao, C. (1978) Groundwater resources study of Tongatapu. Assignment report by WHO consultant. UNDP no. TON/75/004, Regional Office for the Western Pacific.

Lambrakis, N. & Kallergis, G. (2001) Reaction of subsurface coastal aquifers to climate and land use changes in Greece: modeling of groundwater refreshing patterns under natural recharge conditions. *J. Hydrol.* **245**, 19–31.

IPCC (2007) *Climate Change 2007: The Physical Science Basis.* Summary for Policymakers. IPCC WGI Fourth Assessment Report.

Meehl, G. (1996) Vulnerability of freshwater resources to climate change in the tropical Pacific region. *Water, Air Soil Pollut.* **92**, 203–213.

Meehl, G. (1998) Pacific region climate change. *Ocean Coast. Manage.* **37**, 137–147.

Michael, H. A., Mulligan, A. E. & Harvey, C. F. (2005) Seasonal oscillations in water exchange between aquifers and the coastal ocean. *Nature* **436**, 1145–1148.

Oude Essink, G. (2001) Improving fresh groundwater supply-problems and solutions. *Ocean Coast. Manage.* **44**, 429–449.

Pfeifer, D. & Stach, L. (1972) Hydrogeology of the island of Tongatapu, Kingdom of Tonga. *Geol. Jahrbuch,* C4.

Solomon, S. M. & Forbes, D. L. (1999) Coastal hazards and associated management issues on South Pacific Islands—a reconnaissance survey. *Ocean Coast. Manage.* **42**(6), 523–554.

Timmermann, A., Oberhuber, J., Bacher, A., Esch, M., Latif, M. & Roeckner, E. (1999) Increased El Niño frequency in a climate model forced by future greenhouse warming. *Nature* **398**, 694–696.

Tsonis, A., Elsner, J., Hunt, A. & Jagger, T. (2005) Unfolding the relation between temperature and ENSO. *Geophys. Res. Lett.* **32**, L09701, doi:10.1029/2005GL022875.

UNESCO (1999) *Sustainable Criteria for Water Resource Systems*. Cambridge University Press, Cambridge, UK.

van der Velde, M., Green, S. R., Gee, G. W., Vanclooster, M. & Clothier, B. E. (2005) Evaluation of drainage from passive suction and non-suction flux meters in a volcanic clay soil under tropical conditions. *Vadose Zone J.* **4**, 1201–1209.

van der Velde, M. Javaux, M., Vanclooster, M. & Clothier, B. E. (2006) El Niño-Southern Oscillation determines the salinity of the freshwater lens under a coral atoll in the Pacific Ocean. *Geophys. Res. Lett.* **33**, L21403, doi:10.1029/2006GL027748.

van der Velde, M., Green, S. R., Vanclooster, M. & Clothier, B. E. (2007) Sustainable development in small island developing states: agricultural intensification, economical development, and freshwater resources management on the coral atoll of Tongatapu. *Ecol. Economics* **61**, 456–468.

A New Focus on Groundwater–Seawater Interactions
(Proceedings of Symposium HS1001 at IUGG2007, Perugia, July 2007). IAHS Publ. 312, 2007.

Effect of an offshore sinkhole perforation in a coastal confined aquifer on submarine groundwater discharge

SARAH E. FRATESI[1], H. LEONARD VACHER[1] & WARD E. SANFORD[2]

1 *Department of Geology, University of South Florida, 4202 East Fowler Avenue, SCA328, Tampa, Florida 33620, USA*
sfratesi@mail.usf.edu

2 *United States Geological Survey, Mail Stop 431, Reston, Virginia 20192, USA*

Abstract In order to explore submarine groundwater discharge in the vicinity of karst features that penetrate the confining layer of an offshore, partially confined aquifer, we constructed a three-dimensional groundwater model using the SUTRA (Saturated–Unsaturated TRAnsport) variable-density groundwater flow model. We ran a parameter sensitivity analysis, testing the effects of recharge rates, permeabilities of the aquifer and confining layer, and thickness of the confining layer. In all simulations, less than 20% of the freshwater recharge for the entire model exits through the sinkhole. Recirculated seawater usually accounts for 10–30% of the total outflow from the model. Often, the sinkhole lies seaward of the transition zone and acts as a recharge feature for recirculating seawater. The permeability ratio between aquifer and confining layer influences the configuration of the freshwater wedge the most; as confining layer permeability decreases, the wedge lengthens and the fraction of total discharge exiting through the sinkhole increases.

Key words submarine spring; sinkhole; submarine groundwater discharge; karst; freshwater; saltwater; transition zone; coastal aquifer; SUTRA; Florida

INTRODUCTION

Florida may have the greatest potential for submarine groundwater discharge (SGD) of any location in the world due to its high precipitation rates, its long coastline, and the high permeability of the Floridan Aquifer (Zektzer *et al.*, 1973; Swarzenski & Kindinger, 2003). The hydrogeological units of the karstified Floridan Aquifer extend to the edges of the Florida Platform, some 100 km on the northeastern side and twice that distance on the western shelf, and are at least partially confined by the predominantly siliciclastic Hawthorn Formation for most of this area. A freshwater wedge extending offshore in such conditions will discharge water in a diffuse manner across the confining layer, creating concentrated discharge points wherever the confining layer is breached.

On the coastlines of Florida, such a breach is likely to be a sinkhole. In places where the Floridan Aquifer is partially confined, the Hawthorn Formation is less than 100 m thick and perforated by sinkholes and collapse features, the majority of which formed during sea level minima of the late Oligocene and early Miocene when the entire shelf was exposed (Stringfield, 1966). Above sea level, these sinkholes provide a route for recharge to the aquifer. When submerged, they connect the aquifer directly with the overlying seawater and create point sources (or sinks) of SGD (Kohout, 1966; Stringfield & LeGrand, 1969; Swarzenski *et al*., 2003). In breaching the confining layer, Kohout says, the sinkhole acts as an artesian well. Kohout's analogy is echoed by Stringfield & LeGrand (1969, and LeGrand & Stringfield, 1971), who examined the dynamics of freshwater–saltwater systems within karstic conduits that penetrate to the base of the freshwater lens or wedge.

The Crescent Beach Spring just off the coast of Jacksonville, Florida, may be the most spectacular example of a submerged sinkhole spring, ranking among Florida's many first-magnitude springs (Swarzenski *et al*., 2001). However, as Zektzer and others (1973) pointed out, there must be countless smaller submarine springs that may not be easily observed from the sea surface. Such springs may lack integrated conduit systems and still discharge substantially more water than the areas surrounding them.

The purpose of this study was to examine the magnitude of discharge that might be diverted to a sinkhole or other breach in the confining layer of an aquifer discharging freshwater off shore. We wanted to study the parameters that most affect the proportion of discharge that occurs through these structures and their impact on the configuration of the freshwater wedge and its transition zone.

RELATED STUDIES

SGD often consists of meteoric water from land and recirculated seawater, driven by an array of processes operating at different scales, including terrestrial hydraulic gradients, wave and tidal pumping and set-up, and thermal and density gradients (Burnett *et al*., 2003). In this paper, we examine only the SGD driven by a flow of freshwater from land and the recirculation of seawater that accompanies it.

A seafloor discharge face has been included in theoretical models of the coastal freshwater wedge since Hubbert (1940). Glover (1959) calculated the width of the seepage face in his potential-theory model of a coastal aquifer that is unconfined beneath the seafloor. Edelman (1972) presented an analytical solution to the problem of the offshore discharge face of a partially confined aquifer. Kooi & Groen (2001) expanded on Edelman's work, addressing the question of how far groundwater can be carried offshore in a confined aquifer.

Along with terrestrially derived freshwater, the SGD associated with an offshore freshwater wedge includes a certain amount of recirculated seawater. Cooper (1959) and Kohout (1960) described the potential for cyclic flow of seawater into the transition zone and corresponding loss of head in the seawater part of the regime. Stringfield & LeGrand (1969) pointed out that vertical karst features that penetrate to this zone of negative head may have water levels lower than sea level and, if open to the sea, could continuously intake seawater in a cyclic flow. They described such features at Tarpon Springs, Florida, and the island of Cephalonia, Greece (Stringfield & LeGrand, 1969).

The coastal freshwater wedge and the flow system associated with it can also be modified by upconing of the freshwater–saltwater transition zone beneath a freshwater sink (usually conceptualized as a pumping well). Many early models utilized the assumption of a sharp interface and a radially symmetrical cone (Bear & Dagan, 1964; Bennett *et al*., 1968; Haubold, 1975) and are concerned mostly with well contamination.

Numerical modelling is used in later studies to allow for the inclusion of a freshwater–seawater transition zone. Reviews of previous work on upconing beneath pumping wells are included in Reilly & Goodman (1987) and Zhou *et al.* (2005).

DESCRIPTION OF THE MODEL

A simple, three-dimensional aquifer model using the SUTRA (Saturated–Unsaturated TRAnsport) variable-density groundwater flow model (Voss, 1984) allowed us to test the effects of aquifer parameters on the proportion of aquifer discharge coming out of the sinkhole (Fig. 1).

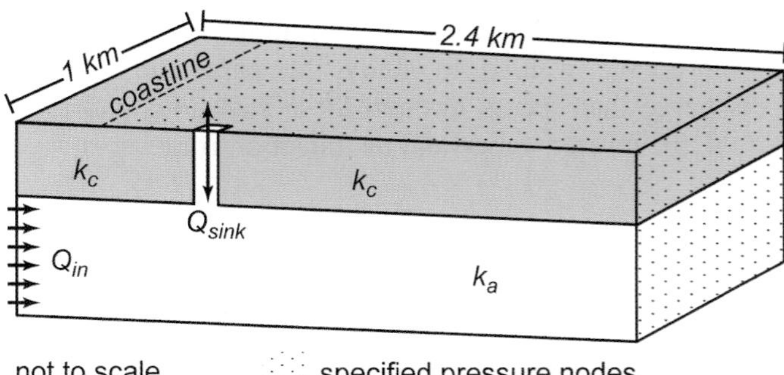

Fig. 1 The model domain, showing boundary conditions and aquifer parameters. Permeability of the confining layer (k_c), aquifer permeability (k_{aq}), freshwater input into the model (Q_{in}), and the discharge (Q_{sink}) through the specified-pressure nodes of the sinkhole.

The aquifer measures 2 km by 2.4 km by 100 m for all simulations and is overlain by a confining layer of 46-m, 66-m, or 86-m thickness. The confining layer was assigned a permeability k_c, except for a 5-by-5 column of nodes representing the sinkhole. The permeabilities of these sinkhole elements are equal to that of the aquifer (k_{aq}). This sinkhole is positioned 440 m from the coast in about 5 m of seawater. We actually ran the model for only half of this model space – the space shown in Fig. 1. We assumed that the model is symmetrical around a vertical plane of symmetry perpendicular to the coastline and passing through the centre of the sinkhole.

Nodes corresponding to the sea floor were assigned specified pressures consistent with those of hydrostatic seawater. Freshwater recharge to the aquifer (Q_{in}) is through specified-source nodes at the landward face of the model. Water discharges through specified-pressure nodes at the top of the confining layer, at the top of the sinkhole (Q_{sink}), and at the seaward end of the model. We ran each simulation to steady state, testing the effects of several recharge rates, the permeability magnitude and ratio of the aquifer (k_{aq}) and confining layer (k_c), and the thickness of the confining layer (b). We constructed a parameter matrix from the following parameters:

(a) Recharge rates: 0.18 m^3/d, 0.15 m^3/d, 0.11 m^3/d, 0.092 m^3/d, and 0.074 m^3/d per metre of coastline.

(b) Permeability of aquifer (k_{aq}): 1.00×10^{-12} m^2, 1.00×10^{-13} m^2.
(c) k_{aq}/k_c ratio: 1, 3, 10, 20, 30, 50, 70, 100, 130, 160, 200, 250, 300, 1000.
(d) Confining layer thickness: 46 m, 66 m, 86 m.

We chose these parameters to fall within acceptable values for a small-scale, coastal flow system. For each parameter set, we ran two simulations: one simulation with the sinkhole in place and a simulation with an uninterrupted confining layer for comparison. The "no sinkhole" cases were essentially two-dimensional models. Of the simulations that we ran, we ended up using about 600 for analysis.

RESULTS

Sinkhole discharge and seawater recirculation

At a steady state, the percentage of freshwater discharging through the sinkhole depends most strongly on the ratio of permeabilities k_{aq}/k_c (Fig. 2(a)), and is at most around 20% of Q_{in} for the parameters that we tested.

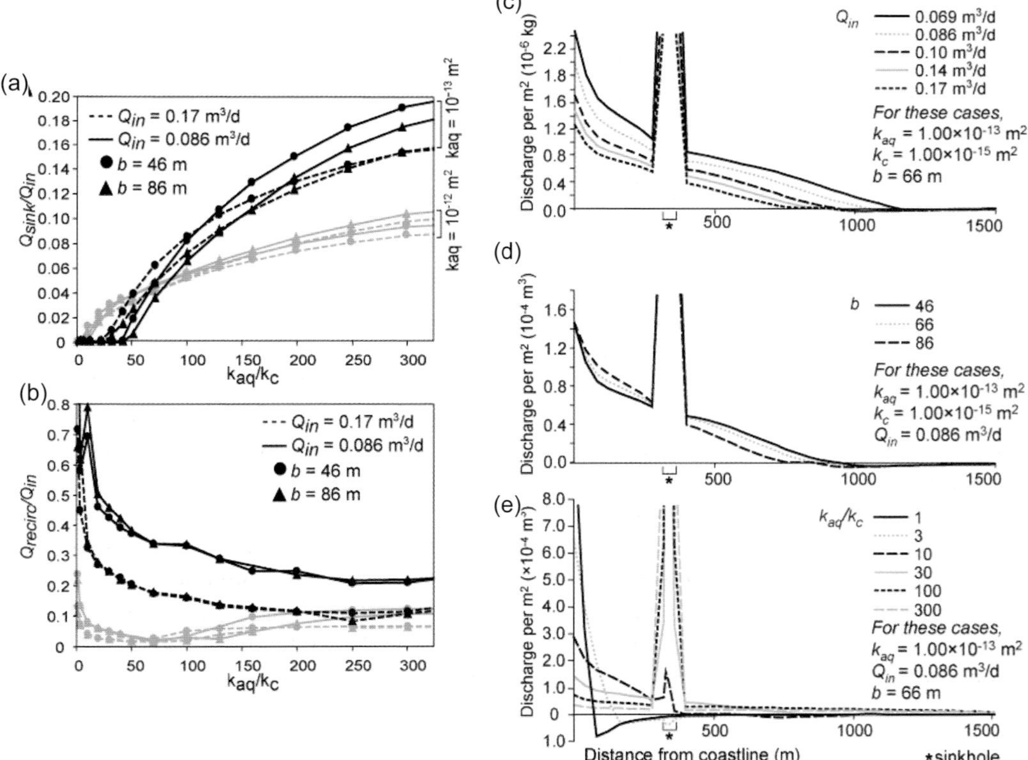

Fig. 2 Results of parameter sensitivity analysis. (a) Proportion of water discharging out of the sinkhole. (b) Amount of recirculated seawater. (c)–(e) Discharge profiles through sinkhole, perpendicular to coastline. (c) The effect of changing freshwater input into the model Q_{in}. (d) The effect of changing aquifer thickness b. (e) The effect of changing the k_{aq}/k_c ratio. This is done by decreasing k_c in six steps from 1.00×10^{-13} m^2 (equal to k_{aq}) to 3.33×10^{-16} m^2 (a k_{aq}/k_c ratio of 300).

Simulations with a higher aquifer permeability k_{aq} generally have a lower discharge through the sinkhole, generally less than 10% of the total freshwater throughput. In most simulations with k_{aq}/k_c ratios of less than 10, the sinkhole discharges nothing.

Total fluxes through the model are often higher than the specified freshwater flux, indicating that seawater is circulating through the model space. The rate of seawater circulation (Q_{recirc}) is inversely proportional to k_{aq}/k_c (Fig. 2(b)), ranging from about 10% to 30% of Q_{in} at higher k_{aq}/k_c ratios.

Discharge profiles

The specific discharge across the sediment-water interface decreases with distance from the coast (Fig. 2(c), (d) and (e)), ending in a very slight negative discharge corresponding with the slow influx of seawater into the aquifer. The profile of discharge perpendicular to the coast is generally stable with a change in Q_{in} or b (Fig. 2(c) and (d)). The k_{aq}/k_c ratio, however, has a far greater effect on the discharge profile. At a k_{aq}/k_c ratio of 1, the freshwater wedge is very short, and the profile has a prominent negative peak corresponding to the intake of seawater (Fig. 2(e)). The freshwater wedge is short enough that virtually all of the recirculated seawater enters the model landward of the sinkhole. Decreasing k_c spreads the discharge out over a larger area and creates focusing of recirculated seawater through the sinkhole. At higher k_{aq}/k_c ratios, the length of the freshwater wedge approaches the length of the model and freshwater starts to discharge out of the outcrop at the end of the model.

Flow regimes

The presence of a confining layer spreads the transition zone out considerably, the parameters of the model determining its shape and position. The magnitude of Q_{in} affects its position (Fig. 3(a)), whereas k_{aq}/k_c completely changes its character (Fig. 3(b)).

The upconing of the concentration contours at the seaward edge of the sinkhole is evident in Fig. 3. Upconing occurs in the upper concentration contours (representing lower fractions of seawater); saltier water remains undisturbed. For $k_{aq}/k_c = 1000$, the concentration contours representing greater than 50% seawater are crowded together near the end of the model.

Flow within the transition zone is largely parallel to the concentration contours (Fig. 3(c)). As expected, the highest velocities occur beneath the sinkhole and within the freshwater portion of the model. Beneath the transition zone, the general flow direction is landward. This flow reverses in the vicinity of the 70%-seawater concentration contour. In this region, flow is imperceptibly slow and vertical in direction.

DISCUSSION

The profiles presented here exhibit a classic freshwater wedge extending offshore with diffuse upward discharge through a confining layer to the sea. Our results are consistent with those of Kooi & Groen (2001), who found that the length of the

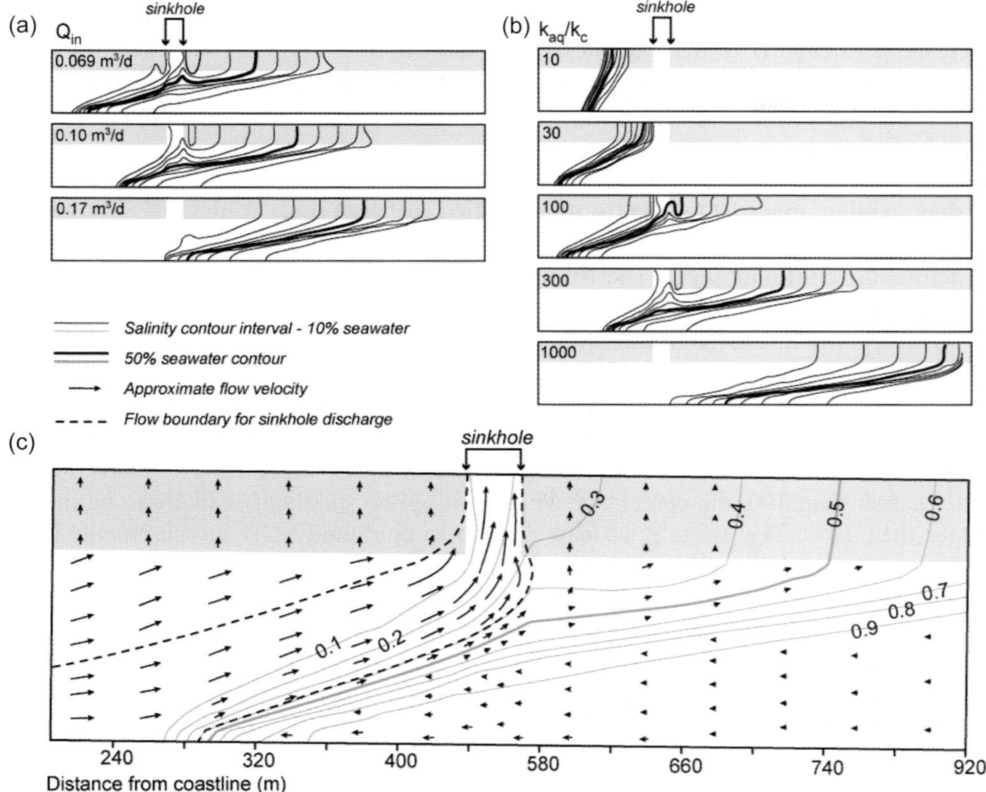

Fig. 3 Salinities and flow regime within the freshwater wedge. (a) and (b) Transition zone profiles through the sinkhole. (c) Flow diagram of profile through centre of the sinkhole. This slice is a no-flow boundary: there is no flow into or out of this section. Except where specified, all of these simulations have Q_{in} = 0.086 m^3/s, k_{aq} = 1.00 × 10^{-12} m^2, k_c = 3.33×10^{-15} m^2, and b = 46.

freshwater wedge depends upon the permeability and thickness of the confining layer and the discharge at the coastline. Generally, thicker and less-permeable confining layers, coupled with high recharge at the coast, produce longer freshwater wedges (Fig. 3(a) and (b); higher k_{aq}/k_c ratio corresponds to less-permeable confining layer).

Kooi & Groen (2001) pointed out that the length of the freshwater wedge is limited by downward dispersive effects at the outflow face, which overcome upward seepage in very long freshwater wedges. It is possible that, like Kooi & Groen's calculated "length of freshwater discharge", our ratio Q_{sink}/Q_{in} may not continue to increase with k_{aq}/k_c but instead may reach a maximum. With this particular model domain, however, the simulations with higher k_{aq}/k_c are too influenced by discharge out of the "outcrop" end of the model for us to be certain that this is the case.

Within our simulations, the sinkhole diverts up to 20% of the freshwater discharge from the freshwater wedge. The discharge through the sinkhole depends mostly on its position relative to the freshwater–saltwater transition zone. In this study, the sinkhole is stationary; the position and shape of the transition zone is largely influenced by the widely varying k_{aq}/k_c ratio. Field values of k_{aq}/k_c may vary by several orders of magnitude in an area such as Florida. We used experimental values that reflect this. In fact, the permeability of the units in the Floridan Aquifer can often be around 10 000 times the permeability of the Hawthorne Formation. The results shown in Fig. 2(a), however, suggest that a percentage greater than 30% would be unlikely, even at very high k_{aq}/k_c ratios.

Recirculated seawater accounts for some fraction of the total flux through the system, adding about 10 to 30% to the specified recharge (Q_{in}) at higher k_{aq}/k_c ratios. Recirculation is higher at lower k_{aq}/k_c ratios which allow seawater to more easily penetrate the confining layer. The freshwater wedge ends landward of the sinkhole in most of these simulations, causing the sinkhole to act as a recharge feature.

The specific discharge of the sinkhole dwarfs that of the diffuse outflow face (Fig. 2(c), (d) and (e)), but the area of diffuse flow is much larger. Without a conduit system feeding the sinkhole discharge, the sinkhole's influence is mostly local, and much of the freshwater in the system will still exit through the diffuse outflow face. However, given that sinkholes tend to exist in clusters and that the k_{aq}/k_c ratio is likely to be much larger than 300, the percentage of Q_{in} exiting the sinkholes is likely to be much larger than 20%. The effect is at least enough to confound SGD measurements in a highly heterogeneous offshore face, especially if spot seepage measurements are integrated across the entire outflow face.

Zhou *et al.* (2005), Johannsen *et al.* (2002) and Oswald *et al.* (2002) reported that in experiments concerning upconing beneath pumping wells, only the part of the transition zone with lower salinities was upconed. The water with higher salinities was not affected by the upconing. We found similar results, with a salinity profile comparable to that of a cased well penetrating the confining layer.

CONCLUSIONS

A sinkhole that perforates a confining layer in a coastal aquifer acts similarly to a cased well into the freshwater wedge, discharging freshwater or pulling in seawater, depending on its position within the flow regime. In the cases we examined, the sinkhole diverts as much as 20% of the freshwater flow through the model. This number is likely to increase with an increase in the ratio of aquifer permeability to confining-layer permeability. At lower k_{aq}/k_c ratios, seawater circulation can double the total water flux through the model compared to specified recharge values. At higher k_{aq}/k_c ratios, the rate of seawater recirculation is an additional 10–30% added to the specified freshwater recharge rate. The lower-salinity concentration contours are upconed at the seaward edge of the sinkhole; however, the sinkhole almost always discharges more than 95% freshwater. Even without a substantial conduit system feeding it, a submarine spring of this type can substantially alter the surrounding flow regime and complicate studies of submarine groundwater discharge.

Acknowledgements We appreciate the assistance of the University of South Florida and the US Geological Survey for funding of this study. We would also like to acknowledge the assistance of two reviewers whose comments greatly improved this paper.

REFERENCES

Bear, J. & Dagan, G. (1964) Some exact solutions of interface problems by means of the hodograph method. *J. Geophys. Res.* **69**, 1563–1572.

Bennett, G. D., Mundorff, M. J. & Hussain, S. A. (1968) Electric-analog studies of brine coning beneath freshwater wells in the Punjab Region, West Pakistan. *US Geol. Survey Water-Supply Paper 1608-J.*

Burnett, W. C., Bokuniewicz, H., Huettel, M., Moore, W. S. & Taniguchi, M. (2003) Groundwater and pore water inputs to the coastal zone. *Biogeochemistry* **66**, 3–33.

Cooper, H. H. Jr (1959) A hypothesis concerning the dynamic balance of fresh water and salt water in a coastal aquifer. *J. Geophys. Res.* **64**, 461–467.

Edelman, J. H. (1972) Groundwater hydraulics of extensive aquifers. *International Institute for Land Reclamation and Drainage Bulletin 13*. Wageningen, The Netherlands.

Glover, R. E. (1959) The pattern of fresh water flow in a coastal aquifer. *J. Geophys. Res.* **64**, 457–459.

Haubold, R. G. (1975) Approximation for steady interface beneath a well pumping freshwater overlying saltwater. *Ground Water* **13**, 254–259.

Hubbert, M. K. (1940) The theory of ground-water motion. *J. Geol.* **48**, 785–944.

Johannsen, K., Kinzelbach, W., Oswald, S. & Wittum, G. (2002) The saltpool benchmark problem – numerical simulation of saltwater upconing in a porous medium. *Adv. Water Resour.* **25**, 335–348.

Kohout, F. A. (1960) Flow pattern of fresh and salt water in the Biscayne Aquifer of the Miami area, Florida. *J. Geophys. Res.* **64**, 461–467.

Kohout, F. A. (1966) Submarine springs: A neglected phenomenon of coastal hydrology. *Hydrology* **26,** 391–413.

Kooi, H. & Groen, J. (2001) Offshore continuation of coastal groundwater systems; predictions using sharp-interface approximations and variable-density flow modelling. *J. Hydrol.* **246**, 19–35.

LeGrand, H. E. & Stringfield, V. T. (1971) Development and distribution of permeability in carbonate aquifers. *Water Resour. Res.* **7**, 1284–1294.

Oswald, S. E., Scheidegger, M. B. & Kinzelbach, W. (2002) Time-dependent measurement of strongly density-dependent flow in a porous medium via nuclear magnetic resonance imaging. *Transp. Porous Med.* **47**, 169–193.

Reilly, T. E. & Goodman, A. S. (1987) Analysis of saltwater upconing beneath a pumping well. *J. Hydrol.* **89**, 169–204.

Stringfield, V. T. (1966) Artesian water in Tertiary limestone in the southeastern US. *US Geol. Survey Professional Paper 517.*

Stringfield V. T. & LeGrand, H. E. (1969) Relation of sea water to fresh water in carbonate rocks in coastal areas, with special reference to Florida, U.S.A., and Cephalonia (Kephallinia), Greece. *J. Hydrol.* **9**, 387–404.

Swarzenski, P. W. & Kindinger, J. L. (2003) Leaky coastal margins: Examples of enhanced coastal groundwater/surface water exchange from Tampa Bay and Crescent Beach submarine Spring, Florida, USA. In: *Coastal Aquifer Management, Monitoring and Modeling and Case Studies* (ed. by A. Cheng & D. Ouazar), 93–112. Lewis Publishers, Boca Raton, Florida, USA.

Swarzenski P. W., Reich, C. D., Spechler, R. M., Kindinger, J. L. & Moore, W. S. (2001) Using multiple geochemical tracers to characterize the hydrogeology of the submarine spring off Crescent Beach, Florida. *Chem. Geol.* **179**, 187–202.

Voss, C. (1984) Finite-element simulation model for saturated-unsaturated fluid density-dependent groundwater flow with energy transport or chemically-reactive single-species solute transport. *US Geol. Survey Water-Resources Investigations Report 84-4369.*

Zektzer, I. S., Ivanov, V. A. & Meskheteli, A. V. (1973) The problem of direct groundwater discharge to the seas. *J. Hydrol.* **20**, 1–36.

Zhou, Q., Bear, J. & Bensabat, J. (2005) Saltwater upconing and decay beneath a well pumping above an interface zone. *Transp. Porous Med.* **61**, 337–363.

A New Focus on Groundwater–Seawater Interactions
(Proceedings of Symposium HS1001 at IUGG2007, Perugia, July 2007). IAHS Publ. 312, 2007.

Numerical modelling to determine freshwater/ saltwater interface configuration in a low-gradient coastal wetland aquifer

ERIC SWAIN & MELINDA WOLFERT
US Geological Survey, Florida Integrated Science Center, 3110 SW 9th Avenue, Fort Lauderdale, Florida 33315, USA
edswain@usgs.gov

Abstract A coupled hydrodynamic surface-water/groundwater model with salinity transport is used to examine the aquifer salinity interface in the coastal wetlands of Everglades National Park in Florida, USA. The hydrology differs from many other coastal areas in that inland water levels are often higher than land surface, the flow gradients are small, and, along parts of the coastline, the wetland is separated from the offshore waters by a natural embankment. Examining the model-simulated aquifer salinities along a transect that cuts the coastal embankment, a small zone of fresh groundwater is seen beneath the embankment, which varies seasonally in size and salinity. The simulated surface-water and groundwater levels suggest that this zone exists because of ponding of surface water at the coastal embankment, creating freshwater underflow to the offshore waters. The seasonal variability in the freshwater zone indicates that it is sensitive to the wetland flows and water levels. The small size of the zone in the simulation indicates that a model with a higher spatial resolution could probably depict the zone more accurately. The coastal ecology is strongly affected by the salinity of the shallow groundwater and the coastal freshwater zone is sensitive to wetland flows and levels. In this environment, predicting the aquifer salinity interface in coastal wetlands is important in examining the effects of changing water deliveries associated with ecosystem restoration efforts.
Key words numerical model; freshwater–saltwater interface; wetlands

INTRODUCTION

Coastal groundwater, under the influence of saltwater intruding from the sea, tends to have a definable interface dividing the fresh and saline zones. The classical conceptual model of the salinity interface in a coastal aquifer is shown in Fig. 1, which shows an intruding saltwater wedge at the base of the aquifer with the upper edge of the interface located a short distance offshore and a freshwater discharge face located between the upper edge and the shore. This situation can be represented by the Ghyben-Herzberg approximation, which assumes a hydrostatic pressure distribution and implicitly assumes flow is only horizontal (Chin, 2000). Horizontal-only flow is not possible, however, because the freshwater that moves downgradient towards the coast must discharge to surface. The Ghyben-Herzberg representation can be modified to include the offshore freshwater discharge face (actual saltwater interface in Fig. 1).

The modified Ghyben-Herzberg representation assumes that the only significant vertical flow occurs at the offshore freshwater discharge face and the inland water table lies below land surface. The conditions of a coastal wetland, however, may be different. For example, inland water levels are higher than land surface with a surface-water flow

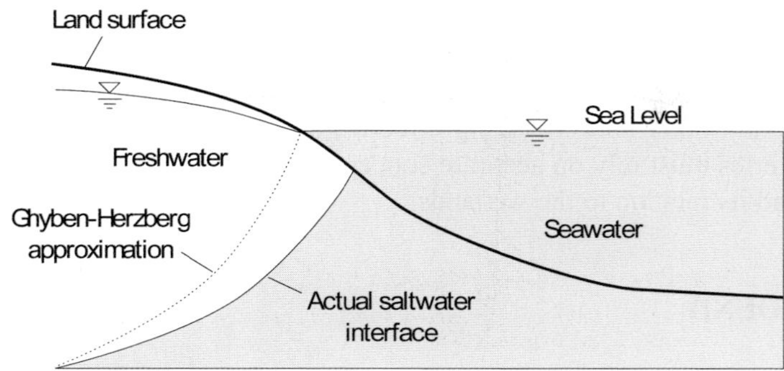

Fig. 1 Saltwater interface in a coastal aquifer where the Ghyzen-Herzberg approximation nearly applies.

connection between the wetlands and the offshore waters. Coastal wetlands often form in "protected areas", meaning that a partial obstruction to surface-water flow, such as a mudbank, shoal, or coastal embankment is present between the coastal wetlands and the offshore area. The surface-water level difference across this coastal embankment can be large relative to the mean gradient, and significant vertical flow is induced in the groundwater. In this case, salinity can still intrude inland from the coastal embankment, but freshwater flows would impound inland, creating fresh groundwater flow through and beneath the embankment. This atypical groundwater salinity-interface configuration affects the quantity and distribution of fresh groundwater discharge offshore.

The coastal wetland in the southern Everglades of Florida, USA, has been the subject of many studies to determine existing and historical conditions, trends, and the effects of restoration efforts. The numerical modelling effort described in this paper involves coupling the two-dimensional hydrodynamic surface-water model SWIFT2D (Schaffranek, 2004) with a three-dimensional groundwater flow model SEAWAT (Guo & Langevin, 2002). Both models simulate variable-density salinity transport. The coupling represents the leakage and salt flux between the surface and groundwater. SWIFT2D computes vertically-integrated two-dimensional forms of the equations of surface-water mass and momentum conservation, and solute transport equations for salt, heat, and other constituents. The code was modified for application to coastal wetlands, such as the Everglades, to allow the input of spatially variable rainfall, computation of evapotranspiration, variation of frictional resistance with depth, and other necessary features (Swain, 2005). SEAWAT combines the three-dimensional groundwater flow model MODFLOW (McDonald & Harbaugh, 1988) with the solute-transport code MT3DMS (Zheng & Wang, 1998) to incorporate the effect of salinity transport. Linking SWIFT2D and SEAWAT to account for leakage and salt flux between the surface water and groundwater is accomplished by constructing a code that calls both models and passes the necessary information between them (Langevin *et al.*, 2005). This coupled code is referred to as Flow and Transport in a Linked Overland/Aquifer Density Dependent System (FTLOADDS).

The purpose of this paper is to describe the results of using the FTLOADDS model to examine the groundwater salinity configuration in an aquifer underlying a coastal wetland. The application of a numerical model supports and helps quantify the proposed salinity distribution. The model also provides a guide to future field measurement locations that can better define the salinity interface. Ecosystem restoration efforts in south Florida concentrate on the timing and volumes of freshwater to the Everglades and other areas. Making effective and meaningful changes to Everglades water deliveries must rely on accurate conceptualizations of the coastal aquifer salinity interface and its relation to the wetlands.

BACKGROUND

The application of FTLOADDS to Everglades National Park is referred to as Tides and Inflows in the Mangroves of the Everglades (TIME). The application simulates the wetland and offshore tidal stage along with the groundwater heads and salinity in both regimes. A robust field data set has been compiled for many hydrological parameters and used for model development and comparison. As a result, the data set supports a complete calibration, and the model has been shown to reproduce field-measured flows, water levels and salinities (Swain *et al.*, 2004; Langevin *et al.*, 2005; Wang *et al.*, 2007). One of the field data-collection projects includes airborne electromagnetic surveying to determine the location of the subsurface saltwater interface (Fitterman & Deszcz-Pan, 2001). The top of this saltwater interface tends to be farther inland than the classical conceptual model.

The TIME application domain consists of about 5250 km^2 of pine uplands, cypress swamps, hardwood hammocks, wetland marsh, wet prairies, lakes, sloughs and rivers composing the Everglades National Park area. The domain of the TIME application is bounded to the north by the Tamiami Trail (US Highway 41), to the west by US Highway 29 and the Gulf of Mexico, to the south by Florida Bay, and to the east by Levee 31N, Levee 31W, and US Highway 1 (Fig. 2).

Several major drainage features, including sloughs and topographic depressions, intersect the approximately 85 × 75 km domain. The largest feature is Shark River Slough, which extends southwest from the northeastern corner of the domain to the west coast shoreline. Taylor Slough is a smaller drainage feature in the southeastern corner of the domain and, together with several coastal creeks, is the main source of runoff to northeastern Florida Bay. The northwestern part of the domain has several additional sloughs and rivers that are connected by the Wilderness Waterway and discharge to the coast. A higher elevation feature along the southeastern coast, the Buttonwood Embankment (Fig. 2), is estimated to be about 15 cm higher than the surrounding marsh (Holmes *et al.*, 2000). An equivalent feature does not exist along the west coast, where the coastal area has less topographic relief.

The climate of southern Florida is characterized by a wet season from May to September and a dry season from October to April. Sixty percent of the total rainfall occurs during this wet season. Daily rainfall patterns during the wet season and dry season are characterized by local, small-scale afternoon showers and frontal patterns, respectively. The area of surface-water inundation in the domain fluctuates greatly between the wet and dry seasons, as does the quantity of coastal freshwater flow.

The highly permeable surficial aquifer system extends over most of the Everglades National Park/Big Cypress National Preserve area and underlies a thin peat layer in

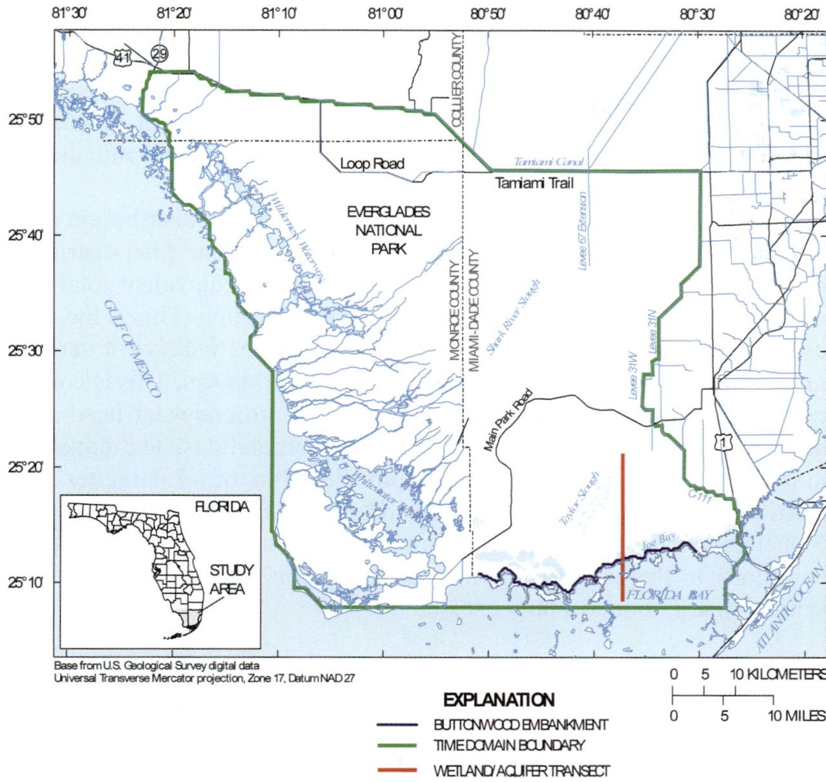

Fig. 2 Location of the Tides and Inflows to the Mangrove Everglades (TIME) domain, geographic features and water management features.

some areas. The surficial aquifer system generally thins towards the west in the study area. Leakage between surface water and groundwater is affected by the surficial peat layer over the majority of the domain (Harvey *et al.*, 2000).

APPLICATION

Because the TIME application involves a hydrodynamic surface-water model, a three-dimensional groundwater model, and the coupling between the two, a large amount of field data is needed; much of which was collected specifically for model development. These data include information pertaining to topography, rainfall, frictional resistance of vegetation, evapotranspiration rates, water level, inland and coastal discharge, and aquifer properties. Model development is described in Swain *et al.* (2004), Wolfert *et al.* (2006) and Wang *et al.* (2007). The primary purpose of the TIME development is to represent the coastal Everglades flow regime for discharges to Florida Bay (Fig. 2). To represent this regime's response to ecosystem restoration efforts, however, TIME must have boundary inflows that represent the restoration changes. These simulated inflows are generated by a large-scale regional model of South Florida developed by the South

Florida Water Management District (Wolfert *et al.*, 2004). Using these inflows, a 10-year simulation for 1990–1999 is made with a 10-minute timestep in the surface-water model and a 1-day timestep in the groundwater model. Leakage and salt flux between the surface water and the groundwater is calculated for each 10-minute surface-water timestep, using the current computed surface-water stage and the latest computed groundwater head.

In order to get a reasonable initial condition for the distribution of salinity, the 10-year simulation was repeated a number of times with the final distribution of each run used as the initial condition for the next. After an equivalent total of 160 years, the saltwater interface reaches a relatively constant location. This is the simulation used to examine the configuration of the saltwater interface, and how it varies over time. The groundwater model grid divides the aquifer into 10 layers. This is considered sufficient resolution to represent the vertical variations in groundwater head and salinity to the accuracy needed for the regional restoration simulations. The uppermost layer of the numerical model aquifer has a bottom elevation of –7.0 metres (North American Vertical Datum of 1988; NAVD88), meaning that, at the coastline, the layer is about 7 metres thick. This model resolution was not designed to represent small-scale variations in the salinity interface configuration, but to study major features of the salinity distribution at a larger scale.

RESULTS

Figure 3 shows computed aquifer salinity for the north–south transect shown in Fig. 2. Two simulation dates were chosen that have substantially different salinities under the coastal embankment. Even with the 7-m vertical discretization, sufficient resolution exists to see the effect of the coastal embankment on the salinity in layer 1. Figure 3 indicates, however, substantial salinity variations at the coast over a distance of several model grid cells, so a higher resolution groundwater simulation should increase the accuracy of the representation and give a better estimate on the actual salinity variation. On both dates shown in Fig. 3, the lower salinity zone under the embankment can be seen with somewhat higher salinities on both sides. This is particularly distinct on 7/30/1999, where salinity at a point over a kilometre inland is on the order of 6 points higher than under the embankment. This distribution differs somewhat from the salinity distribution of 11/18/1995, which corresponds to several months after a particularly wet season, with a larger freshwater zone under the embankment and lower salinity on the inland side. The simulation indicates that the surface-water hydrology causes this phenomenon, which can be loosely visualized as a freshwater "bubble" at the coastline. Other transects in an east–west direction, cutting the western Everglades coast (Fig. 2) did not indicate any such freshwater bubble.

The surface-water stage and groundwater head averaged over the entire simulation along the flow transect are shown in Fig. 4, along with the land elevation. The temporal variability of the surface water produces a more irregular average level than that of the groundwater. It can be seen that the surface-water stage is higher than groundwater head inland from the coastal embankment and lower than groundwater head offshore. Whereas the ground-water gradient is more uniformly downward, the surface-water stage drops abruptly at the coastal embankment.

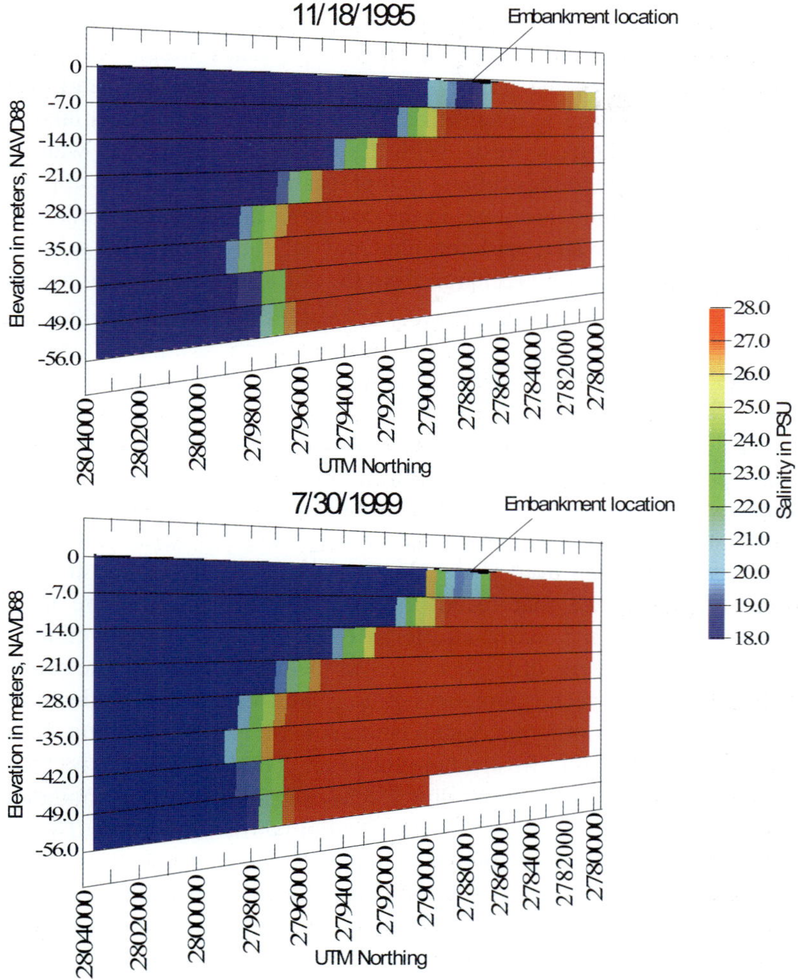

Fig. 3 Computed aquifer salinity showing the effect of the coastal embankment. Transect location shown in Fig. 2.

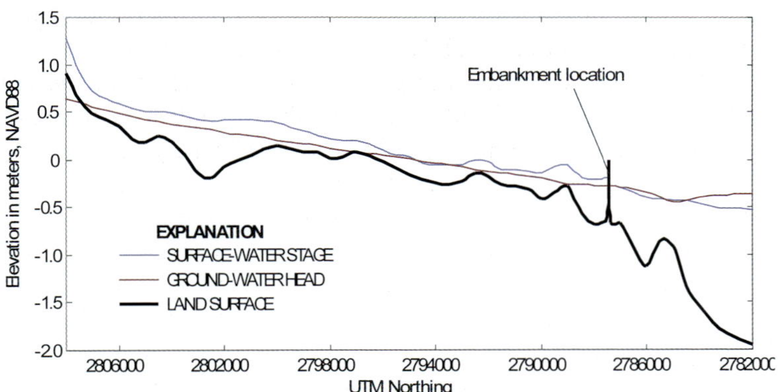

Fig. 4 Average surface-water stage and groundwater head showing effect of coastal embankment.

Further examination of Fig. 4 does not indicate an inland zone where average groundwater head is notably higher than the surface water. At approximately UTM Northing 2794000 in the transect, the average groundwater head is slightly higher. This location in Fig. 3 is inland from where the interface meets land surface, and the groundwater is relatively fresh. With the yearly fluctuations in water levels, groundwater can discharge to the wetland at this location; however, this appears to be a minor effect. The model indicates that, in the coastal Everglades, the wetlands predominantly supply the groundwater, and the tendency of the top of the saltwater interface to intrude inland is offset by recharge from the surface water ponded near the coast.

DISCUSSION

The TIME application is uniquely qualified to simulate a coastal wetland with underlying aquifer. As well as representing groundwater flow and salinity transport three-dimensionally, the wetlands and offshore waters are simulated by a fully hydrodynamic two-dimensional code. In addition, prodigious field measurements of multiple parameters have been used to develop and calibrate the model. It is thus with a high degree of confidence that results of the TIME application are interpreted for the saltwater interface in a coastal aquifer affected by an overlying wetland.

The simulated zone of fresher water in the transect cutting the southern Everglades coast, which is not seen in transects cutting the western Everglades coast, supports the concept that the southern coastal embankment is a controlling feature. The most apparent cause of the freshwater zone simulated in Fig. 3 is the damming and channelization of surface-water flow by the embankment. The simulation indicates that the saltwater interface resembles the situation shown in Fig. 5. The coastal embankment effect is the main phenomenon that differentiates this coastal wetland situation from the classical conceptual model of the salinity interface in a coastal aquifer.

Simulation results indicate that the presence of a coastal embankment produces an aquifer salinity distribution that can be roughly visualized in three zones, as depicted in Fig. 5. The first zone is inland where the surficial groundwater is fresh, and groundwater can discharge to the surface. The simulation indicates predominant recharge of the groundwater from the wetland in this zone, but exchange occurs in both directions. The second zone is at the coast, and the higher salinity groundwater at the top of the saltwater interface is just inland from a "bubble" of fresher groundwater that lies

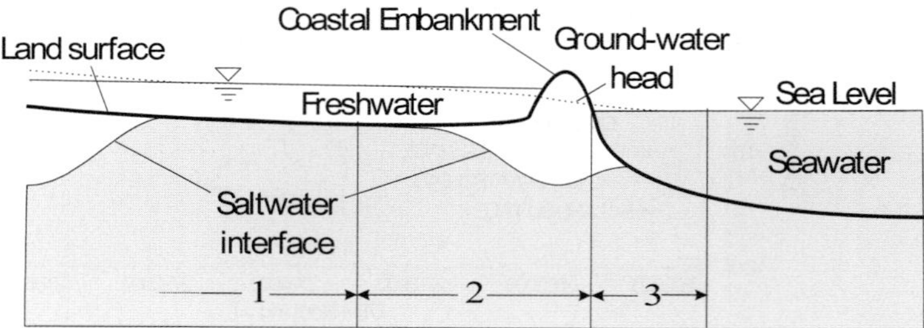

Fig. 5 Saltwater interface in the coastal aquifer where the wetland lies inland of a coastal embankment.

under the coastal embankment. This bubble is apparently induced by recharge from ponded surface water at the embankment. The third zone is offshore, where groundwater becomes more saline and discharges to the sea. The salinity variations induced by the coastal embankment are observed only in the uppermost layer of the model, which is approximately 7 m thick, so a higher resolution model would most likely produce a more accurate picture of the extent and salinity of this phenomenon.

This study indicates a new conceptual model of the configuration of the saltwater interface in coastal wetland environments. The implications affect the development and restoration of coastal wetlands and the associated ecology. The coastal mangroves in the Everglades are significantly affected by the salinity of the shallow groundwater, and changes in the mangrove communities affect other flora and fauna in the area. The TIME simulation indicates that somewhat fresher groundwater along the coastal embankment may be a natural feature of the embankment hydrology, and a feature that is sensitive to the wetland water levels. The location of the saltwater and freshwater marsh can change according to this coastal salinity. Thus, a change in upstream water deliveries can potentially have a complex effect on the configuration of the saltwater interface, salinity of the shallow groundwater, and hence the marsh communities.

REFERENCES

Chin, D. A. (2000) *Water-Resources Engineering*. Prentice-Hall, New Jersey, USA.

Fitterman, D. V. & Deszcz-Pan, M. (2001) Saltwater intrusion in Everglades National Park, Florida measured by airborne electromagnetic surveys. In: *First International Conference on Saltwater Intrusion and Coastal Aquifers-Monitoring, Modeling, and Management* (SWICA-M3), (Essaouira, Morocco). Laboratoire d'Analyse des Systémes Hydrauliques (LASH), Rabat, Morocco.

Guo, Weixing & Langevin, C. D. (2002) User's Guide to SEAWAT: A Computer Program for Simulation of Three-Dimensional Variable-Density Ground-Water Flow. *US Geological Survey Techniques of Water-Resources Investigations, Book 6, Chapter A7.*

Harvey, J. W., Jackson, J. M., Mooney, R. H. & Choi, J. (2000) Interactions between ground water and surface water in Taylor Slough and vicinity, Everglades National Park, south Florida: study methods and appendixes. *US Geological Survey Open-File Report 00-483.*

Holmes, C. W., Robbins, J., Halley, R. B., Bothner, M., Tenbrink, M. & Marot, M. (2000) Sediment dynamics of Florida Bay mud banks on a decadal time scale: US Geological Survey Program on the South Florida Ecosystem: 2000. Proc. Greater Everglades Ecosystem Restoration (GEER) Conference (December 2000). In: *US Geological Survey Open-File Report 00-449.*

Langevin, C. D., Swain, E. D. & Wolfert, M. A. (2005) Simulation of integrated surface-water/ground-water flow and salinity for a coastal wetland and adjacent estuary. *J. Hydrol.* **314**, 212–234.

McDonald, M. G. & Harbaugh, A. W. (1988) A modular three-dimensional finite-difference ground-water flow model. *US Geological Survey Techniques of Water Resources Investigations, Book 6, Chapter A1.*

Schaffranek, R. W. (2004) Simulation of surface-water integrated flow and transport in two dimensions: SWIFT2D user's manual. *US Geological Survey Techniques and Methods, Book 6, Chapter B1.*

Swain, E. D. (2005) A model for simulation of surface-water integrated flow and transport in two dimensions: user's guide for application to coastal wetlands. *US Geological Survey Open-File Report 2005-1033.*

Swain, E. D., Wolfert, M. A., Bales, J. D. & Goodwin, C. R. (2004) Two-dimensional hydrodynamic simulation of surface-water flow and transport to Florida Bay through the Southern Inland and Coastal Systems (SICS). *US Geological Survey Water-Resources Investigations Report 03-4287.*

Wang, J. D., Swain, E. D., Wolfert, M. A., Langevin, C. D., James, D. E. & Telis, P. A. (2007) Applications of Flow and Transport in a Linked Overland/Aquifer Density Dependent System (FTLOADDS) to simulate flow, salinity, and surface-water stage in the southern Everglades, Florida. *US Geological Survey Scientific Investigations Report 2007-5010.*

Wolfert, M. A., Langevin, C. D. & Swain, E. D. (2004) Assigning boundary conditions to the Southern Inland and Coastal Systems (SICS) model using results from the South Florida Water Management Model (SFWMM). *US Geological Survey Open-File Report 2004-1195.*

Wolfert, M. A., Swain, E. D. & Wang, J. D. (2006) Utilizing the TIME model to simulate Comprehensive Everglades Restoration Plan (CERP) scenarios for Florida Bay and Florida Keys Feasibility Study (FBFKFS). In: *2006 Greater Everglades Ecosystem Restoration Conference* (Lake Buena Vista, Florida, June, 2006), p. 249.

Zheng, C. & Wang, P. P. (1998) MT3DMS, A modular three-dimensional multispecies transport model for simulation of advection, dispersion and chemical reactions of contaminants in groundwater systems. *Vicksburg, Mississippi Waterways Experiment Station, US Army Corps of Engineers.*

Simulation of submarine groundwater discharge salinity and temperature variations: implications for remote detection

ALYSSA M. DAUSMAN[1,2], CHRISTIAN D. LANGEVIN[1] & MICHAEL C. SUKOP[2]

1 *US Geological Survey-Florida Integrated Science Center, 3110 SW 9th Avenue, Fort Lauderdale, Florida 33315, USA*
adausman@usgs.gov

2 *Florida International University-Department of Earth Sciences, Miami, Florida 33199, USA*

Abstract A hydrological analysis using a numerical simulation was done to identify the transient response of the salinity and temperature of submarine groundwater discharge (SGD) and utilize the results to guide data collection. Results indicate that the amount of SGD fluctuates depending on the ocean stage and geology, with the greatest amount of SGD delivered at low tide when the aquifer is in direct hydraulic contact with the ocean. The salinity of SGD remains lower than the ocean throughout the year; however, the salinity difference between the aquifer and ocean is inversely proportional to the ocean stage. The temperature difference between the ocean and SGD fluctuates seasonally, with the greatest temperature differences occurring in summer and winter. The outcome of this research reveals that numerical modelling could potentially be used to guide data collection including aerial surveys using electromagnetic (EM) resistivity and thermal imagery.

Key words submarine groundwater discharge

INTRODUCTION

In coastal areas, submarine groundwater discharge (SGD) is a topic of increasing concern to marine scientists, primarily because terrestrially derived groundwater can release nutrients and other potentially harmful contaminants into ecologically sensitive marine environments (Finkl & Krupa, 2003; Taniguchi *et al.*, 2002). It is difficult to quantify SGD, however, because spatial variations in recharge, aquifer hydraulic properties, tides, and other complicating factors can substantially affect patterns and rates of groundwater flow to the ocean. The presence of geological heterogeneity, for example, can complicate the measurement of SGD (Bokuniewicz *et al.*, 2003).

Improved aerial survey techniques that utilize electromagnetic (EM) and thermal imagery provide a direct means to map zones of increased SGD (Weiss *et al.*, 2005). EM methods can be used to detect subsurface salinity differences, whereas thermal imagery can detect small water temperature variations at the sea surface (greater than 0.08°C in calm seas). Therefore, it is possible to detect SGD with an aerial survey if the salinity or temperature difference between SGD and the ocean is relatively large. The best time to conduct an aerial survey is when salinity and temperature differences between SGD and surface water are the greatest.

Variable-density numerical models are often used for management purposes to simulate groundwater flow in coastal aquifers, including saltwater intrusion and SGD. A deep (~3 km) model of North Carolina was developed to investigate fresh and saline SGD, as well as geothermal convection in a coastal setting (Wilson, 2005). Langevin (2003) used a variable-density flow and transport model to quantify rates of fresh SGD into Biscayne Bay in southern Florida. Smith & Zawadski (2003) calibrated a two-dimensional cross-sectional model to seepage measurements in the northeastern Gulf of Mexico. Previous research with data collection and modelling has also shown that seasonality can affect the amount of SGD resulting from water-table changes (Michael *et al.*, 2005). However, few numerical modelling studies, if any, have been designed specifically to guide data collection efforts.

The purpose of this paper is to demonstrate how a numerical model can be used to guide data collection efforts in studies of SGD. A variable-density numerical model was developed using generalized hydrological conditions for southeastern Florida to determine when and where SGD rates are the greatest and to quantify expected temperature and salinity differences between groundwater and ocean water. The model is unique in that it represents SGD characteristics, including salinity and temperature, at short-term (tidal) and longer-term (seasonal) time scales. Results from the model are used to provide insight into optimal times to perform aerial surveys.

METHODS

To evaluate transient variations in the salinity and temperature of SGD, a 2-D cross-sectional model was developed using idealized hydrogeological conditions and hydraulic parameters common to southeastern Florida (Dausman & Langevin, 2005) (Figs 1 and 2). The numerical model is intended to represent generalized flow conditions in the Biscayne aquifer in southeastern Florida rather than one specific location. The data used to develop the model were obtained from various places in southeastern Florida including Miami-Dade, Palm Beach, and Broward Counties (Fig. 1). The 2-D vertical cross-sectional model was aligned along a west-to-east groundwater-flow line perpendicular to the coastline. The model represents a 1-year period, from 1 January 1998 to 31 December 1998, using hourly stress periods. Consecutive simulations were run using head, salinity, and temperature from the previous model run until salinities and temperatures achieved dynamic equilibrium.

The SEAWAT code (Langevin *et al.*, 2003), which is a combined version of MODFLOW (McDonald & Harbaugh, 1988) and MT3DMS (Zheng, 1990), simulates variable-density flow driven by changes in salinity. A modified version of this code was used in this study to simulate density-dependent flow caused by changes in salinity, as well as simulate changes in temperature (Thorne *et al.*, 2006).

The 2-D model grid contains 152 columns and 14 layers; each cell is 150 by 7.5 m. The active cells represent the Biscayne aquifer (Table 1), which is underlain by a low-permeability unit (represented with no-flow cells). The specified head cells along the sea floor represent the ocean. The ocean water levels and temperature values vary hourly according to measured data collected during 1998. The salinity value is specified at 35 parts per thousand (ppt). General-head boundary (GHB) cells on the

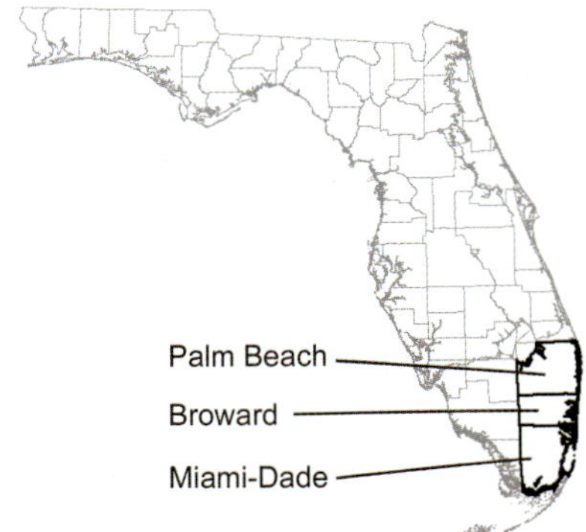

Fig. 1 Map of Florida with the three counties containing the Biscayne aquifer.

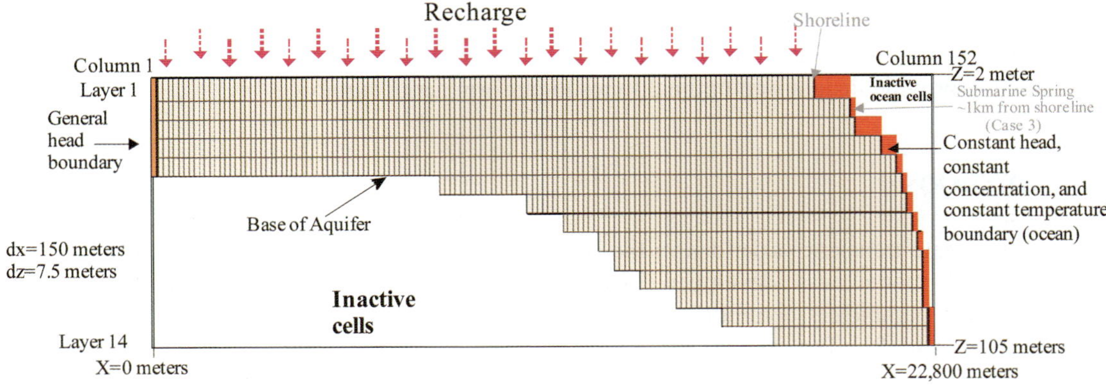

Fig. 2 Boundary conditions and finite difference model grid.

Table 1 Aquifer parameters used in the variable-density cross-sectional models (from Dausman & Langevin, 2005).

Parameter (units)	Parameter definition	Value
K_H (m/d)	Horizontal hydraulic conductivity	1150
K_v (m/d)	Vertical hydraulic conductivity	150
α_L (m)	Longitudinal dispersivity	3.0
α_T (m)	Transverse dispersivity	0.3
S_y (dimensionless)	Specific yield	0.1
S (dimensionless)	Storage	10^{-5}
n (dimensionless)	Porosity	0.1
D* (m^2/d)	Bulk thermal diffusivity	0.67996
μ (kg/(m d))	Equivalent freshwater viscosity	86.4
ρ_f (kg/m^3)	Density of fresh water	1000
ρ_s (kg/m^3)	Density of sea water	1025
$\frac{d\rho}{dC}$ (dimensionless)	Density change with concentration	0.7
$\frac{d\mu}{dC}$ (m^2/d)	Viscosity change with concentration	10^{-6}

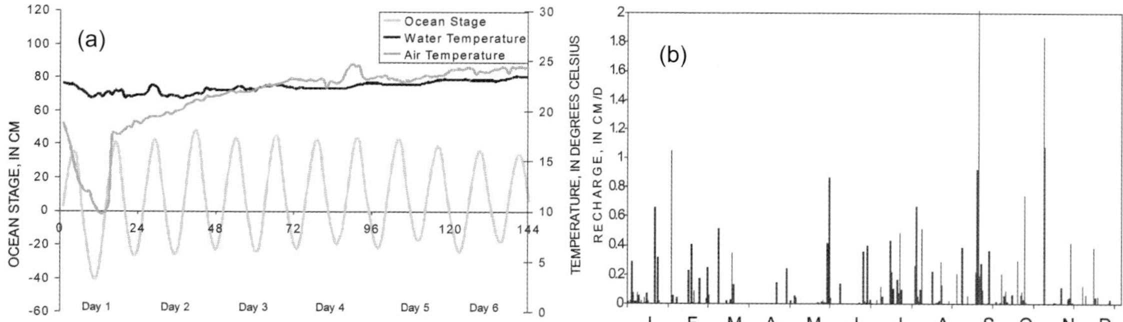

Fig. 3 (a) The first 6 days of the stage in the ocean boundary, temperature of the ocean boundary, and air temperature used in the recharge and general head boundary cells. (b) Model recharge.

western model boundary were used to represent hydraulic connection with the Florida Everglades. Assigned GHB heads, salinities, and temperatures were based on observed conditions. Net recharge, approximated as 15% of rainfall, changes daily; the temperature of recharge changes hourly according to measured air temperatures (Fig. 3(a), (b)).

Three scenarios were simulated to quantify different types of hydraulic connection with the ocean. In the first case, the Biscayne aquifer is in direct hydraulic connection with the ocean (Fig. 4). In Case 2, a low hydraulic conductivity layer is assigned along the sea floor (representing low-permeability marine sediments) to reduce the hydraulic connection with the ocean (Fig. 4). This low-permeability layer is included in the model by setting the vertical and horizontal hydraulic conductivities of the constant head cells (CHD) to a value of 0.05 m/d. In this approach, the thickness of the marine sediments is about one half of a cell thickness of each CHD cell. The storage of the low permeability layer is assumed to be the same as that of the aquifer, 10^{-5}. We did not test the effects of different storage values. However, the decrease in hydraulic conductivity of the low permeability layer results in a lower hydraulic diffusivity causing a damping of the ocean tides. Case 3 is identical to Case 2, except for the presence of a spring 1 km from the shoreline (column 137) (Figs 2 and 4). At this location, the Biscayne aquifer is in direct hydraulic connection with the ocean.

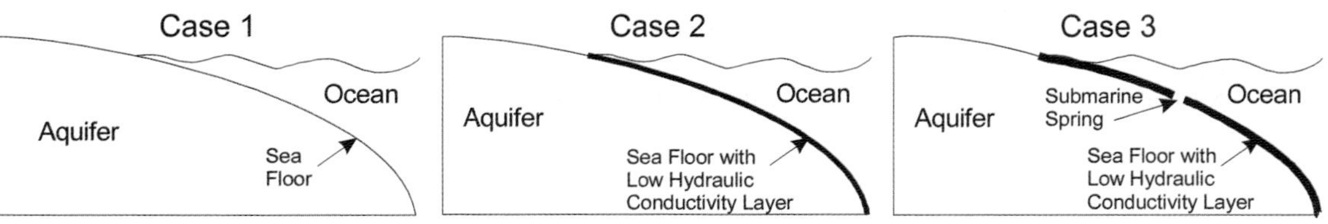

Fig. 4 Conceptual models of hydraulic connection between the aquifer and the ocean.

RESULTS

Simulation results include groundwater discharge rates (and salinities and temperatures) for each hourly stress period and for each model cell. These discharge rates were combined or averaged in different ways to highlight the differences between the three cases. Rates to individual model cells are expressed in units of m/d. When discharge rates are combined, they are expressed in units of m^3/d per meter of shoreline (expressed as m^2/d), or in the case of a cumulative discharge, as m^3 per meter of shoreline.

Figure 5(a) shows the total SGD (calculated by summing the discharge for the ocean cells) as a function of time for the first 7 days of the simulation. The highest discharge rates (positive indicating flow from the aquifer into the ocean) occur during low tide; the lowest rates occur during high tide. As expected, the minimum and maximum discharges are a function of the hydraulic connection between the aquifer and the ocean. Case 1 has the highest discharge rates, ranging between about 150 and –150 m^2/d. Case 2 has the lowest discharge rates, ranging between 50 and –50 m^2/d. Because Case 3 contains hydraulic features of both cases 1 and 2, the discharge rates range between cases 1 and 2. The combined discharge totals for the entire simulation are shown in Fig. 5(b). The largest discharge occurs in Case 1, followed by cases 2 and 3, respectively. A mass balance analysis revealed that approximately the same amount of freshwater is discharged to the ocean for all three models; however, more seawater is recirculated in Case 1, resulting in a larger total.

Fresh recharge is applied to the top of the model. About 25% of that fresh recharge flows to the west and discharges to the western boundary. The remaining 75% discharges into the ocean. The type of hydraulic connection has a large effect on where SGD occurs in relation to the shoreline. The average annual rate of SGD was calculated for each ocean constant-head cell (Fig. 6). In Case 1, most of the SGD is into the first ocean cell. Case 1 also shows large components of recirculated seawater, observed as negative SGD rates farther offshore. The low hydraulic conductivity layer in Case 2 impedes SGD, forcing terrestrially-derived groundwater to discharge as far as 1.8 km from the shoreline. For Case 3, most of the SGD is concentrated into the cell representing the submarine spring; however, relatively low discharge rates are observed near the shoreline.

As shown in Fig. 6, SGD rates fluctuate between positive and negative values during a tidal cycle making it difficult to determine a net discharge rate. For this reason, the cumulative discharge (in m^3 per meter of shoreline) as a function of time was calculated for the first model cell representing the ocean (Fig. 7(a)). The final value of about 6,000 m^3/m of shoreline over the course of 1 year is equivalent to an average SGD rate of about 0.11 m/d (calculated from the 150-m cell width). Also

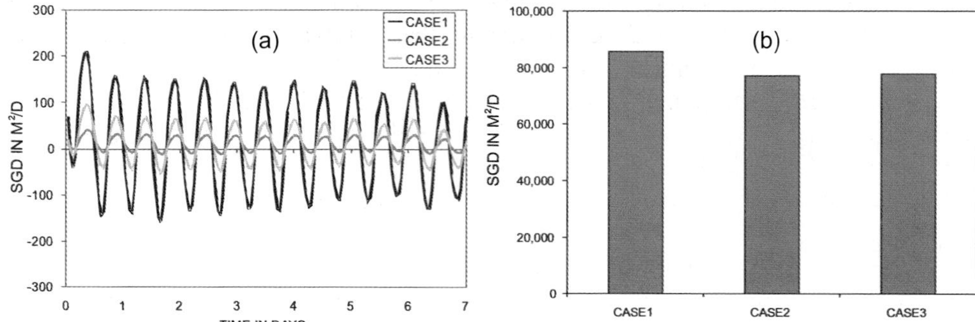

Fig. 5 (a) The sum of SGD for each case over time. The first 7 days of the model are shown. (b) The total amount of SGD for each case.

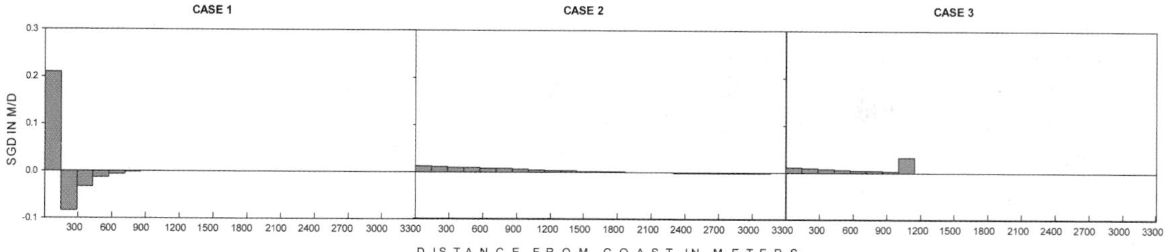

Fig. 6 Model results showing the annual average discharge with distance from the coast.

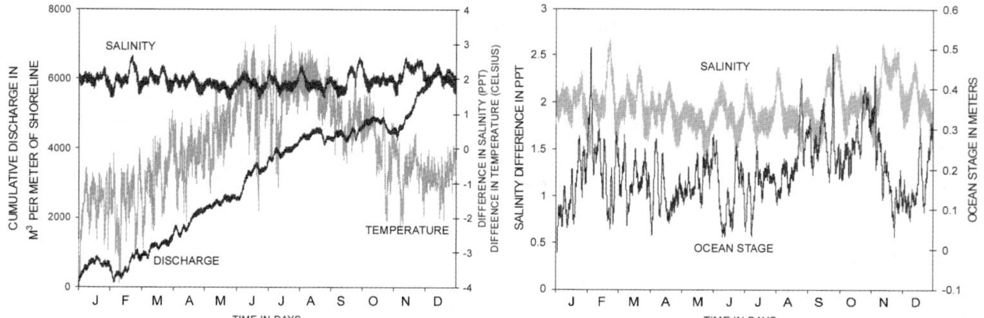

Fig. 7 (a) Cumulative discharge, the difference in salinity and the difference in temperature with time at the first cell representing the ocean for Case 1. (b) The difference in salinity and the ocean stage with time at the coast for Case 1.

shown in Fig. 7(a) are the temperature and salinity differences between the ocean and underlying aquifer cell at the shoreline. The average salinity difference is only about 2 ppt, indicating that most of the SGD for this case has near-seawater salinities. Fluctuations in salinity difference are primarily caused by fluctuations in ocean stage as shown by an inverse correlation (Fig. 7(b)); a higher ocean stage corresponds with a smaller difference in salinity between the aquifer and ocean.

The difference in temperature between the ocean and SGD fluctuates seasonally, with warmer groundwater discharging into the colder saline ocean in the winter (Figs 7(a) and 8(a)). During late summer, cooler groundwater discharges into the warmer ocean waters (Fig. 8(b)). Temperature differences between groundwater and the ocean are minimal during the spring and fall (Fig. 8(c)). In southern Florida, the ocean and rainfall are both warmer (~28°C, assuming that rainfall temperature equals air temperature) during summer than during winter (~22°C). Groundwater that is discharged to the coast remains at about 25°C, which is a mixture of the older warmer and cooler waters from previous seasons. Because the ocean water temperature varies from about 22°C in winter to 28°C in summer, these are periods of greatest temperature difference between ocean waters and older mixed aquifer waters that are discharged to the coast.

Model results at low tide (when SGD rates and salinity differences are large) and in winter (when temperature differences between the ocean and aquifer are the largest) are shown in Fig. 9 for the three cases. Temperature and salinity differences are shown here with the SGD rates to give an indication of whether the discharge would be

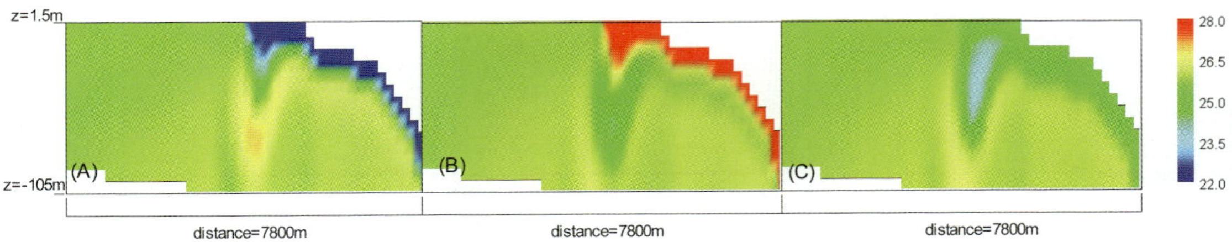

Fig. 8 Temperature (°C) for Case 1 during (A) winter, (B) summer and (C) spring.

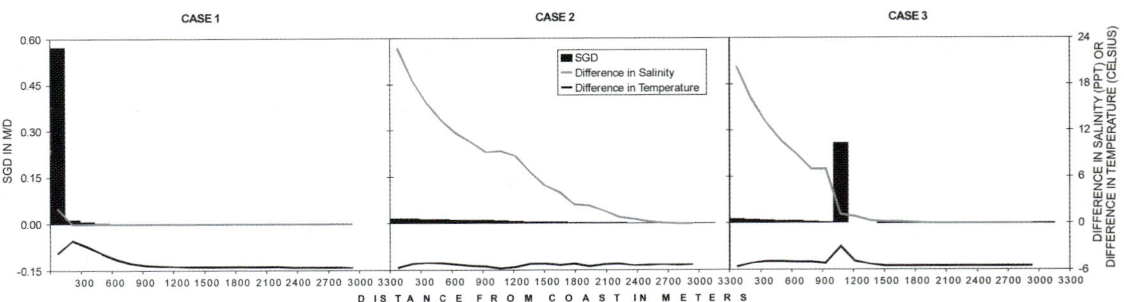

Fig. 9 Discharge, difference in salinity and difference in temperature *vs* distance from the coast, at low tide in winter.

detectable from an aerial survey (i.e. do the largest temperature and salinity differences correspond with the locations of largest discharge?). For Case 1, where most of the discharge is at the shoreline, the salinity difference is about 2 ppt and the temperature difference is about 3°C. In Case 2, the fresh SGD is not concentrated near the shoreline, but discharged over a 1-km outflow face with distance from the coast. The low hydraulic conductivity layer in Case 2 impedes the flow of SGD and reduces the mixing of the aquifer and ocean waters. The reduced mixing results in a salinity difference between the ocean and aquifer of from 6 to 23 ppt across the outflow face, with the greatest difference near the shoreline (Fig. 9). However, differences in temperature between groundwater and the ocean in Case 2 stay relatively consistent at ~6°C across the outflow face. Case 3 shows a relatively large amount of SGD at the submarine spring, where fresher groundwater is allowed to discharge to the spring. The increased mixing at the spring results in a decrease in the salinity difference between the ocean and aquifer to approximately 3 ppt at the main point of discharge. The temperature difference between the aquifer and ocean at the spring is approximately 3°C (Fig. 9).

DISCUSSION

Variable-density groundwater flow models can be used to understand the complex nature of a coastal groundwater system. This analysis illustrates how the salinity, temperature and rate of SGD are affected by the hydraulic character of the coastal system from the geological setting to ocean characteristics. The results from this type of analysis can be used to help guide data collection efforts for studies of SGD.

Aerial surveys using EM resistivity are used to detect fresh SGD to the ocean. An EM resistivity survey has the ability to detect salinity variations at depth. During all times of the year, fresher groundwater is discharged to the ocean. However, the salinity difference between the aquifer and ocean is affected by the hourly and monthly ocean stage variations. The modelling for this research reveals that aerial EM surveys are best conducted at low tide and when ocean stages are at a relative low. This is when the difference in salinities between the aquifer and ocean are greatest (Figs 7(b) and 9). Therefore, the results from an aerial survey could potentially be improved by including expected ocean stage variations for the specific target area as part of the data collection plan.

An aerial survey using EM resistivity may or may not be successful depending upon the geology of the aquifer. An aquifer in direct hydraulic connection to the ocean, such as Case 1, may not have large enough salinity differences between the aquifer and ocean for an aerial survey to detect (a difference of only 2 ppt at the shoreline, decreasing to 0 ppt with distance from the shore). Case 2 contains a low hydraulic conductivity layer of sediments across the ocean floor. Although the sediments impede SGD and increase the zone of discharge over a larger area of the ocean floor, the difference in salinity between the ocean and aquifer waters is relatively high in this environment (up to 23 ppt), enabling likely detection by EM resistivity (Fig. 9). If there is a fracture in the low conductivity layer, resulting in a submarine spring, as in Case 3, the SGD across the ocean floor and at the spring may be high enough for detection by the survey as long as the salinity differences between the aquifer and ocean are relatively high, as in Fig. 9.

Data collection methods utilizing thermal imagery to quantify SGD are successful in calm waters where the temperature difference between the SGD and ocean waters is greater than 0.08°C at the ocean surface. Thermal imagery cannot detect temperature variations below the surface; therefore, a larger temperature difference between the SGD and the ocean and a relatively large amount of SGD is necessary for detection. Thermal images should be collected at low tide when SGD is relatively high and at a location where SGD rates would be large enough to affect the ocean surface temperature. In southern Florida, thermal imagery surveys are best conducted during the late summer or late winter when temperature differences between SGD and the ocean waters are the largest ~3°C (Figs 7(a) and 8). During the spring and fall, SGD is about the same temperature as the ocean and would probably not be detected.

The success of a survey using thermal imagery can also be affected by the geology of the aquifer. When the aquifer is in direct hydraulic connection to the ocean, as in Case 1, a narrow zone of high discharge exists at the shoreline close to the surface. In this environment, the temperature difference of ~3°C between the SGD and the ocean would probably be detected by an aerial survey using thermal imagery (Fig. 9). If there is a low conductivity layer of marine sediments producing a wide zone of discharge across the sea floor and a decreased amount of SGD, thermal imagery will likely not work. Even though the temperature difference is ~6°C, the lack of concentrated discharge reaching the ocean surface would make SGD undetectable (Fig. 9). However, if there is a submarine spring in this environment, it would likely be identified by thermal imagery as long as the spring exists in shallow waters, there is an increased amount of SGD (i.e. at low tide), and the temperature difference is greater than 0.08°C (Fig. 9). If the submarine spring is too deep, thermal imagery will likely not detect the spring.

CONCLUSIONS

Numerical modelling can help enhance and guide SGD data collection efforts by providing estimates of expected salinity and temperature variations. Aerial surveys using EM resistivity are more appropriate for some coastal environments, whereas, thermal imagery is likely more successful in others. Results from this research indicate that modelling of a coastal environment prior to data collection can reveal the potential for success using either types of aerial surveys. This investigation also reveals that ignoring short-term temporal changes on the system, such as tides, could hinder modelling efforts to guide data collection. Although the results presented here appear to be representative for south Florida, the general conclusions may also be valid for other environments with similar hydrogeological and ocean characteristics.

REFERENCES

Bokuniewicz, H., Buddemeier, R., Maxwell, B. & Smith, C. (2003) The typological approach to submarine ground-water discharge (SGD). *Biogeochemistry* **66**, 145–158.

Dausman, A. & Langevin, C. D. (2005) Movement of the saltwater interface in the surficial aquifer system in response to hydrologic stresses and water-management practices, Broward County, Florida. *US Geol. Survey Scientific Investigations Report 2004-5256*.

Finkl, C. W. & Krupa, S. L. (2003) Environmental impacts of coastal-plain activities on sandy beach systems: hazards, perception and mitigation. *J. Coastal Res.* **35**, 132–150.

Langevin, C. D. (2003) Simulation of submarine ground water discharge to a marine estuary: Biscayne Bay, Florida. *Ground Water* **41**(6), 758–771.

Langevin, C. D., Shoemaker, W. B. & Guo, W. (2003) MODFLOW-2000: the US Geological Survey modular ground-water model. Documentation of the SEAWAT-2000 version with the variable-density flow process (VDF) and the integrated MT3DMS transport process (IMT). *US Geol. Survey Open-File Report 03-426*.

McDonald, M. G. & Harbaugh, A. W. (1988) A modular three-dimensional finite-difference ground-water flow model. *US Geol. Survey Techniques of Water Resources Investigations*, Book 6.

Michael, H. A., Mulligan, A. E. & Harvey, C. F. (2005) Seasonal oscillations in water exchange between aquifers and the coastal ocean. *Nature* **436**(6054), 1145-1148.

Smith, L. & Zawadski, W. (2003) A hydrologeologic model of submarine groundwater discharge: Florida intercomparison experiment. *Biogeochemistry* **66**, 95–110.

Taniguchi, M., Burnett, W. C., Cable, J. E. & Turner, J. V. (2002) Investigation of submarine groundwater discharge. *Hydrol. Processes* **16**, 2115–2129.

Thorne, D., Langevin, C. D. & Sukop, M. C. (2006) Addition of simultaneous heat and solute transport and variable fluid viscosity to SEAWAT. *Computers and Geosciences* **32**, 1758–1768.

Weiss, M., Kruse, S., Burnett, W. C., Chanton, J., Greenwood, W., Murray, M., Peterson, R. & Swarzenski, P. (2005) Evaluation of geophysical and thermal methods for detecting submarine groundwater discharge (SGD) in the Suwannee River Estuary. American Geophysical Union, Fall Meeting 2005, abstract #H43F-0554.

Wilson, A. M. (2005) Fresh and saline groundwater discharge to the ocean: a regional perspective. *Water Resour. Res.* **41**(2), 1–11.

Zheng, Chunmiao (1990) MT3D: A modular three-dimensional transport model for simulation of advection, dispersion and chemical reactions of contaminants in groundwater systems. Report to the US Environmental Protection Agency, Ada, Oklahoma, USA.

Submarine groundwater discharge under extreme rainfall events

EUNHEE LEE, YUNJUNG HYUN & KANG-KUN LEE
School of Earth and Environmental Sciences, Seoul National University, Korea
egg10@snu.ac.kr

Abstract In coastal areas, it is recognized that seasonal variation of precipitation rates affects the hydraulic gradient of groundwater and thus submarine groundwater discharge (SGD) rates. In this study, we estimated the total outflux rate through the seepage face and seabed from the coastal aquifer by using the numerical code, FEFLOW. In particular, we focused on the effect of a localized pulse-type precipitation on SGD flux pattern. The calculated SGD flux with time-varying recharge rate shows a quite different pattern from the one with constant recharge rate and each type of the recharge yields a unique pattern of SGD. The results imply the dynamic boundary condition along the land side is significant and that SGD can be miscalculated when incorporating constant precipitation or recharge rate.

Key words submarine groundwater discharge (SGD); heavy rainfall effect; FEFLOW

INTRODUCTION

Submarine groundwater discharge (SGD) is one of the main processes which transfer contaminants and nutrients from inland to the sea; therefore, it has a significant effect on the estuarine environment and ecosystem. It being difficult to observe its existence and effects on the environment, scientific interest toward SGD has not occurred until recently. Even though numerous studies have been performed to identify SGD for several years, many components of it still exist that cannot be explained scientifically. In particular, recent studies of SGD mainly focus on the effect of dynamic oceanic conditions such as tide, waves and current, and show insufficiency in reflecting the dynamic hydrological conditions of inland areas (Taniguchi, 2002; Kim *et al.*, 2003; Li *et al.*, 2005). Michael *et al.* (2005) show the impact of seasonal oscillation of inland recharge rate with a sinusoidal shape to the SGD flux, but such a gradual change of recharge is not enough to represent rapid changes due to storms or heavy rainfall events. So in this study, we focus on the localized effect of a heavy rain event on SGD. Then, conducting the simulation for calculating outflux through the aquifer for several years with actual precipitation data, we investigated seasonal and annual variations of SGD.

METHOD

Conceptual model

A two dimensional vertical section of an unconfined coastal aquifer was constructed (Fig. 1). Then, a prescribed flux boundary condition, which considers the temporal

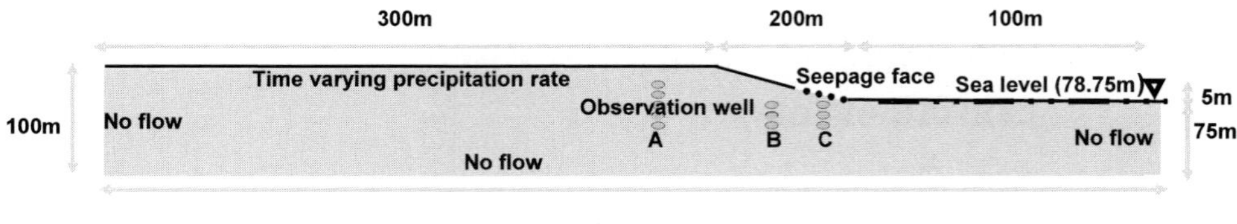

Fig. 1 Schematic representation of domain.

change of precipitation rate, is set on the top boundary and the time-varying SGD flux rate is calculated at each time. Also, three monitoring wells (A, B, C) are installed to investigate the variation of hydraulic head and salinity. Two different kinds of recharge rate are assigned: one is a sinusoidal curved type which emphasizes seasonal climate variation, and the other is a pulse type which focuses on the heavy rain effect in summer. For the heavy rainfall condition model, we pick the two days with the highest precipitation rate in a year. We also keep the average of the recharge rate the same as each other for the comparison. Finally, monthly averaged recharge rates for the last 10 years in Korea are assigned to the model and the SGD flux is calculated to demonstrate the relationship between the intensity of precipitation rate and SGD flux.

Governing equations

FEFLOW (Diersch, 2005) is used for solving coupled equations of variably saturated density-dependent flow and solute transport. Balance equations of fluid mass and momentum are given by:

$$[S_o \cdot s(\psi) + \varepsilon \cdot C(\psi)]\partial h / \partial t + \nabla \cdot \mathbf{q} = Q_h \tag{1}$$

$$\mathbf{q} = -K_r(s)\mathbf{K} \cdot \left[\nabla h + (\rho - \rho_o)/\rho_o \mathbf{e}\right] \tag{2}$$

where S_o is the specific storage coefficient, $C(\psi)$ is the relationship for the saturation dependent moisture capacity, $s(\psi)$is the empirical relationships for the capillary pressure head- saturation, ε is the porosity, h is the hydraulic head, t is the time, $\mathbf{q}$ is the Darcy flux vector, Q_h is the fluid source, $K_r(s)$ is the relative conductivity, $\mathbf{K}$ is the tensor of hydraulic conductivity, ρ is the fluid density, ρ_o is the reference fluid density and $\mathbf{e}$ is the gravitational unit vector.

And, the conservation equation for solute mass is given by:

$$s(\psi)\frac{\partial C}{\partial t} + \mathbf{q} \cdot \nabla C - \nabla \cdot [(\varepsilon \cdot s(\psi) D_d \mathbf{I} + \mathbf{D}) \cdot \nabla C] = 0 \tag{3}$$

where C is the concentration of solute in the fluid phase, D_d is the molecular diffusion coefficient, $\mathbf{I}$ is the identity tensor, $\mathbf{D}$ is the mechanical dispersion tensor.

The following equations are constitutive relations for the nonlinear coupling processes.

$$h = p / \rho_o g + z \tag{4}$$

$$\rho = \rho_o[1 + \alpha(C - C_o)] \tag{5}$$

where p is the fluid pressure, g is the gravitational coefficient, z is the elevation head, α is the fluid density difference ratio, C_o is the reference concentration of solute.

RESULTS

Hydrograph of SGD with different types of recharge rate

Figure 2 illustrates the relationship between recharge rate and SGD rate under a sinusoidal curved recharge rate. About 50 days of time lag between the maximum (and minimum) recharge rate and the maximum (and minimum) SGD rate exists. It is known that seasonal variation of recharge rate mainly controls the height of seasonal water table change and thus, the location of the freshwater–saltwater interface. This seasonal shift of interface makes the amount of recirculated seawater change with time (Michael *et al.*, 2005).

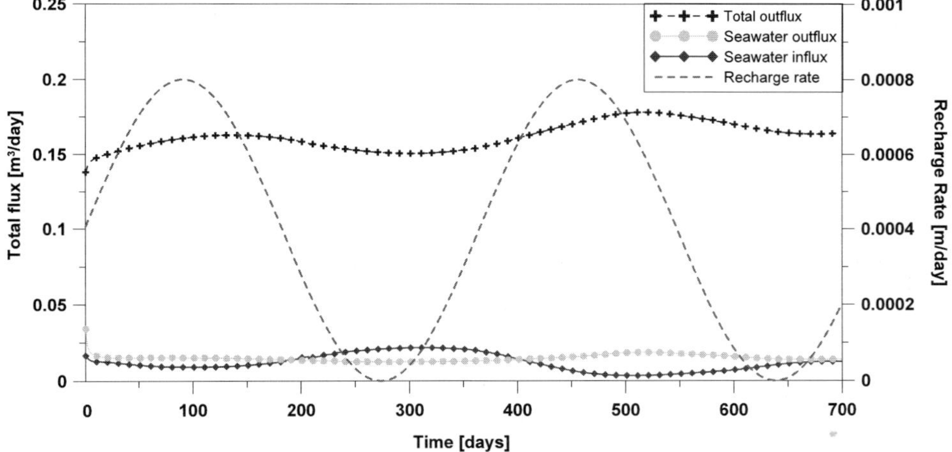

Fig. 2 Hydrograph of SGD with sinusoidal recharge rate.

Figure 3(a) and (b) illustrates the hydrograph of SGD when heavy rainfall events are included. For a pulse-type recharge event, the time lag between maximum precipitation and maximum SGD is about 2–3 days which is much shorter than that for the sinusoidal recharge. The high intensity of rainfall flux near the seepage face may cause this short lag. Modelling results indicate that a sudden increase of precipitation causes an increase of freshwater and seawater outflux to the ocean, while seawater influx through the seafloor decreases after the specific event. The increase of seawater outflux probably occurred not only as the freshwater–seawater interface moved backward to the seaside but also because of dispersive flushing due to the increased freshwater flux.

Figure 4(a) and (b) shows the equivalent hydraulic head and relative concentration of salinity change with time in the monitoring wells. The effect of heavy rainfall on the salinity and hydraulic head can also be observed.

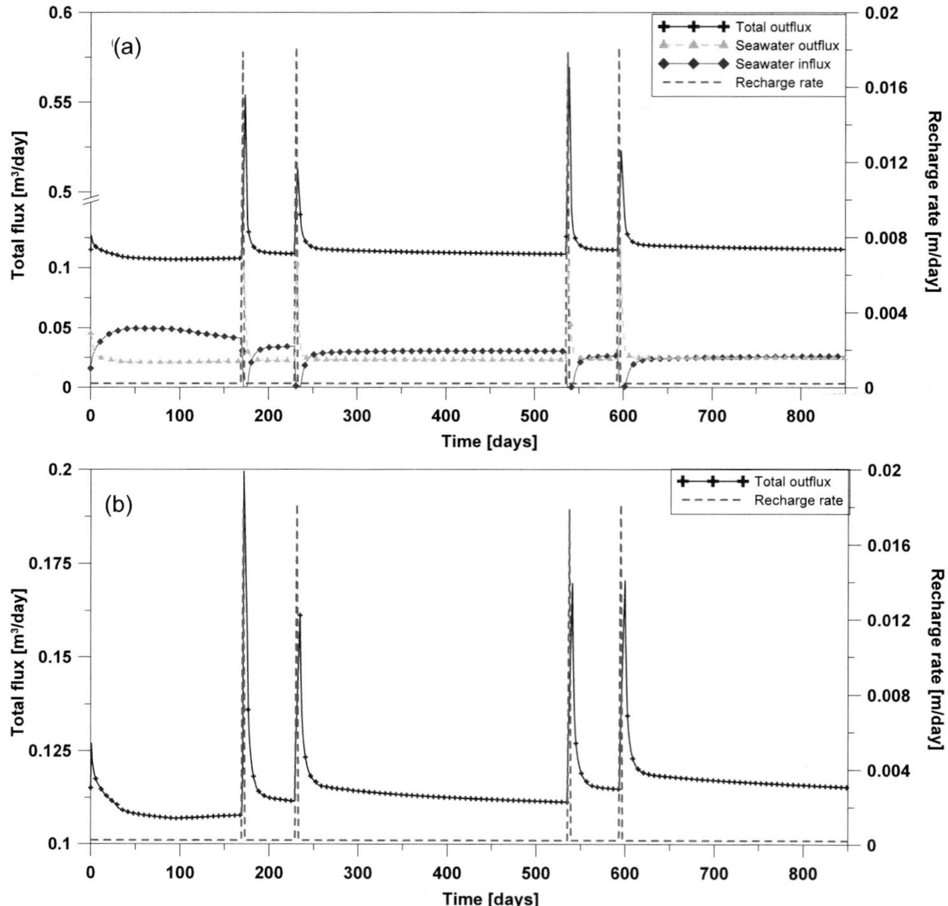

Fig. 3 (a) Hydrograph of SGD with pulse type recharge rate. (b) An enlarged partial section of total outflux through the aquifer, increase of SGD after the heavy rainfall event is observed.

Long-term variation of SGD

Figure 5 shows the long-term relationship between the precipitation rate and the SGD flux. The simulation result indicates time-varying precipitation and its intensity has a formidable effect on SGD variation. The modelling results also indicate that there exist different patterns of SGD flux every year. This annual change of outflux represents the effect of precipitation rate of each year. Also, the volume of outflux is influenced by the recharge rate from the previous years. Figure 5 shows this effect in which the higher recharge rate between days 210 and 240 leads to the upper shift of the SGD flux till about 1000 day. This phenomenon is more evident in the latter part of the graph.

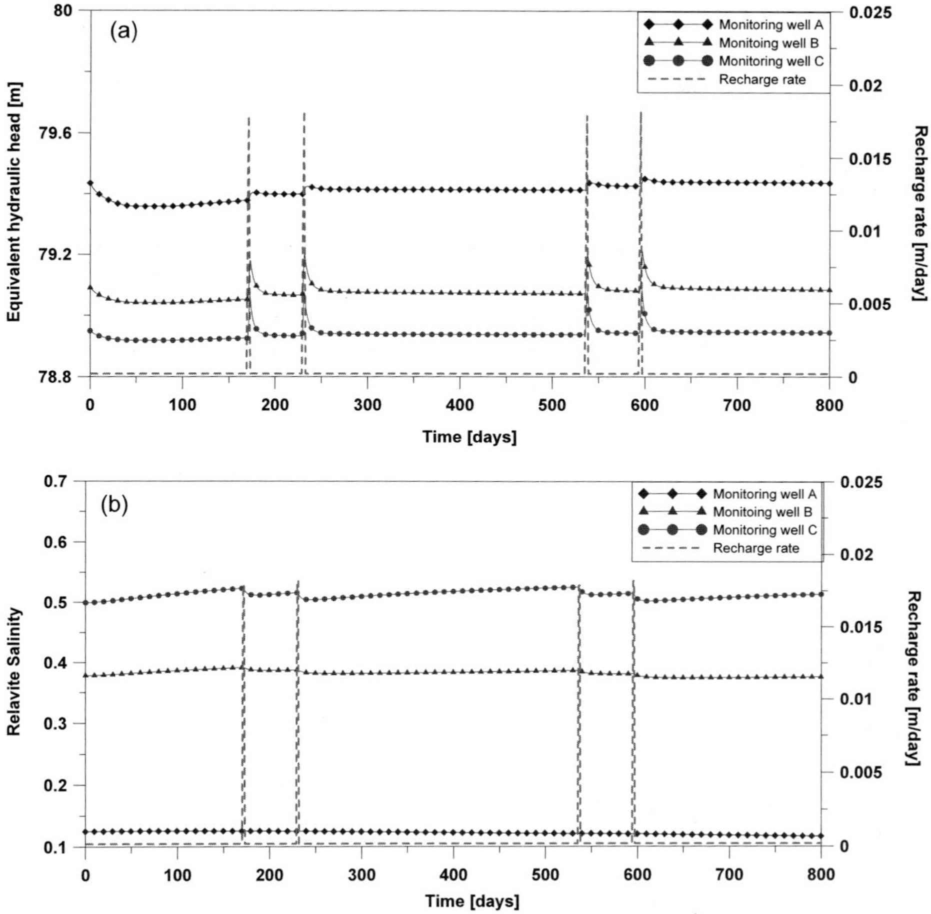

Fig. 4 (a) Equivalent hydraulic head and recharge rate. (b) Relative salinity and rechar ge rate. Monitoring depth from wells A, B and C are 50 m, 24 m, and 18 m below the surface respectively.

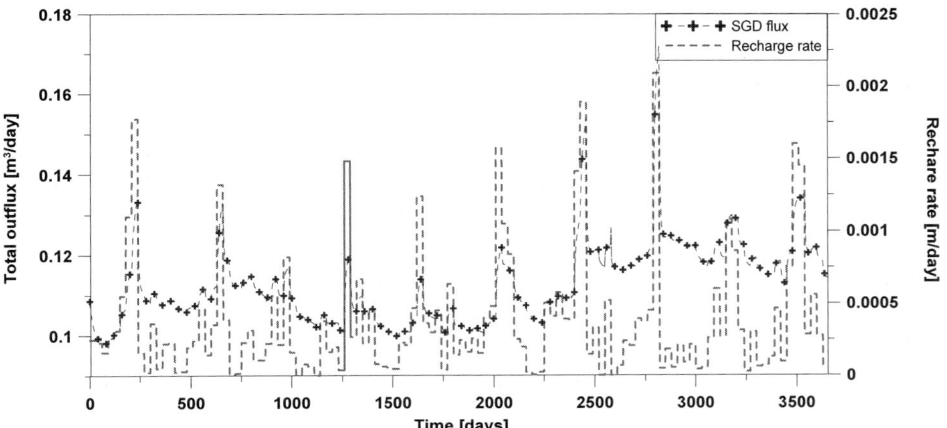

Fig. 5 Hydrograph of SGD with long-term precipitation rate variation.

DISCUSSION AND CONCLUSION

Our simulation results show that different patterns of recharge rate make unique patterns of SGD. When a gradual change of seasonal climate is emphasized, the outflux shows similar patterns to the recharge rate with a lag time. This phenomenon is caused by the oscillation of hydraulic head and interface. But when the heavy rain effect in the summer is emphasized, a periodic change of SGD cannot be observed. Rather, its variation is followed by the high intensity of precipitation for the long period. From this result, we can infer that unusually frequent storms or heavy rainfall events give rise to the increased outflux rate and flux of dissolved nutrients and contaminants to the sea, leading to unexpected eutrophication and contamination of the estuary environment. This research demonstrates that not only the change of oceanic conditions but also the change of precipitation rate should be considered when quantifying SGD flux to the ocean in transient condition.

Acknowledgements This study was supported by the BK21 project of the Korean Government and Advanced Environment Biotechnology Research Center at POSTECH.

REFERENCES

Diersh, H. J. G. (2005) FEFLOW finite element subsurface flow and transport simulation system, reference manual. WASY, Inst. for Water Resour. Plann. and Syst. Res., Berlin, Germany.

Kim, G., Lee, K. K., Park, K. S., Hwang, D. W. & Yang, H. S. (2003) Large submarine groundwater discharge (SGD) from a volcanic island. *Geophys. Res. Lett.* **30**(21), doi:10.1029/2003GL018378.

Li, L., Barry, D. A., Stagnitti, F. & Parlange, J. U. (1999) Submarine groundwater discharge and associated chemical input into a coastal sea. *Water Resour. Res.* **35**(11), 3253–3260.

Michael, H. A., Mulligan, A. E. & Harvey, C. F. (2005) Seasonal oscillations in water exchange between aquifers and the coastal ocean. *Nature* **436**, doi:10.1038/nature03935.

Taniguchi, M. (2002) Tidal effect on submarine groundwater discharge into the ocean. *Geophys. Res. Lett.* **29**(12), doi:10.1029/2002GL014987

Impact of a water diversion project on the groundwater environment of Xiamen Island

LIU ZHENGHUA[1,2], HUANG HAO[2], LIU JIANLI[3], WANG JINKENG[2] & CHEN BIN[2]

1 *School of Life Science, Xiamen University, South Siming Road, Xiamen 361005, China*
lzh_xm@126.com

2 *The Third Institute of Oceanography, SOA, 178A Daxue Road, Xiamen 361005, China*

3 *Institute of Soil Science, CAS, No. 71 East Beijing Road, Nanjing 210008, China*

Abstract The water in the Yuandang Lake of Xiamen Island has been seriously polluted by human activities. In order to alleviate water pollution, two schemes have been proposed to divert seawater, one east-to-west and the other west-to-east, both dividing Xiamen Island and posing great risks to the groundwater. From 2002, water has been diverted from west to east, but it has not significantly improved the lake-water quality; the east to west water diversion theme is under planning. In this study, a groundwater model was set up to further understand the local groundwater–seawater interaction, and to assess the risk of seawater intrusion. The finite-element method was adopted to analyse groundwater flow and seawater transport. We conclude that under annual average precipitation, seawater invasion will become stable one year after the project is completed. In low flow years, the seawater fringe will be close to the lakeside, resulting in seawater recharging the lake water through groundwater flow.

Key words water diversion project; seawater intrusion; finite element method; Xiamen Island

INTRODUCTION

The study of seawater invasion and its effects on the groundwater environment is attracting more and more researchers. Seawater intrusion simulation has advanced from the theory to experiment, and it is developing into the process of numerical simulation. Badon Ghyben and Herzberg brought forward the fresh–salt water interface theory, and they proposed the famous formula of Ghyben-Herzberg for freshwater–saltwater interface calculation in 1889 and 1901, separately. So we are already more than 100 years into the study of seawater intrusion. Custodio comprehensively analysed the seawater invasion problem in his book *Groundwater Problems in Coastal Areas,* and this work established the foundation for the study of the fresh–salt water interface. In the past 10 years, with the enhancement of computing capability, numerical simulation has become the mainstream research approach. Zheng (1994) proposed the of HMOC (Hybrid method of characteristic) model for the calculation of conversion weight by using a three-dimensional finite-difference method. Yuan (2001) recommend a mathematical model for evaluation of the after effects of seawater invasion and its prevention, and he applied his method in the area of Shandong Province in China.

Xiamen Island is located in the southeast of China (Fig. 1). In recent years, the economy of Xiamen Island has developed rapidly, but the ecological environment in this island has deteriorated, and this has caused the degradation of the ecosystem service. The most serious problem is water pollution, such the pollution of the main water body, Yuandang Lake and Songbai Lake. In order to solve this problem, and restore the ecosystem service function of the waterfront area, the possibility of using seawater to dilute the sewages was considered. But this scheme may cause seawater intrusion to the groundwater along the project line. In order to understand the extent of the seawater intrusion, a Finite Element Method (FEM) is used to aid decision-making. First, we analyse the hydrological and geological condition of the study area. Second, through a mathematical analogue, we analyse the movement of the phreatic water and transport of salt, and thus understand the movement of groundwater and solutes of the island. Third, regarding the problem of the water diversion, we analyse its effects on the groundwater environment of the island.

WATER DIVERSION PROJECT FROM EAST TO WEST OF XIAMEN ISLAND

The total length of the water diversion project from the east to west of Xiamen Island is planned to be about 14.7 km, which begins from the tide-receive mouth in the East Sea area, and ends at the point of the west dyke located at the Yuandang Lake (Fig. 1). The conduit consists of open ditches and culverts, and results in a 4-km long seawater canal in the east part of Xiamen Island which directs flow to the geometric centre of the island, and then connects with the Jiangtou Park lake. The flow of diversion water will be 60–90 m^3/s, and a total of 1 200 000–1 340 000 m^3/d after the flood tide twice a day. This can completely improve the water quality of Yuandang Lake and the surrounding environment, and build up a travelling route and environmental protection demonstration site. After this project, the tourism and leisure places and inhabited region along the two canal sides will be constructed, and this will stimulate the economic development of this region.

Fig. 1 The location of Xiamen Island and the project position.

WATER BODIES WITHIN THE STUDY AREA

There are two main water bodies along the line of the water diversion project: Yuandang Lake and Songbai Lake (Fig. 1). Improvement of water quality is one of the main purposes of this project. There are 1.5 km^2 of water in Yuandang Lake, and the drainage area is 37.1 km^2, which accounts for 30% of this area of Xiamen Island. Yundang Lake was originally a harbour in the West Sea area of Xiamen, and it was later enclosed and cultivated. It is now composed of a building area and by a saltwater lake. Yuandang Lake may be divided into three parts: the Trunk canal (0.151 km^2), the inland lake (0.518 km^2), and the outer lake (0.820 km^2). The outer lake links the West Sea area (300 m away) with a tidal culvert. Usually, the water level can be adjusted between 0.00 and 0.70 m a.m.s.l. (above mean sea level) by tidewater coming from the West Sea area. The volume of Yuandang Lake is 3.03×10^6 m^3 when the lake water level is 0.00 m a.m.s.l. Currently, Yuandang Lake is not only a landscape feature at the centre of Xiamen island, but also has an important role in flood prevention of the urban area of Xiamen (about 37.1 km^2).

Songbai Lake lies upstream of Yuandang Lake, and the coverage of water is only 0.11 km^2, with a capacity of 160×10^3 m^3. Though Songbai Lake and Yundang Lake are both fed through the trunk canal, the lake water level of the Songbai is 0.7 m higher than that of Yuandang Lake because of the topography. Thus, the water of Songbai Lake is difficult to exchange with Yuandang Lake. Three pump stations were set up to pump seawater out from Yuandang Lake and divert it to Songbai Lake at a rate of 90×10^3 m^3/d, which thus formed an artificial "little circulation" exchange system between the water bodies.

The source area for the project, Zhongzai Gulf, is currently isolated from the outer seawater by a coastal dyke. After the water diversion project is finished, the sea wall will be removed and Zhongzai Bay will be linked with the outer seawater.

The introduction of seawater may have strong influences on the water quality and quantity of the groundwater and surface water near the project. In order to study the process of seawater invasion and quantify its extent when the water diversion project is completed, we adopted a model that used the Finite Element Method.

GEOLOGICAL AND HYDROLOGICAL CONDITIONS AND THEIR CONCEPTUALIZATION

In order to simulate the influence (especially on the project line) of the seawater diversion project from east to west, we simplified the geological and hydrological conditions, and on the basis of the simplification we have simulated the flow of freshwater and seawater.

The hydrogeological setting was divided into the following three zones based on the topography and water storage properties: diluvian deposits (including the marine deposit area), sedimentary deposits, and bedrock (Fig. 2). With pumping experiments, the unit-drawdown pumping rates of the three zones were determined to be 0.085–5.019, 0.05–0.561 and 0.004–0.161 L s^{-1} m^{-1} respectively. Through the experimental data we know that the diluvian area contains relatively abundant water and the sedimentary deposits and bedrock area contain very little water.

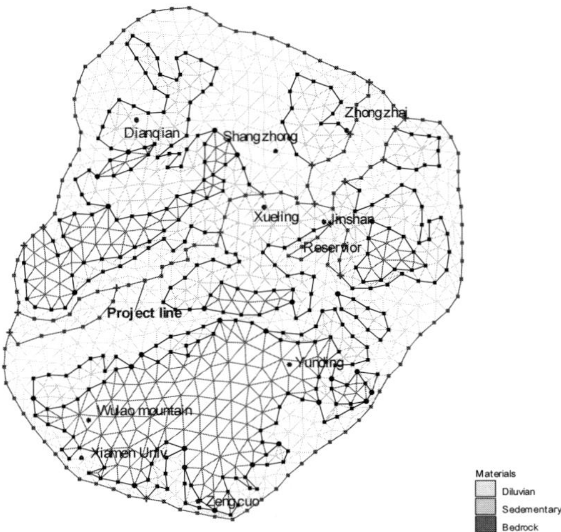

Fig. 2 The geological and hydrological condition conceptualization of Xiamen Island.

The groundwater supply, movement process, and discharge and their relationship are relatively simple in the study area. The precipitation (Table 1) is the major source of supply, although in some locations groundwater gets supply from surface water bodies. Supply and discharge may also occur between different water-contained terraces.

Groundwater flow in Xiamen Island may be described as a three-dimensional unconfined aquifer system. As a simplification, the peripheral sea area is considered to be a constant-head boundary. Groundwater in the coastal regions is in close contact with seawater, and experiences both recharge from and discharge to the sea.

Table 1 Average annual precipitation adopted in the simulation (1952–1990).

Month	1	2	3	4	5	6	7	8	9	10	11	12
Precipitation (mm)	36.3	74.2	91.6	139.5	162.6	191.6	147.0	148.7	115.1	25.9	32.5	23.2
Intensity (mm/d)	4.4	6.5	6.6	10.1	10.0	1.9	14.7	13.3	13.1	6.5	6.9	4.6

NUMERICAL SIMULATION OF SEAWATER INVASION

Based on the mass conservation and continuity equation, we have the three-dimensional saturated groundwater-flow equation:

$$\frac{\rho}{\rho_0} F \frac{\partial h}{\partial t} = \nabla \cdot \left[K \cdot \left(\nabla h + \frac{\rho}{\rho_0} \nabla z \right) \right] + \frac{\rho^*}{\rho_0} q \tag{1}$$

where F is the coefficient of storage, h is hydraulic head, t is time, K is the conductivity tensor, q is the source/sink flow rate, ρ is the water density at a concentration of C, ρ_0 is the density of water , and ρ^* is the density of source water. Similarly, the salt transport equation can be written as:

$$\theta\frac{\partial C}{\partial t}+\rho_b\frac{\partial S}{\partial t}+V\cdot\nabla C-\nabla\cdot(\theta D\cdot\nabla C)=-\left(\alpha'\frac{\partial h}{\partial t}+\lambda\right)(\theta C+\rho_b S)-$$
$$(\theta K_w C+\rho_b K_s S)+m-\frac{\rho^*}{\rho}qC+\left(F\frac{\partial h}{\partial t}+\frac{\rho_0}{\rho}V\cdot\nabla\left(\frac{\rho}{\rho_0}\right)-\frac{\partial\theta}{\partial t}\right)C \tag{2}$$

where θ is the water content, ρ_b is the density of the porous media, C is the solute concentration, S is the density of absorption, V is the permeability velocity, D is the dispersion coefficient tensor, α' is the compression ratio of the porous medium, λ is the constant of attenuation, m is the mass of injected water, q is the flux of infused water, K_w is the coefficient of degradation of the dissolved phase, and K_s is the coefficient of degradation of absorption phase. The boundary conditions can be written as:

$$h=h_1(x,y,z,t)\quad (x,y,z)\in\Gamma_1 \tag{3}$$

$$C=C_1(x,y,z,t)\quad (x,y,z)\in\Gamma_2 \tag{4}$$

where h_1 and C_1 are known functions relating to spatial and temporal factors, and Γ_1 and Γ_2 are prescribed hydraulic head and concentration boundary conditions, respectively. In this study $h_1=4$ m and $C_1=29$ g/L. The initial conditions are:

$$h=h(x,y,z,0)\quad (x,y,z)\in\Omega \tag{5}$$

$$C=C(x,y,z,0)\quad (x,y,z)\in\Omega \tag{6}$$

where Ω is the study area.

THE POSSIBILITY AND EXTENT OF SEAWATER INTRUSION

In order to estimate long-term seawater intrusion to the groundwater within the island, FEMWATER (Lin, 1996) is utilized to simulate the groundwater flow field and the distribution of salinity after the project operation. The results are shown in Figs 3 and 4. Figure 3 is the flow field one year after the project has been completed. It represents the equilibrium of the groundwater supply from the precipitation and the groundwater discharge to the ocean from inside the island.

Analysis of the groundwater velocity field indicated that, because the project line is low and flat in topography, and the permeability of the terrace is weak, there is low possibility that the project line can be supplied by adjacent places. We conclude that there is only a low risk of a large-scale invasion of seawater because of the topography and geological conditions along the line of the project. The same conclusion can be drawn from the distribution of the flow field one year after project completion (Fig. 5).

Figure 5 shows the range of seawater intrusion (salinity >0.1g/L) one year after the project completion. We conclude that the salinity will change noticeably along the project in the range of 150–800 m, with an average of 400 m, and the total area affected area will be 13.67 km^2 (including the Zhongzai Gulf).

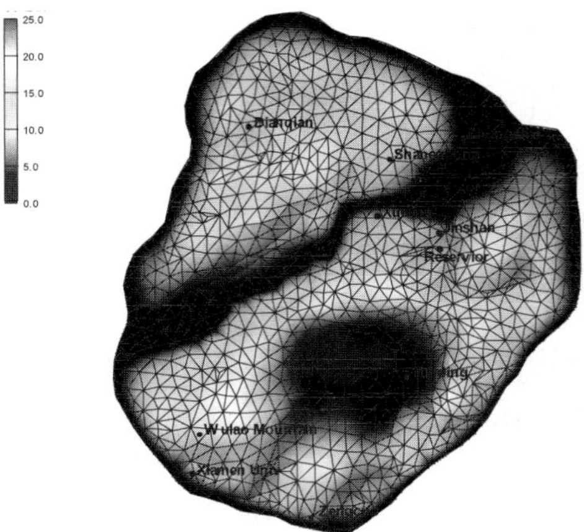

Fig. 3 The groundwater levels after one year of operation (m).

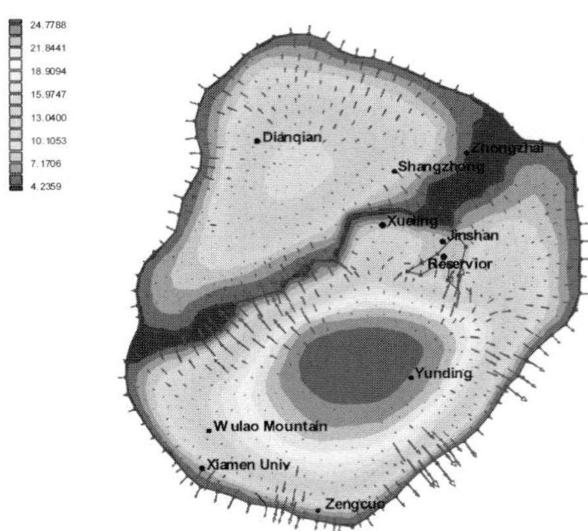

Fig. 4 The groundwater velocity fields after one year of operation.

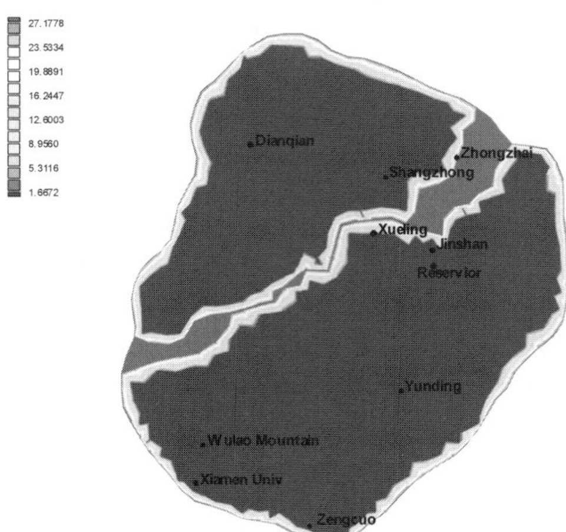

Fig. 5 The spatial extent of seawater intrusion after one year of operation (g/L).

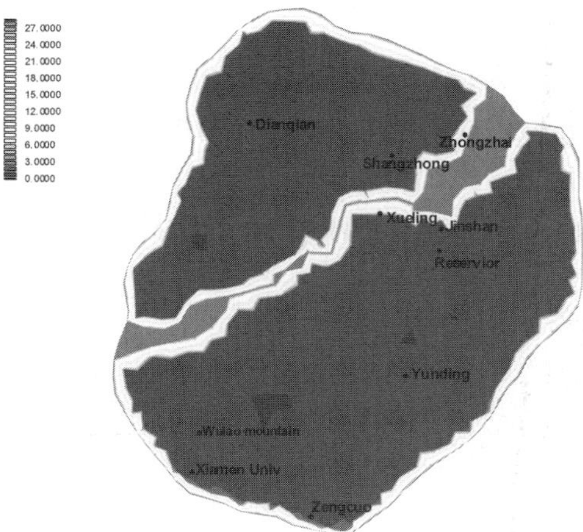

Fig. 6 The spatial extent of sea water intrusion after 10 years of operation (g/L).

The groundwater flow field and salinity distribution up to 10 years after the project operation were also simulated. The result indicates that the range of seawater intrusion has not noticeably expanded (Fig. 6). It may be concluded that the range of the seawater intrusion is almost stable one year after the project is completed. However, in years with inadequate precipitation, when the peripheral groundwater level falls, the distance that the seawater will invade will increase accordingly.

As shown in Fig. 5, the seawater intrusion will reach the Hubian Reservoir (major freshwater supply of the island) after one year of operation. If the precipitation is low and the water level of the reservoir recedes, seawater may enter the reservoir, and cause the salinity of the reservoir to rise.

Acknowledgements The work described in this paper was supported by the Sciences and Technology Programme of FuJian Province (2006F3078) and the National Special Programme, China (no. 908-02-04-08)., and also supported by the Xiamen Road & Bridge Group Co. Ltd.

REFERENCES

Cheng Jianmin (2003) Automatic generation method of triangle mesh with inner specific restriction. *J. Yangtze River Scientific Research Institute* **20**(04), 39–43.

Du Qungui & Liu Shaohong (2004) 2D Finite element adaptive mesh generation based on geometric features - Part I Detailed algorithm description. *J. Computer-Aided Design & Computer Graphics* **7**(3), 85–90.

Lin, H. C., Yeh, G. T., Cheng, J. R., Cheng, H. P. & Jones, N. L. (1996) FEMWATER: A three-dimensional finite element computer model for simulating density-dependent flow and transport in variably saturated media. US Army Engineers Waterways Experiment Station Technical Report.

Moor, Y. H., Stoessell, B. K. & Easley, D. H. (1992) Fresh-water/sea-water relationship within a groundwater flow system, northeastern coast of the Yucatan Peninsula. *Ground Water* **30**(3), 343–350.

Pinder, G. F. & Cooper, H. H. (1970) A numerical technique for calculating the transient position of the saltwater front. *Water Resour. Res.* **6**(4), 875–883.

Zheng, C. (1990) MT3D: A modular three-dimensional transport model for simulation of advection, dispersion and chemical reactions of contaminants in groundwater systems. Report to the US Environmental Protection Agency. Robert S. Kerr Environmental Research Laboratory, Ada, Oklahoma, USA.

Three-dimensional numerical simulation of density-dependent groundwater flow and salt transport due to groundwater pumping in a heterogeneous and true anisotropic coastal aquifer system

JU-HYUN PARK, CHAN-SUNG OH & JUN-MO KIM

School of Earth and Environmental Sciences, Seoul National University, Seoul 151-742, Korea
junmokim@snu.ac.kr

Abstract A series of three-dimensional numerical simulations using a multidimensional hydrodynamic dispersion numerical model is performed to analyse seawater intrusion under groundwater pumping in an unsaturated fractured porous coastal aquifer system, which is heterogeneous and true anisotropic. The numerical simulation results show that such heterogeneity and true anisotropy have significant effects on spatial and temporal distributions of density-dependent groundwater flow and salt transport. Therefore, it may be concluded that both heterogeneity and true anisotropy must be properly considered when more rigorous and reasonable predictions of long-term density-dependent groundwater flow and salt transport induced by groundwater pumping are to be obtained for the optimal management of coastal groundwater resources.

Key words fractured porous coastal aquifer system; heterogeneity; true anisotropy; groundwater pumping; seawater intrusion; hydrodynamic dispersion; numerical simulation

INTRODUCTION

Groundwater has been pumped indiscreetly and excessively from coastal aquifers for various human activities as the population has grown throughout the world, especially in Asia. As a result, extensive coastal groundwater depletion and salinization have occurred under intensive seawater intrusion, and thus the sustainability of coastal groundwater resources has become a significant issue in coastal areas and even in the adjacent inland areas.

Various hydrodynamic dispersion numerical models have been developed on the basis of the fully coupled governing equations for groundwater flow and solute transport and then used as useful tools to simulate seawater intrusion and related phenomena for the management of coastal groundwater resources (see Bear *et al.*, 1999; Cheng & Ouazar, 2004). However, numerical simulations of seawater intrusion in heterogeneous and true anisotropic fractured porous coastal aquifer systems, which contain complex spatial distributions of geological media (rock masses), and each of the geological media has joints or joint sets of various orientations, have not yet been performed seriously.

The objectives of this study are to simulate density-dependent groundwater flow and salt transport due to groundwater pumping in an unsaturated heterogeneous and true anisotropic coastal aquifer system using a multidimensional hydrodynamic dispersion numerical model and to evaluate the effects of heterogeneity and true anisotropy on such coupled groundwater flow and salt transport phenomena. From a practical point of view, such a quantitative understanding of the effects of heterogeneity and true anisotropy may suggest improved guidelines for planning, designing, assessing, and modifying groundwater pumping schemes for the optimal management of coastal groundwater resources.

NUMERICAL MODEL

The hydrodynamic dispersion numerical model used in this study is COFAT3D (Kim & Yeh, 2004), which has been developed from 3DFEMFAT (Yeh *et al.*, 1994). This numerical model is a general multidimensional hybrid Lagrangian-Eulerian finite element model and can simulate density-dependent groundwater flow and solute transport in saturated–unsaturated porous, fractured, and fractured porous geological media, which are heterogeneous and true anisotropic. This numerical model uses the adaptive finite difference time-stepping scheme for transient problems, the incremental Picard method for coupled nonlinear problems, and one of four conventional and preconditioned conjugate gradient (PCG) iterative methods for matrix solutions. In this study, the linearized groundwater flow and solute transport matrix equations are solved sequentially by the incomplete Cholesky LU decomposed preconditioned conjugate gradient (ICPCG) iterative method, while the convergence criteria for pressure head and seawater-normalized solute concentration are set equal to 10^{-3} m and 10^{-3} for nonlinear iterations and 10^{-4} m and 10^{-4} for linear iterations.

NUMERICAL SIMULATIONS

Coastal aquifer system

The coastal aquifer system considered in this study is located on the western coast of Korea and is composed of Quaternary alluvial layers underlain by Precambrian gneiss and Cretaceous quartz monzonite, rhyolitic tuff, Kyokpori Formation (conglomerate, sandstone, and shale), and rhyolite with a major fault as shown in Fig. 1 (Korea Institute of Geoscience and Mineral Resources, 1997; Kim, 2000). Such geological rock masses (media) and faults do not only have irregular thicknesses and boundaries but also contain numerous joints and bedding planes of various orientations. Thus the coastal aquifer system is hydrogeologically heterogeneous and true anisotropic.

Groundwater has been pumped indiscreetly from the overlying alluvial layers and even from the underlying bedrocks for various agricultural and municipal activities since the late 1990s. As a result, seawater intrusion has intensified, and it has caused extensive groundwater depletion and salinization. In order to predict and analyse the seawater intrusion under groundwater pumping in such a heterogeneous and true anisotropic coastal aquifer system, a series of three-dimensional numerical simulations of density-dependent groundwater flow and salt transport has often been requested.

The material properties of the geological media and fault, which compose the coastal aquifer system, are obtained from a variety of geological surveys and hydrogeological tests and are summarized in Table 1, and the characteristics of the joint

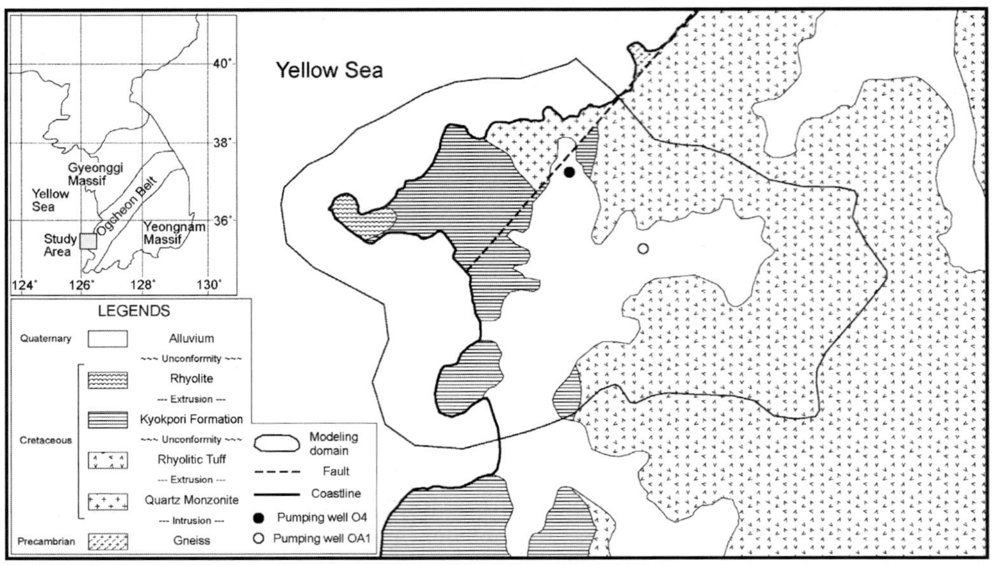

Fig. 1 Location and geological maps of the study area. The horizontal extent of the coastal aquifer system modelled in this study is outlined by a thin solid line. The pumping wells O4 and OA1 are marked by the full and open circles, respectively.

Table 1 Material properties of the geological media and fault.

Property	Quartz monzonite	Rhyolitic tuff	Kyokpori Formation (Unit 1)	Kyokpori Formation (Unit 2)	Rhyolite	Weathered zone	Alluvium and marine sediment	Fault
Porosity	1.60×10^{-2}	8.50×10^{-2}	2.60×10^{-2}	2.60×10^{-2}	1.00×10^{-1}	1.90×10^{-1}	4.30×10^{-1}	1.90×10^{-1}
Sat. hydraulic conductivity (m/s)	2.50×10^{-12}	2.16×10^{-10}	1.10×10^{-6}	1.01×10^{-6}	4.00×10^{-7}	2.10×10^{-5}	2.88×10^{-4}	6.01×10^{-7}
Dry bulk density (kg/m^3)	2.61×10^{3}	2.40×10^{3}	2.58×10^{3}	2.58×10^{3}	2.31×10^{3}	2.15×10^{3}	1.52×10^{3}	2.15×10^{3}
Longitudinal dispersivity (m)	15.80	15.80	15.80	15.80	15.80	31.60	31.60	31.60
Transverse dispersivity (m)	1.58	1.58	1.58	1.58	1.58	3.16	3.16	3.16
Residual water saturation	8.64×10^{-3}	4.59×10^{-2}	8.60×10^{-3}	8.60×10^{-3}	6.93×10^{-2}	7.35×10^{-2}	1.05×10^{-1}	8.94×10^{-2}
Unsaturated hydraulic parameters of van Genuchten's (1980) model								
α_v (1/m)	0.50	0.50	2.00	2.00	2.70	7.50	14.50	1.60
n_v	1.09	1.09	1.41	1.41	1.23	1.89	2.68	1.37

sets in the geological media and fault are summarized in Table 2 (Kim, 2006). The tortuosity is set equal to 0.41 assuming that the solid particles which compose the geologic media are spherical in shape. The compressibility of water is set equal to 4.40×10^{-10} m^2/N, the density of fresh water is set equal to 1000 kg/m^3, and the dynamic viscosity of water is set equal to 1.00×10^{-3} kg m s^{-1}. The density of seawater is set equal to 1025 kg/m^3; that is, the density difference ratio is set equal to 0.025. The molecular diffusion coefficient of salt in water is set equal to 1.68×10^{-9} m^2/s (Freeze & Cherry, 1979; Domenico & Schwartz, 1990; Fetter, 1994).

Table 2 Characteristics of the joint sets in the geological media and fault.

Geologic media	Joint set	Strike	Dip	Aperture (m)	Spacing (m)
Quartz monzonite	Set 1	N76°W	4°W	3.00×10^{-5}	3.75×10^{-1}
	Set 2	N52°W	64°W	6.00×10^{-5}	2.64×10^{-1}
	Set 3	N88°W	88°E	6.00×10^{-5}	2.18×10^{-1}
Rhyolitic tuff	Set 1	N9°E	10°W	3.00×10^{-5}	2.03×10^{-1}
	Set 2	N25°W	80°E	6.00×10^{-5}	2.32×10^{-1}
	Set 3	N72°E	82°E	6.00×10^{-5}	1.92×10^{-1}
Kyokpori Formation (Unit 1)	Matrix	N47°E	14°W		
	Set 1 (bedding plane)	N47°E	14°W	3.00×10^{-5}	2.70×10^{-1}
	Set 2	N10°E	79°W	6.00×10^{-5}	3.38×10^{-1}
	Set 3	N84°W	85°W	6.00×10^{-5}	2.71×10^{-1}
Kyokpori Formation (Unit 2)	Matrix	N35°E	7°W		
	Set 1 (bedding plane)	N35°E	7°W	3.00×10^{-5}	2.20×10^{-1}
	Set 2	N49°E	90°	6.00×10^{-5}	3.10×10^{-1}
	Set 3	N58°W	85°W	6.00×10^{-5}	6.00×10^{-1}
	Set 4	N16°E	72°W	6.00×10^{-5}	5.32×10^{-1}
Rhyolite	Set 1	N84°E	82°W	5.00×10^{-5}	1.25×10^{-1}
	Set 2	N10°E	90°	5.00×10^{-5}	1.20×10^{-1}
Fault	Matrix	N49°E	90°		
	Set 1	N49°E	90°	1.00×10^{-4}	1.25×10^{0}
	Set 2	N49°E	90°	2.00×10^{-4}	2.00×10^{-1}
	Set 3	N49°E	90°	2.00×10^{-4}	6.67×10^{-2}

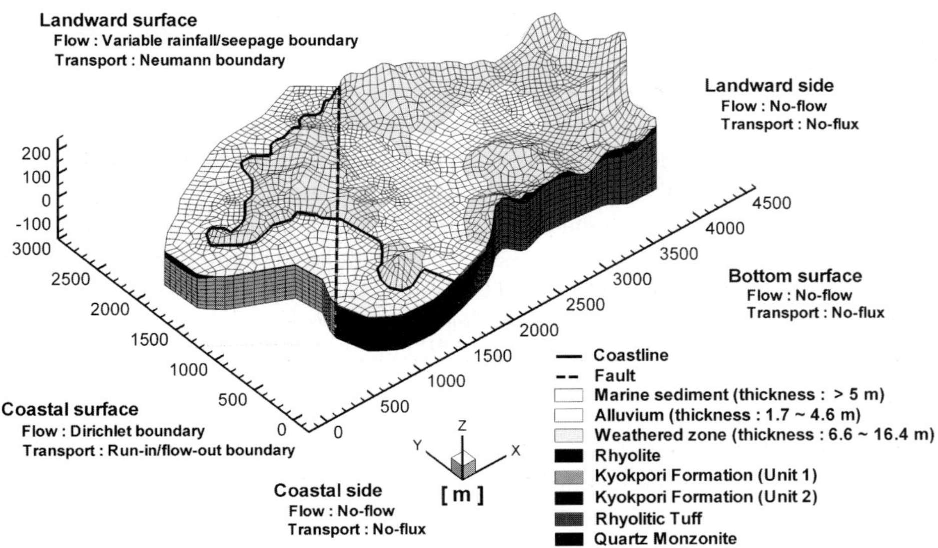

Fig. 2 Finite element mesh and boundary conditions of the coastal aquifer system used in the numerical simulations. The x, y and z axes are aligned toward the east, north, and zenith, respectively.

The horizontal landward and coastal side boundaries of the coastal aquifer system modelled in this study are determined considering onshore mountain ridges and offshore topographies beyond the coastline, respectively, and are outlined by a thin solid line as shown in Fig. 1. The bottom boundary of the coastal aquifer system is horizontally flat and is located at a depth of 150 m b.s.l. The coastal aquifer system with these horizontal and bottom boundaries is discretized into 53 314 irregular hexahedral elements with 58 240 nodes as shown in Fig. 2. The boundary conditions for the coastal aquifer system are also summarized in Fig. 2. The net annual average rainfall rate is set equal to 20% of the annual average rainfall rate of 1209 mm/year considering evaporation and transpiration (Korea Meteorological Administration, 1971–2000).

RESULTS AND ANALYSES

In order to obtain initial steady-state spatial distributions of density-dependent groundwater flow and salt transport in the coastal aquifer system before groundwater pumping, a steady-state numerical simulation is performed first, and its results are illustrated in Fig. 3. Figure 3 clearly shows that the above-mentioned heterogeneity and true anisotropy, especially the major fault, as well as the topography and unsaturated zone, have significant effects on the spatial distributions of hydraulic head, seawater-normalized salt concentration, groundwater flow flux, and seawater-normalized salt transport flux before groundwater pumping. As shown in Fig. 3, the major fault acts as a highly permeable conduit along it but behaves as a less permeable barrier across it. On the other hand, as shown in Fig. 3(d), the measured and simulated transition (or mixing) zones (in other words, interfaces or salt water fronts) between fresh water and salt water before groundwater pumping are reasonably well matched recognizing that there is no calibration procedure involved in this steady-state numerical simulation. Therefore, the results from this steady-state numerical simulation are expected to represent the field behaviour closely and adequately. Here, the measured transition zones are inferred from an electrical resistivity (ER) sounding (Line 1) and an electrical conductivity (EC) logging (Line 2).

In order to obtain spatial and temporal distributions of density-dependent groundwater flow and salt transport in the coastal aquifer system during groundwater pumping at two different pumping wells O4 and OA1 (see Fig. 1) and two different pumping rates (100 and 400 m^3/day), four different transient-state numerical simulations are then performed using their initial steady-state spatial distributions (i.e. Fig. 3) as initial conditions, and their results are illustrated in Figs 4–9 and are compared with Fig. 3(a)–(f), respectively. Figures 4–9 clearly show that the above-mentioned heterogeneity and true anisotropy, especially the major fault, as well as the topography and unsaturated zone, have significant effects on the spatial distributions of hydraulic head, seawater-normalized salt concentration, groundwater flow flux, and seawater-normalized salt transport flux during groundwater pumping. Such effects of heterogeneity and true anisotropy become more prominent when groundwater is pumped from Well O4 since it is closer to the major fault than Well OA1. It is also certain that groundwater depletion and salinization become intensive and extensive as the pumping rate increases at either pumping well. As shown in Fig. 10, the hydraulic

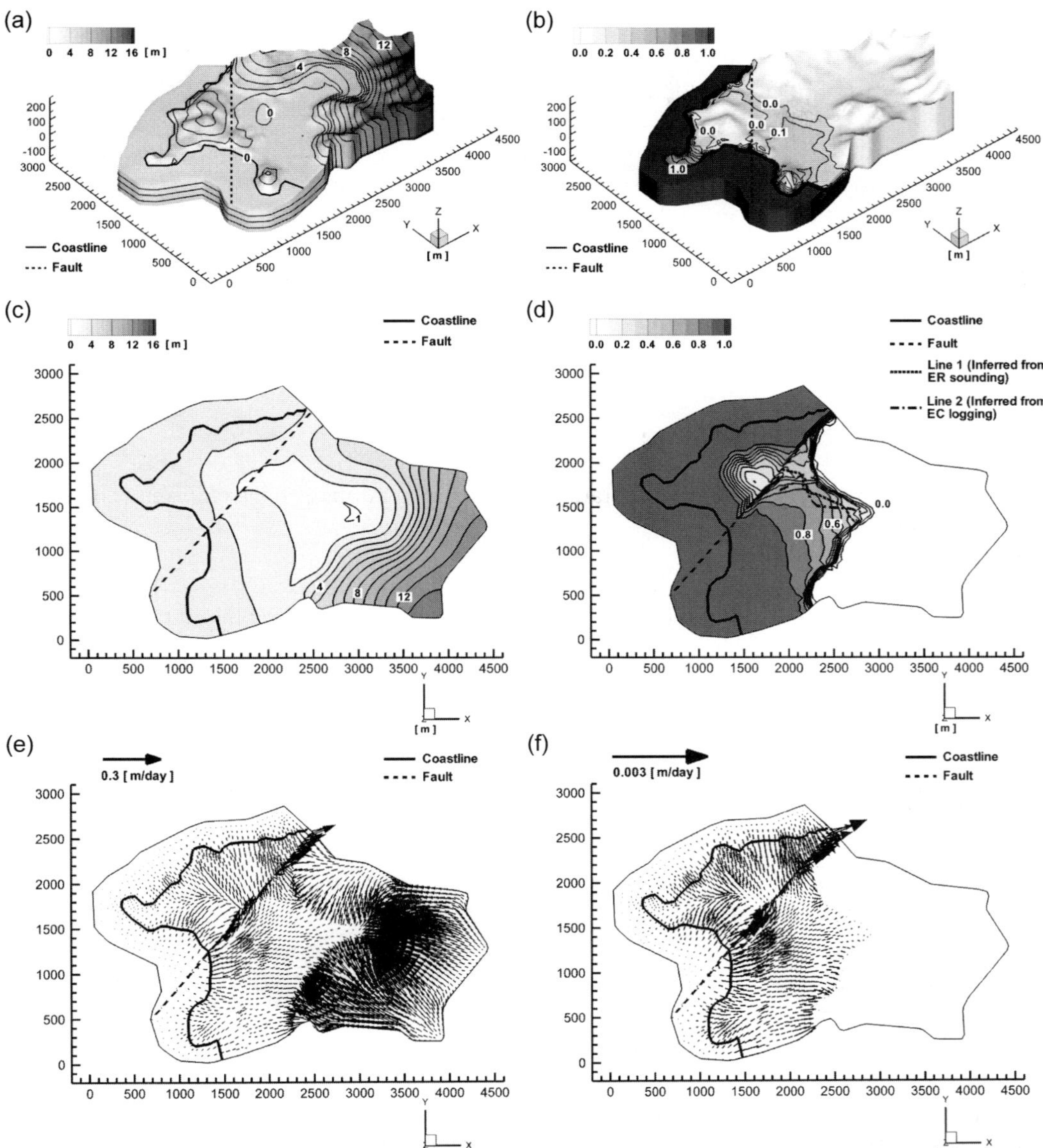

Fig. 3 Initial steady-state spatial distributions of: (a) hydraulic head (m), (b) normalized salt concentration, (c) hydraulic head (m) at the bottom boundary, (d) normalized salt concentration at the bottom boundary, (e) groundwater flow flux (m/day) at the bottom boundary, and (f) normalized salt transport flux (m/day) at the bottom boundary before groundwater pumping in the coastal aquifer system. The contour intervals of hydraulic head and normalized salt concentration are 1 m and 0.1, respectively.

head reaches its final steady-state condition much faster (in about 1 day) than the seawater-normalized salt concentration (after more than 50 years) at either pumping well. This results from salt transport under hydrodynamic dispersion as well as advection. In addition, the seawater-normalized salt concentration at Well O4 still

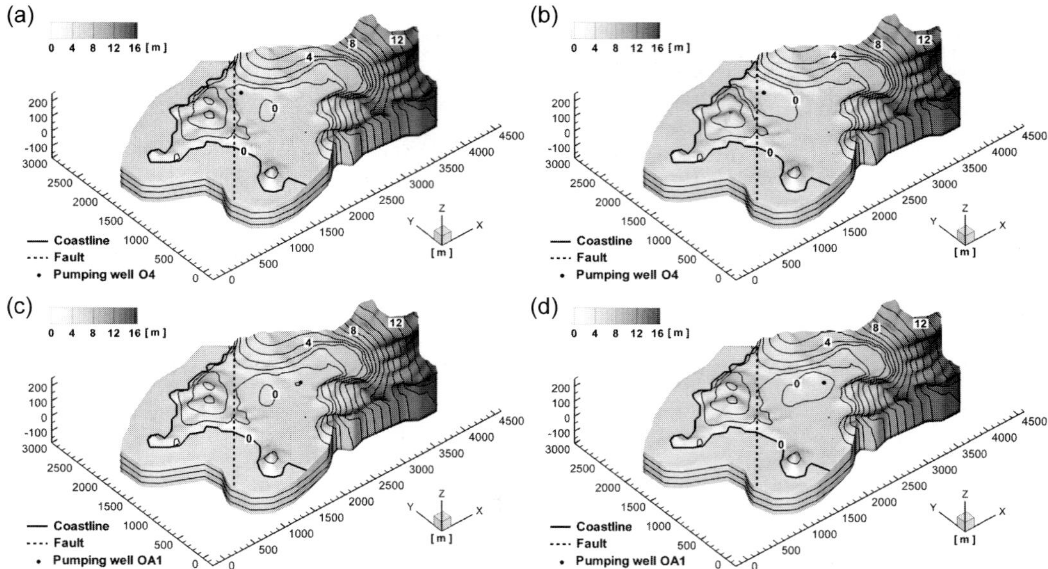

Fig. 4 Final steady-state spatial distributions of hydraulic head (m) during groundwater pumping at rates of: (a) 100 m^3/day at Well O4, (b) 400 m^3/day at Well O4, (c) 100 m^3/day at Well OA1, and (d) 400 m^3/day at Well OA1 in the coastal aquifer system. The contour interval of hydraulic head is 1 m.

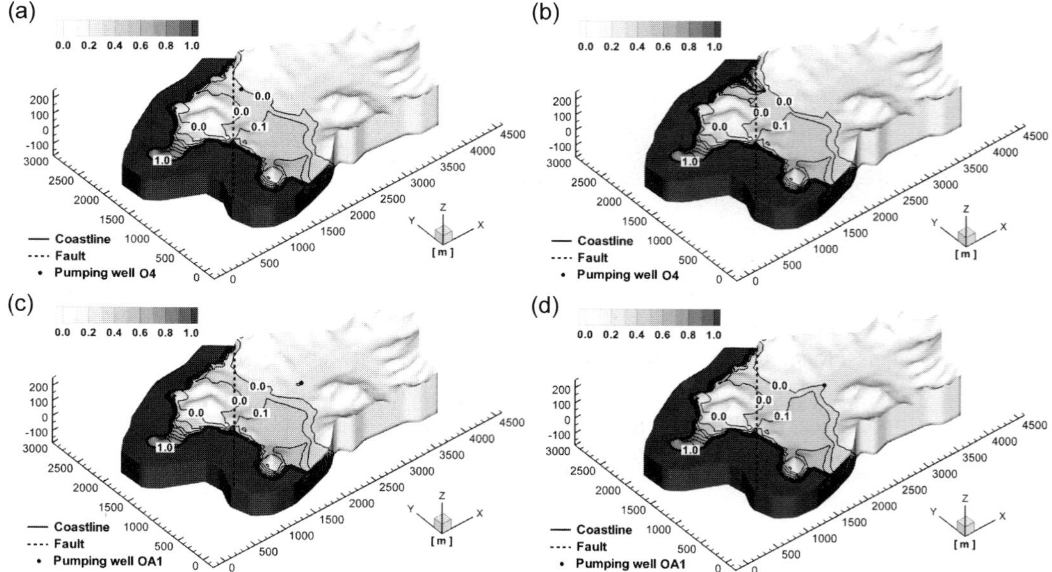

Fig. 5 Final steady-state spatial distributions of normalized salt concentration during groundwater pumping at rates of: (a) 100 m^3/day at Well O4, (b) 400 m^3/day at Well O4, (c) 100 m^3/day at Well OA1, and (d) 400 m^3/day at Well OA1 in the coastal aquifer system. The contour interval of normalized salt concentration is 0.1.

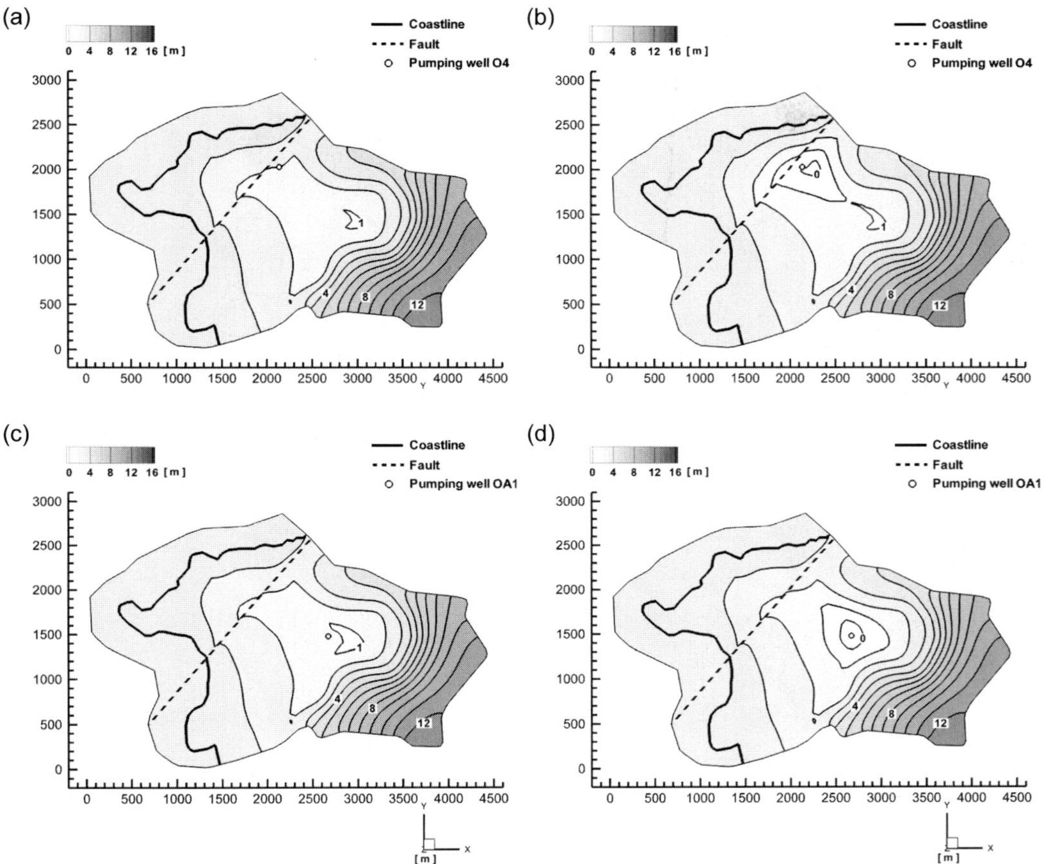

Fig. 6 Final steady-state spatial distributions of hydraulic head (m) at the bottom boundary during groundwater pumping at rates of: (a) 100 m^3/day at Well O4, (b) 400 m^3/day at Well O4, (c) 100 m^3/day at Well OA1, and (d) 400 m^3/day at Well OA1 in the coastal aquifer system. The contour interval of hydraulic head is 1 m.

increases even after that at Well OA1 reaches its final steady-state condition. It strongly suggests that the major fault influences and controls more highly density-dependent groundwater flow and salt transport when groundwater is pumped from Well O4 as it is closer to the major fault than Well OA1. As shown in Figs 4–9, the major fault acts as a highly permeable conduit along it but behaves as a less permeable barrier across it. As a result, it retards seawater intrusion toward Well O4.

CONCLUSIONS

Using a multidimensional hydrodynamic dispersion numerical model, density-dependent groundwater and salt transport before and during groundwater pumping in an unsaturated fractured porous coastal aquifer system, which is heterogeneous and true anisotropic, were simulated, validated, predicted, and analysed. The coastal aquifer system is composed of Quaternary alluvial layers underlain by Precambrian

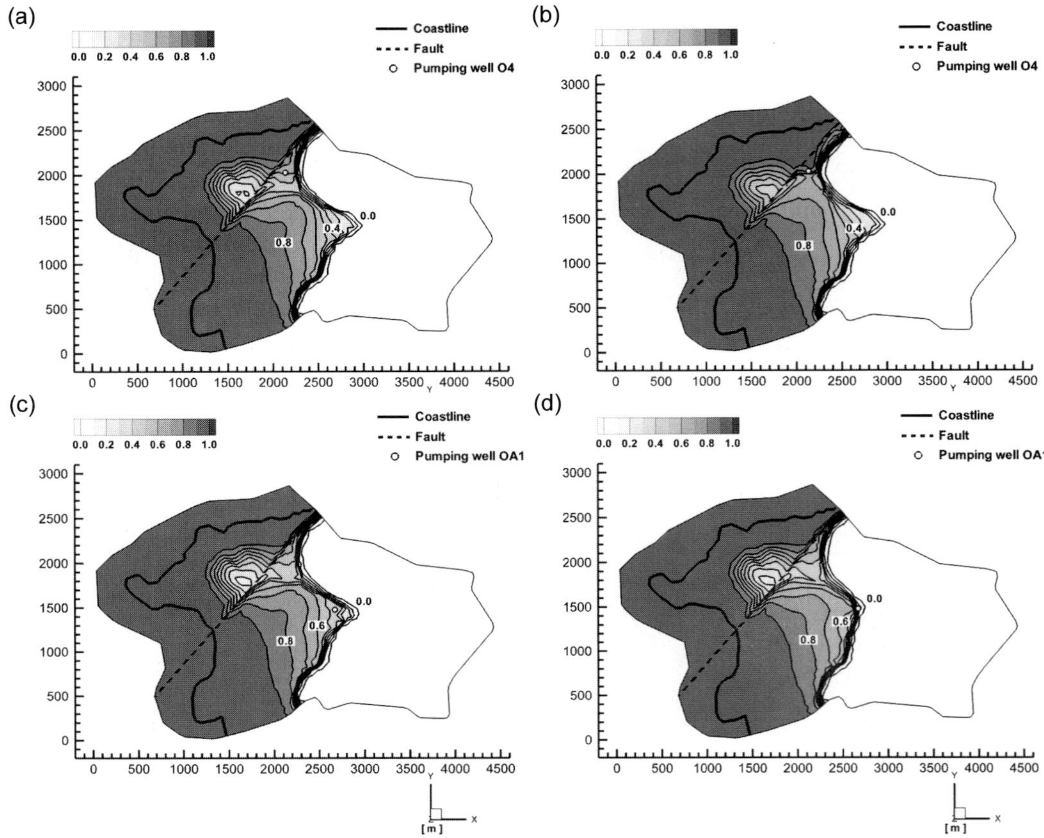

Fig. 7 Final steady-state spatial distributions of normalized salt concentration at the bottom boundary during groundwater pumping at rates of: (a) 100 m^3/day at Well O4, (b) 400 m^3/day at Well O4, (c) 100 m^3/day at Well OA1, and (d) 400 m^3/day at Well OA1 in the coastal aquifer system. The contour interval of normalized salt concentration is 0.1.

gneiss and Cretaceous quartz monzonite, rhyolitic tuff, Kyokpori Formation (conglomerate, sandstone, and shale), and rhyolite with a major fault. A steady-state numerical simulation was performed first to obtain initial steady-state spatial distributions of density-dependent groundwater flow and salt transport before groundwater pumping, and its results were illustrated and validated reasonably with respect to the measured transition zones between fresh water and salt water. Four different transient-state numerical simulations were then performed to obtain spatial and temporal distributions of density-dependent groundwater flow and salt transport during groundwater pumping at two different pumping wells and two different pumping rates, and their results were illustrated and analysed. These steady and transient-state numerical simulation results show that such heterogeneity and true anisotropy have significant effects on spatial and temporal distributions of density-dependent groundwater flow and salt transport. Therefore, it may be concluded that both heterogeneity and true anisotropy cannot always be ignored if they are observed

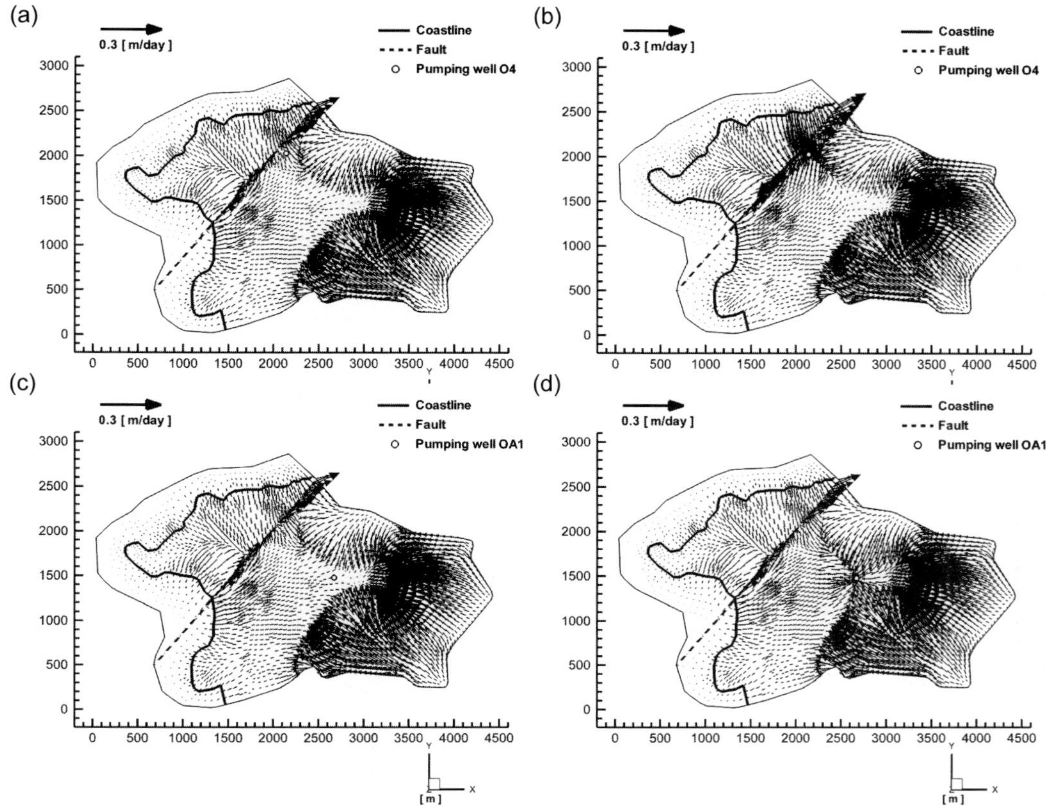

Fig. 8 Final steady-state spatial distributions of groundwater flow flux (m/day) at the bottom boundary during groundwater pumping at rates of: (a) 100 m^3/day at Well O4, (b) 400 m^3/day at Well O4, (c) 100 m^3/day at Well OA1, and (d) 400 m^3/day at Well OA1 in the coastal aquifer system.

in actual aquifer system, and thus they must be properly considered when more rigorous and reasonable predictions of long-term density-dependent groundwater flow and salt transport induced by groundwater pumping are to be obtained for the optimal management of coastal groundwater resources. Further numerical studies of various geologic and hydrogeological settings and field applications are recommended to arrive at more general conclusions concerning the effects of heterogeneity and true anisotropy on three-dimensional density-dependent groundwater flow and salt transport due to groundwater pumping in unsaturated fractured porous aquifer systems.

Acknowledgements This work was supported by the Sustainable Water Resources Research Center of the 21st Century Frontier Research and Development Program, Ministry of Science and Technology, Korea. This work was also supported in part by the Brain Korea 21 Project, Ministry of Education and Human Resources Development, Korea.

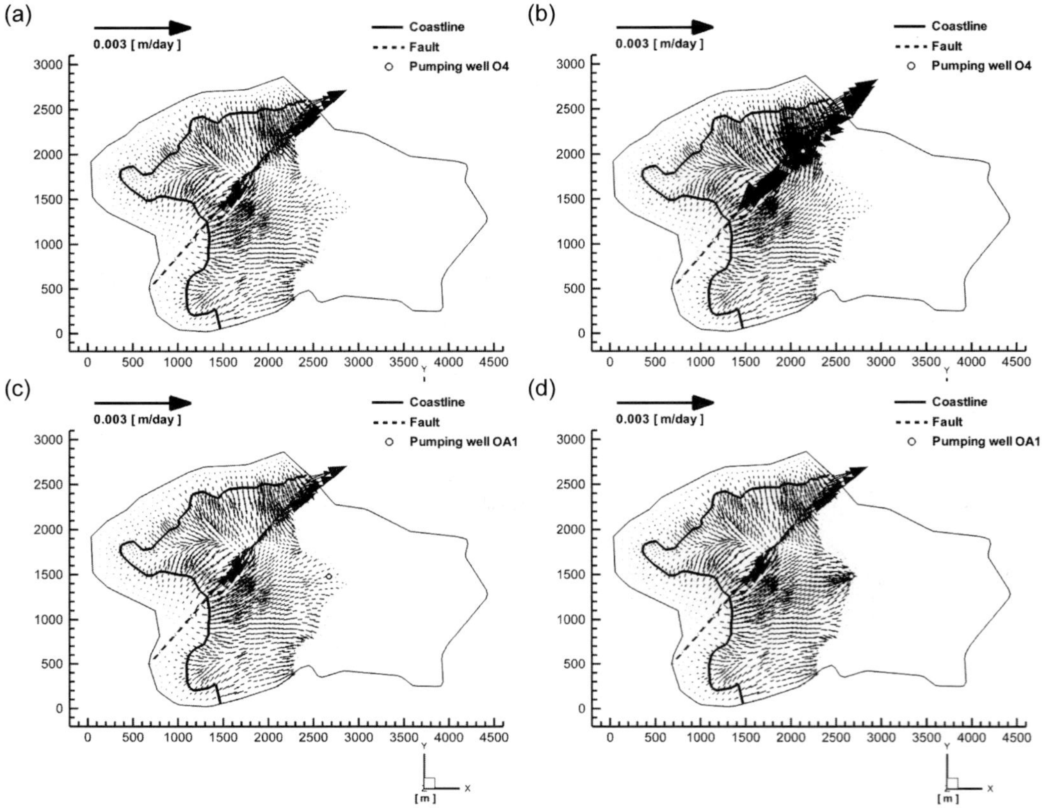

Fig. 9 Final steady-state spatial distributions of normalized salt transport flux (m/day) at the bottom boundary during groundwater pumping at rates of: (a) 100 m^3/day at Well O4, (b) 400 m^3/day at Well O4, (c) 100 m^3/day at Well OA1, and (d) 400 m^3/day at Well OA1 in the coastal aquifer system.

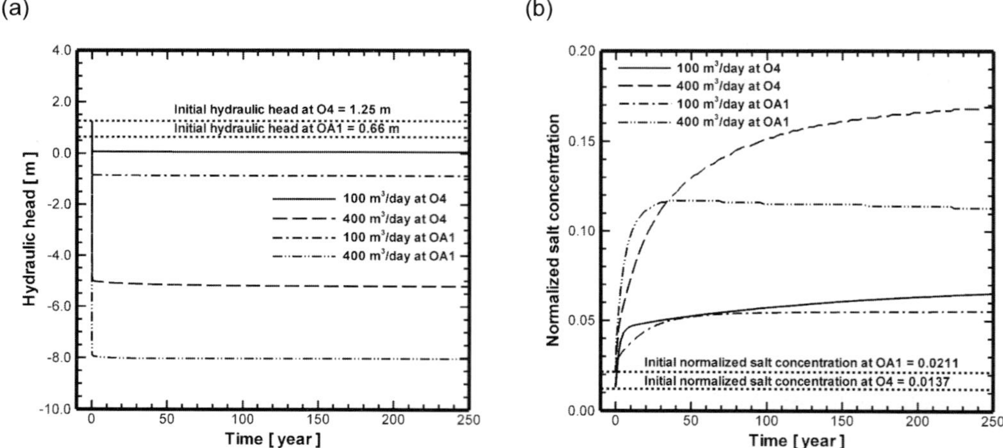

Fig. 10 Temporal changes of: (a) hydraulic head (m) and (b) normalized salt concentration at Wells O4 and OA1 during groundwater pumping at rates of 100 and 400 m^3/day in the coastal aquifer system.

REFERENCES

Bear, J., Cheng, A. H. D., Sorek, S., Ouazar, D. & Herrera, I. (1999) *Seawater Intrusion in Coastal Aquifers: Concepts, Methods, and Practices*. Kluwer Academic Publishers, Norwell, Massachusetts, USA.

Cheng, A. H. D. & Ouazar, D. (2004) *Coastal Aquifer Management: Monitoring, Modeling, and Case Studies*. Lewis Publishers, Boca Raton, Florida, USA.

Domenico, P. A. & Schwartz, F. W. (1990) *Physical and Chemical Hydrogeology*. John Wiley and Sons, New York, USA.

Fetter, C. W. (1994) *Applied Hydrogeology*, third edn. Prentice Hall, Upper Saddle River, New Jersey, USA.

Freeze, R. A. & Cherry, J. A. (1979) *Groundwater*. Prentice Hall, Englewood Cliffs, New Jersey, USA.

Kim, J. M. (2006) Development and application of a hydrodynamic dispersion numerical model for coastal groundwater management. Final Report, Geological and Groundwater Engineering Laboratory, Seoul National University, Seoul, Korea.

Kim, S. B. (2000) Sedimentary processes and environments of the Kyokpori Formation (Cretaceous), SW Korea. PhD Thesis, Seoul National University, Seoul, Korea.

Kim, J. M. & Yeh, G. T. (2004) COFAT3D: a finite element model for fully coupled groundwater flow and solute transport in three-dimensional saturated-unsaturated porous and fractured media, version 1.0. Technical Report, Geological and Groundwater Engineering Laboratory, Seoul National University, Seoul, Korea.

Korea Institute of Geoscience and Mineral Resources (1997) *Geological Map of Kwangju Area (1:250000)*. Korea Institute of Geoscience and Mineral Resources, Daejeon, Korea.

Korea Meteorological Administration (1971–2000) Climatological data of Kunsan, Korea. Annual Report. Korea Meteorological Administration, Seoul, Korea (published annually and available from http://www.kma.go.kr).

van Genuchten, M. Th. (1980) A closed-form equation for predicting the hydraulic conductivity of unsaturated soils. *Soil Sci. Soc. Am. J.* **44**(5), 892–898.

Yeh, G. T., Cheng, J. R. & Cheng, H. P. (1994) 3DFEMFAT: Users' manual of a 3-dimensional finite element model of density-dependent flow and transport through saturated-unsaturated media, version 2.0. Technical Report, Department of Civil and Environmental Engineering, Pennsylvania State University, University Park, Pennsylvania, USA.

Numerical modelling of saltwater–freshwater interaction in the Walawe River basin, Sri Lanka

PRIYANTHA RANJAN[1], SO KAZAMA[1] & MASAKI SAWAMOTO[2]

1 *Graduate School of Environmental Studies, Tohoku University, 6-6-06, Aramaki Aza Aoba, Aoba yama, Sendai 980-8579, Japan*
ranjan@kaigan.civil.tohoku.ac.jp

2 *Department of Civil Engineering, Graduate School of Engineering, Tohoku University, 6-6-06, Aramaki Aza Aoba, Aoba yama, Sendai 980-8579, Japan*

Abstract A finite difference model that simulates freshwater and saltwater flow separated by a sharp interface has been applied to estimate the salinity intrusion in the lower part of the Walawe River basin to the southern coastal aquifer, Sri Lanka. The effect of hydrogeological factors on the dynamics of the freshwater–saltwater interface has been considered through storage coefficients, porosity and hydraulic conductivity. The paper concludes that hydraulic conductivity is the main hydrogeological factor affecting the movement of the freshwater–saltwater interface, and the saltwater intrusion is more sensitive to groundwater recharge than hydrogeological properties. Therefore, the model was calibrated by adjusting the hydraulic conductivity to match the observed salinity profile in the southern coastal aquifer. Simulation results compare well with the observed long-term salinity profile suggesting that the numerical model can be used to successfully simulate the salinity profile in the area.

Key words salinity intrusion; coastal groundwater resources; hydrogeology; southern coastal aquifer Sri Lanka

INTRODUCTION

Use of coastal aquifers as operational reservoirs in water resources systems requires the development of tools that make it possible to predict the behaviour of the aquifer under different conditions. Estimations of the freshwater–saltwater interface either in steady or transient conditions have become necessary in designing and planning of groundwater systems in coastal areas. Quantitative understanding of the patterns of movement and mixing between freshwater and saltwater, and the factors that influence these processes, are necessary to manage the coastal groundwater resources (Ranjan *et al.*, 2006).

In nature a freshwater–saltwater interface seldom remains stationary. Changes in aquifer stresses result in the movement of the interface. The main objectives of this study are to use a numerical model to understand the behaviour of the freshwater–saltwater interface and to evaluate the effects of different hydrogeological settings by applying the model to the lower part of the Walawe River basin in the southern coastal aquifer in Sri Lanka, which faces scarcity of fresh groundwater resources due to saline intrusion.

Modelling of the dynamics of the freshwater–saltwater interface

Many models have been developed to represent and to study the problem of saltwater intrusion. They range from relatively simple analytical solutions to complex numerical models. The first concept of the freshwater–saltwater interface, now widely cited as the Ghyben-Herzberg principle (Reilly & Goodman, 1985), is based on the hydrostatic equilibrium between fresh and saline water. After the introduction of the Ghyben-Herzberg principle, several analytical and numerical solutions were developed to describe the phenomenon. The movement of fresh groundwater and saltwater in coastal aquifer systems has been studied using two different approaches (Reilly & Goodman, 1985). In the first approach, freshwater and saltwater are assumed completely immiscible and a sharp interface exists between these two phases. In the other approach, the freshwater and saltwater are allowed to mix in response to flow and dispersion mechanisms within the aquifer.

The sharp interface models which solve the coupled freshwater and saltwater flow equations have been developed with different numerical techniques (Shamir & Dagan, 1971; Vappicha & Nagaraja, 1976). A finite element solution with an indirect toe tracking technique was presented by Wilson & Costa (1982). Polo & Ramis (1983) discussed an unconditionally convergent finite difference approach to solve the sharp interface problem. A sharp interface model which solves the coupled freshwater and saltwater flow equations has been developed and it was successfully applied to evaluate multilayered aquifer systems (Essaid, 1986, 1990).

MATHEMATICAL DEVELOPMENT OF SHARP INTERFACE MODEL

Sharp interface models couple the freshwater and saltwater flow equations based on the continuity of flux and pressure. In this approach, together with the Dupuit approximation, for each flow domain the equation of continuity may be integrated over the vertical direction to produce the following system of differential equations (Bear *et al.*, 1999):

$$\frac{\partial}{\partial x}\left[K_{fx}\left(h^f-h^i\right)\frac{\partial h^f}{\partial x}\right]+\frac{\partial}{\partial y}\left[K_{fy}\left(h^f-h^i\right)\frac{\partial h^f}{\partial y}\right]+q_f=S_f\frac{\partial h^f}{\partial t}-\theta\left[(1+\delta)\frac{\partial h^s}{\partial t}-\delta\frac{\partial h^f}{\partial t}\right]+\alpha\theta\frac{\partial h^f}{\partial t} \qquad (1)$$

$$\frac{\partial}{\partial x}\left[K_{sx}\left(h^i-z^b\right)\frac{\partial h^s}{\partial x}\right]+\frac{\partial}{\partial y}\left[K_{sy}\left(h^i-z^b\right)\frac{\partial h^s}{\partial y}\right]+q_s=S_s\frac{\partial h^s}{\partial t}+\theta\left[(1+\delta)\frac{\partial h^s}{\partial t}-\delta\frac{\partial h^f}{\partial t}\right] \qquad (2)$$

The location of the interface elevation is given by:

$$h^i=\frac{\rho_s}{\rho_s-\rho_f}h^s-\frac{\rho_f}{\rho_s-\rho_f}h^f \qquad (3)$$

where ρ_f and ρ_s are specific weight in fresh and salt water, respectively, h^f and h^s are the piezometric heads of freshwater and saltwater regions, q_f and q_s are the flow rates in the fresh and salt water, respectively, and K_f and K_s represent the hydraulic conductivity in the fresh and salt water regions. Storage coefficients in the fresh and salt water regions are given by S_f and S_s respectively, θ is the porosity of the aquifer media. $\alpha = 1$ for an unconfined aquifer and $\alpha = 0$ for a confined aquifer.

Numerical scheme

Except for very simple systems, analytical solutions of those two coupled nonlinear partial differential equations are rarely possible. A numerical method must be employed to obtain approximate solutions. From equations (1) and (2), it is possible to derive a numerical model using implicit finite difference techniques. The continuous system described by the above two equations are replaced by a finite set of discrete points in space and time, and the partial derivatives are replaced by terms calculated from the differences in both freshwater and saltwater head values at these points. Spatial discretization is achieved using a block centred finite difference grid which allows for variable grid spacing. Fortran 77 computer code has been used for the modelling.

EVALUATION OF THE EFFECT OF HYDROGEOLOGICAL FACTORS

To investigate the effect of hydrogeological factors on the dynamics of the freshwater–saltwater flow systems, a 2 km × 2 km horizontal strip of an unconfined aquifer has been simulated by changing the hydrogeological properties. The effect of the specific storage was evaluated by increasing the storage coefficient by orders of magnitude. It shows that the changes in storage coefficient do not affect the location of the interface. The system responds in almost the same manner for different specific storage values. Porosity is the other factor which illustrates the storage of the aquifer. To investigate the effect of porosity on the behaviour of the flow system, porosity was changed from 0.1 to 0.4 in increments of 0.1 and the variation of the freshwater–saltwater interface has been checked. The change in porosity does not lead to a change in the position of the interface, but it leads to a change in the time period to achieve the steady state of the interface. Figure 1 shows the time taken to achieve the steady state interface at 500 m from the coastline. Reduction in porosity accelerates the movement of the interface and it drives the system to steady state over a shorter time period.

Hydraulic conductivity is the other factor which affects the change of the fresh-water–saltwater interface. The hydraulic conductivity was changed over the range of

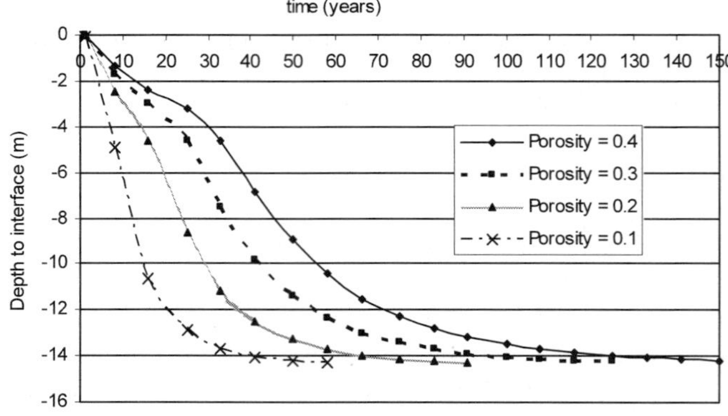

Fig. 1 Effect of porosity on the steady state of the interface.

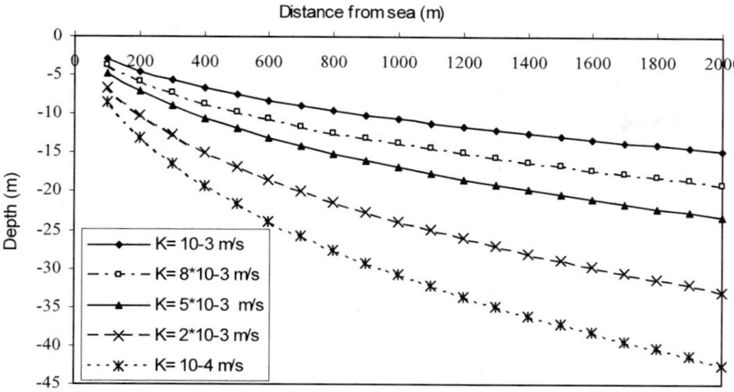

Fig. 2 Change in the interface with hydraulic conductivity (K).

10^{-3} m/s to 10^{-4} m/s. Figure 2 explains that the changes in hydraulic conductivity have an impact on the steady position of the interface. A change in hydraulic conductivity changes the transmissivity and it affects the head gradients necessary to maintain the freshwater flux. This process showed that the model is more sensitive with respect to changes in hydraulic conductivity than other hydrogeological factors.

Simulation of the effect of recharge

The effect of recharge on the dynamics of the freshwater–saltwater interface can be understood most readily by considering a simple, finite groundwater flow system in which all discharge flows to the ocean. Different groundwater recharge values have been simulated without changing the hydrogeological parameters to observe the change of interface. Figure 3 shows the variation of the steady interface with groundwater recharge. It shows that higher recharge can reduce saltwater intrusion. Saltwater will intrude farther inland unless the additional recharge can push the seawater equilibrium surface seaward.

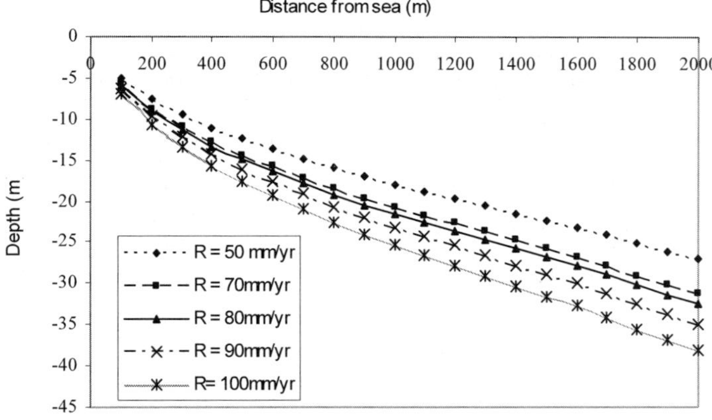

Fig. 3 Change in the interface with groundwater recharge.

APPLICATION TO SOUTHERN COASTAL AQUIFER

To compare the simulated interface with the observed salinity profiles, field data has been collected from the lower part of the Walawe River basin located in the southern coastal aquifer which is in the southern part of Sri Lanka (Fig. 4). The coastal plain, covering a major part of southern Sri Lanka, and has elevations less than 6 m above mean sea level (AMSL), parallel to the coast (Engineering Consultants, 1995). The width of the coastal plain generally ranges from 2 to 10 km. Coastal alluvial soils as well as laterites cover the area parallel to the coast. It includes river sediments and fine to medium green quartzite sand and beach sands (Statkraft Groner, 2000). Groundwater within the area is constrained by the unconsolidated alluvial and deltaic sediments, which were deposited by the main rivers and their distributaries (Kulathunga, 1998).

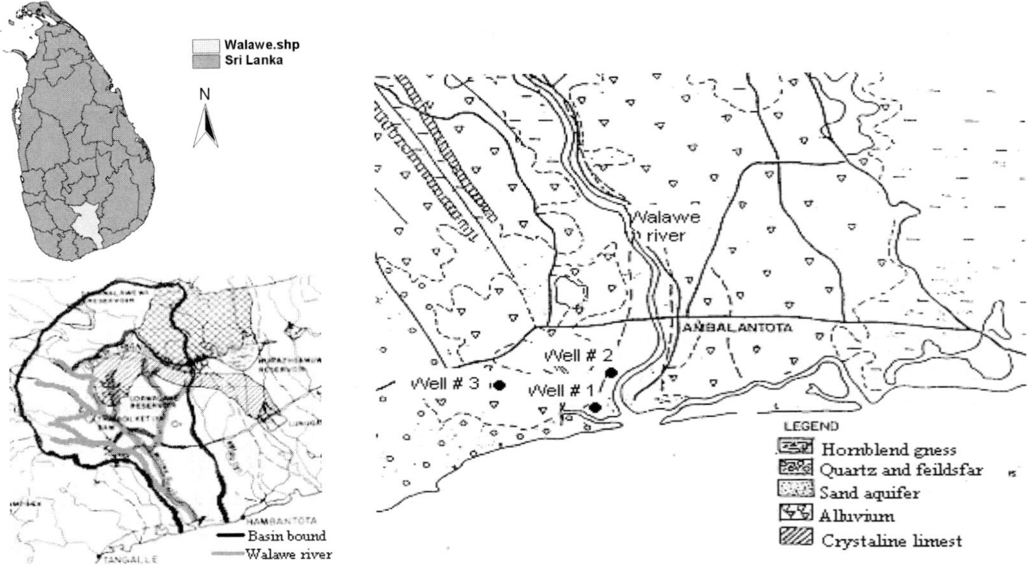

Fig. 4 Observation well locations in the lower part of Walawe River basin.

Calibration and verification of the model using field observation

Three observation wells have been used to measure the change in salinity profile in the area. The observation wells selected were at different distances from the coastline. As the salinity profile shows frequent changes in salinity intrusion, the observation wells were selected to a distance of 2 km from the coastline for the calibration and verification of simulated results. Well #1 is around 400 m from the coast. Well #2 and well #3 are 1 km and 2 km landward from the coast, respectively. Groundwater salinity was measured using the salinity meter, WQC-24. Weekly salinity measurements were measured during the period of investigation (September 2003–November 2004). Observation of a sharp interface is rarely encountered in the field. Literature studies show that the 50% salinity contour is the best representation of the sharp interface between saltwater and freshwater regions (Mahesha & Nagaraja, 1996; Scot & Stephen, 1998; Masciopinto, 2006). Therefore, we assume the 50% seawater salinity contour as equivalent to the sharp interface and depth to the freshwater–saltwater sharp interface was estimated using the vertical salinity profiles of observation wells.

Model calibration

The study area has been simulated as a single-layer unconfined aquifer, with the bottom defined by the position of the granite bedrock. The sharp interface model developed has assumed homogeneous aquifer properties. Since the southern coastal aquifer is not under any major stress such as extensive pumping, it has been assumed as steady until the change in groundwater recharge disturbs the system. On the sea side model boundary, a hydrostatic pressure was imposed where pressure is zero at the sea surface. The average annual groundwater recharge in the Walawe River basin has been estimated as 80 mm/year (Ranjan *et al.*, 2002). This has been used as a realistic value to represent the groundwater recharge in Walawe basin and the simulation runs were conducted with recharge value of 80 mm/year to generate the steady state condition. To calibrate the model for the initial steady state condition, the average elevation of the observed freshwater–saltwater interface in the first three months of the observation period was considered. The average depth to the sharp interface was estimated as 13.5 m at well #1; 21.5 m at well #2; and 34 m at well #3.

For the calibration process, a wide range of values for each parameter have to be tested to estimate the most suitable value. Since the hydraulic conductivity is the main hydrogeological factor affecting the movement of the salinity interface, the model was calibrated by adjusting hydraulic conductivity values to match the simulated steady interface location and the observed interface. The hydraulic conductivity varies across the area. The estimated hydraulic conductivity varies over a large range. The hydraulic conductivity of the southern coastal aquifer has been estimated using the available pumping test results and bore-hole data (Jayaweera, 2001; JICA, 2003). Aquifer materials and their deposition were observed from bore-hole data and relevant hydraulic conductivities were assigned. Since the coastal plain consists of river sediments and coastal alluvium, hydraulic conductivity is estimated to vary between 10^{-4} m/s to 10^{-3} m/s (Freeze & Cherry, 1979). Varying the aquifer properties, several model runs were carried out to find the most reliable location for the steady state salinity intrusion profile. The best fitting profile with the field observation data is shown in Fig. 5.

Verification of the model

The calibrated model has been used to verify the short-term changes in the freshwater–saltwater interface. Output from the steady state results was used as the initial condition to simulate the effects of realistic short-time changes in groundwater recharge on the system. During the model verification period, model coefficients obtained from the calibration process were kept as constants.

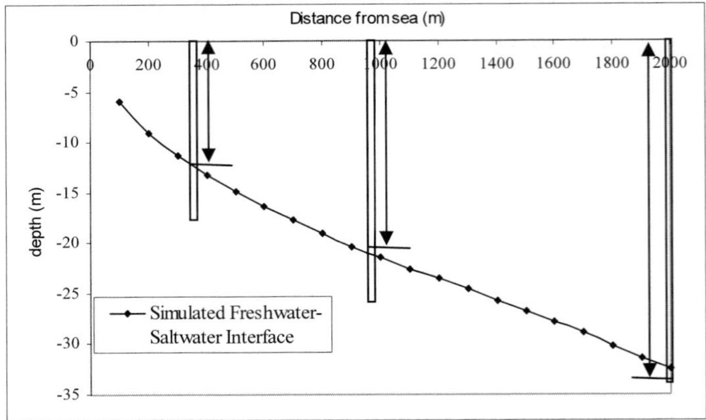

Fig. 5 The best fitting profile of the field observation with calibrated model results.

Since groundwater recharge is the main stress for short-term variations of the freshwater–saltwater interface, seasonal changes in groundwater recharge were assigned to the model. Monthly average groundwater recharge has been estimated using the water balance approach with meteorological data obtained from International Water Management Institute (IWMI) database for the Walawe benchmark basin. Estimated groundwater recharge shows that Walawe basin recharges in two seasons per year; the southwest monsoon season in October–December and the northeast monsoon season in April–June. Groundwater recharge is zero in the other months.

Continuous short-term observations (weekly) are available for well #1 and well #2. These observed weekly average salinity profiles from January 2004 to November 2004 have been used to verify the calibrated model. Figure 6 shows the comparison of weekly variation of the sharp freshwater–saltwater interface in observation wells #1 and #2 and the simulated monthly averaged interface using the sharp interface model.

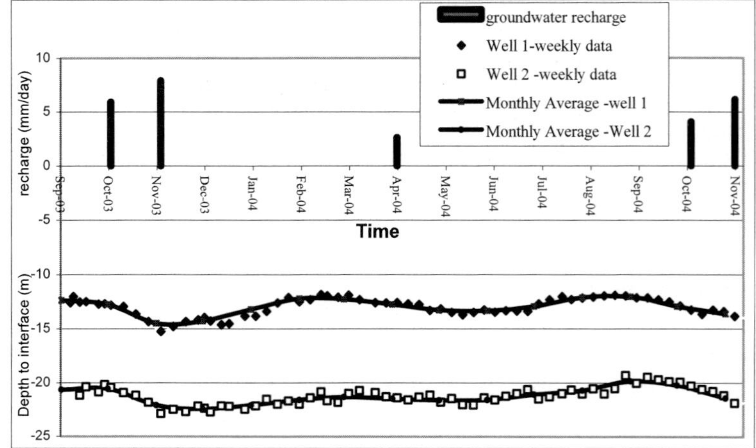

Fig. 6 Weekly variation of the sharp freshwater–saltwater interface.

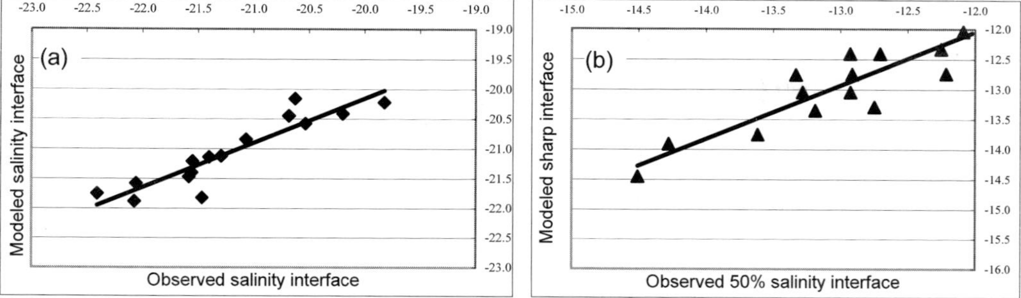

Fig. 7 Comparison of observed and modelled interface: (a) at well #1, (b) at well #2.

The correlation between simulated and monthly averaged observations is shown in Fig. 7. Correlation coefficients between model results and field observations for wells #1 and #2 were 0.86 to 0.77, respectively. These correlations verify a good agreement between model results and field observations.

This kind of simulations shows that numerical models can be used to reproduce the salinity profile in complex hydrogeological settings. It also indicates that any groundwater development activity in the southern coastal aquifer needs to be carefully planned with remedial measures in order to prevent the further intrusion of seawater.

CONCLUSIONS

The accurate solutions of the coupled, nonlinear partial differential flow equations require the utilization of a numerical technique. The sharp interface model developed has been applied to the lower part of the Walawe River basin in the southern coastal aquifer, Sri Lanka. Evaluation of the effect of hydrogeological factors on the dynamics of the freshwater–saltwater interface concluded that hydraulic conductivity is the main hydrogeological factor affecting the movement of the salinity interface. The simulation for groundwater recharge shows that saltwater intrusion is more sensitive to groundwater recharge than hydrogeological properties. Higher values of hydraulic conductivity facilitate intrusion of seawater, whereas increased recharge has the opposite effect. The numerical model was calibrated by adjusting the hydraulic conductivity to match the observed salinity profile in the southern coastal aquifer. The simulation has been carried out for September 2003 to November 2004, considering the changes in groundwater recharge. The observed short-term salinity profiles have been compared with the simulation results. The short-term variation of the salinity interface according to the changes in groundwater recharge shows that the comparison is in reasonable agreement. We conclude that the developed model is able to simulate the changes of the salinity profiles realistically and can be used to address the counter measures to prevent the salinity intrusion in Walawe River basin.

Acknowledgement Authors would like to acknowledge the support received for the field observations from Mr Mahesh Dilroy, University of Ruhuna, Sri Lanka.

REFERENCES

Bear, J., Cheng, A. H. D., Sorek, S., Ouazar, D. & Herrera, I. (1999) *Seawater Intrusion in Coastal Aquifers – Concepts, Methods and Practices*. Kluwer Academic Publishers, The Netherlands.

Engineering Consultants Ltd (1995) Water Resources Inventory Study for Southern Area of Sri Lanka, vol. 1, Main Report.

Essaid, H. I. (1986) A comparison of the coupled freshwater–saltwater flow and the Ghyben-Herzberg sharp interface approaches to modeling of transient behavior in coastal aquifer systems. *J. Hydrol.* **86**, 169–183.

Essaid, H. I. (1990) A multilayered sharp interface model of coupled freshwater and saltwater flow in coastal systems: model development and application. *Water Resour. Res.* **26**(7), 1431–1455.

Freeze, R. A. & Cherry, J. A. (1979) *Groundwater*. Prentice Hall International, London, UK.

Jayaweera, U. (2001) Alluvial aquifer studies at lower part of the Walawe River Basin. Report to Water Resources Board, Sri Lanka.

JICA (2003) The study on comprehensive groundwater resources development for Hambantotaand Monaragala districts in Sri Lanka. Interim report for Water Resources Board; Ministry of Irrigation and Water Resources Management, Sri Lanka.

Kulathunga, N. (1998) Hydrology of a metamorphic terrain: a case study from Hambantota, Sri Lanka. *J. Geol. Soc. Sri Lanka* **1**, 62–71.

Mahesha, A. & Nagaraja, S. H. (1996) Effect of natural recharge on sea water intrusion in coastal aquifers. *J. Hydrol.* **174**(3-4), 211–220.

Masciopinto, C. (2006) Simulation of coastal groundwater remediation: the case of Nardò fractured aquifer in Southern Italy. *Environ. Modelling & Software* **21**(1), 85–97.

Polo, J. F. & Ramis, F. J. R. (1983) Simulation of salt–fresh water interface model. *Water Resour. Res.* **19**(1), 61–68.

Ranjan, S. P. (2002) Assessment of groundwater resources in Walawe Basin, Sri Lanka. Master Thesis, Asian Institute of Technology, Thailand.

Ranjan, S. P., Kazama, S. & Sawamoto, M. (2006) Effects of climate change on coastal fresh groundwater resources, *Global Environ. Change* **16**, 388–399.

Reilly, T. E. & Goodman, A. S. (1985) Quantitative analysis of fresh–salt water relationship in groundwater systems. *J. Hydrol.* **80**, 125–149.

Scot, K. I. & Stephen, B. G. (1998) Estimation of the depth to the freshwater/salt-water interface from vertical head gradients in wells in coastal and island aquifers. *Hydrogeol. J.* **6**, 365–373.

Shamir, U. & Dagan, G. (1971) Motion of saltwater interface in a coastal aquifers: a numerical solution. *Water Resour. Res.* 7(3), 644–657.

Statkraft Groner (2000) Third Water Supply and Sanitation Project, Water Resources Management, Annex D, Groundwater in Walawe Ganga Basin.

Vappicha, V. N. & Nagaraja, S. H. (1976) An approximate solution for the transient interface in a coastal aquifer. *J. Hydrol.* **31**, 161–173.

Wilson, J. L. & Costa, A. (1982) Finite element simulation of salt–fresh water interface with indirect toe tracking, *Water Resour. Res.* **18**(4), 1069–1081.

Seawater intrusion in the coastal aquifer of Wadi Ham, UAE

MOHSEN SHERIF[1] & ANVAR KACIMOV[2]

1 *Civil and Environmental Engineering Department, College of Engineering, UAE University, PO Box 17555, Al Ain, UAE*
msherif@uaeu.ac.ae

2 *Dept of Soils, Water and Agricultural Engineering, College of Agriculture, Sultan Qaboos University, Muscat, Sultanate of Oman*

Abstract A transport model was developed to simulate the seawater intrusion in the aquifer system along the Kalbha and Fujairah coast of the United Arab Emirates. The model was used to simulate the salinity levels of the groundwater of Wadi Ham aquifer and its variation in time and space from January 1994 to March 2005. The area covering the coast of Gulf of Oman in the study domain was taken as a constant concentration boundary with an average salinity (TDS) value of 35 000 mg/L. The effect of artificial recharge on seawater intrusion was evaluated. The results of the simulation indicated that the seawater intrusion is affected by the dry and wet conditions. During the dry years, the velocity vectors are directed from the Gulf of Oman to the aquifer causing severe intrusion problems. During the wet years when rainfall is relatively high and groundwater recharge is encountered from the ponding area of Wadi Ham dam, the velocity vectors are reversed.

Key words numerical modelling; seawater intrusion; recharge; UAE

INTRODUCTION

In arid and semi-arid regions, groundwater constitutes the main, if not the only, source of natural freshwater. Many aquifers around the globe are located in costal areas and are thus vulnerable to the seawater intrusion phenomenon due to the direct hydraulic contact between freshwater in aquifers and the seawater. The growth of population in coastal areas and the conjugate increase in human, agricultural and industrial activities have imposed an increasing demand for freshwater in such areas. This increase in water demand is often covered by extensive pumping of fresh groundwater, causing subsequent lowering of the water table (or piezometric head) and upsetting the dynamic balance between freshwater and saline water bodies. The classical result of such a development is seawater intrusion. This phenomenon has been reported in many countries over the last few decades.

Seawater intrusion phenomena have been reported, with different degree, in almost all coastal aquifers around the globe. The problem is more severe in arid and semi-arid regions where the groundwater constitutes the main freshwater resource. A 3% mixing of seawater with the freshwater in a coastal aquifer would render the freshwater resource unsuitable for human consumption. The shape and degree of the seawater intrusion in a coastal aquifer depend on several factors. Some of these factors are natural and cannot be controlled, while others are manmade and could, thus, be managed. Groundwater resources in coastal aquifers should be carefully exploited to maintain the dynamic balance between the fresh and saline water bodies.

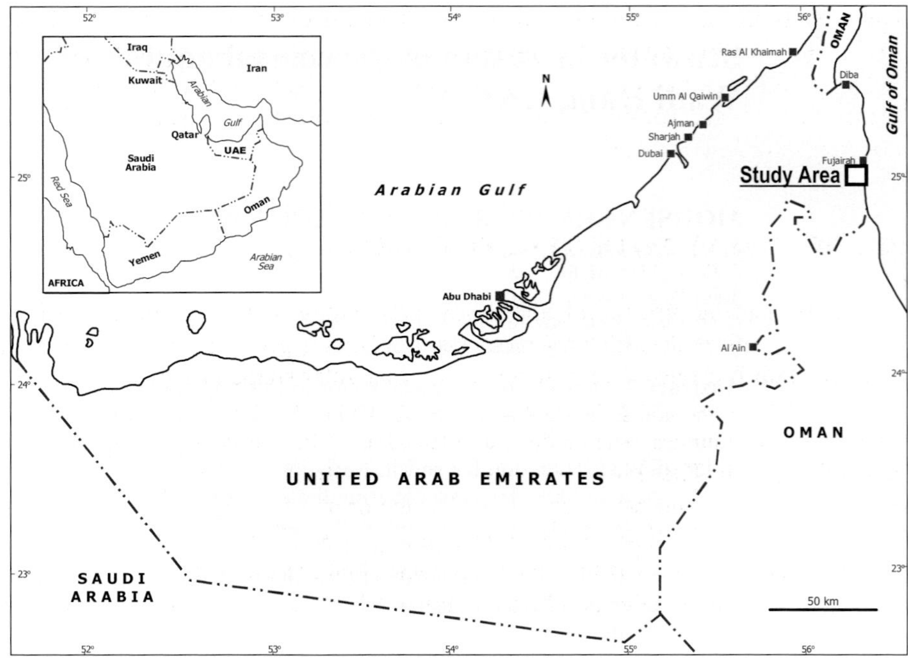

Fig. 1 Location map of the United Arab Emirates.

The United Arab Emirates (UAE) lies in the southeastern part of the Arabian peninsula between latitudes 22°40' and 26°00' N and longitudes 51°00' and 56°00' E. It is bounded to the north by the Arabian Gulf, on the east by the Sultanate of Oman and the Gulf of Oman and on the south and the west by the Kingdom of Saudi Arabia (Fig. 1). The total area of the UAE is about 83 600 km^2. The UAE is divided into two distinct zones: the larger low-lying zone and the smaller mountainous zone. Ninety percent of the country's area is covered by the low-lying zone (Al Hammadi, 2003). The low-lying zone ranges in altitude from sea level up to 300 metres. Its major part is characterized by the presence of sand dunes which rise gradually from the coastal plain reaching their highest elevation of 250 m above sea level (a.m.s.l.). Along the coast of the Arabian Gulf, the low-lying land is punctuated by ancient raised beaches and isolated hills which may reach up to 40 m a.m.s.l. in some locations (Baghdady, 1998).

Renewable water resources in the UAE are very limited. No surface water in the form of rivers or lakes is available. The rainfall is very scarce, random and infrequent. The average rainfall is about 110 mm/year. However, this annual rainfall is mostly encountered in a few events. Records indicate that the average number of rainfall events per year is five or less. These rainfall events are generally characterized by their short durations and heavy intensities. Such rainfall characteristics are quite consistent with regions classified as drought areas.

This paper is devoted to the simulation and assessment of seawater intrusion in the coastal aquifer of Wadi Ham, UAE. Visual MODFLOW (McDonald & Harbaugh, 1988; Waterloo Hydrgeologic Inc., 2000) was used to calibrate and simulate groundwater levels and flow patterns in Wadi Ham under different conditions. MT3D (Zheng, 1990) was used to simulate the solute transport and assess the impacts of groundwater recharge on the degree of seawater intrusion. It is concluded that rainfall and recharge events have a significant effect on the groundwater flow pattern and hence the seawater intrusion problem.

PHYSICAL SETTING OF THE STUDY AREA

A remote sensing image for the study area along with the location of existing monitoring wells and available geophysical sections is given in Fig. 2. The valley floor of Wadi Ham is a flat-gravelly plain with a triangular shape broadening to the sea and draining the surrounding mountains. It rises from sea level at Fujairah to approximately 100 m above sea level to the northwest. A few hills are scattered in different parts of the wadi. These hills subdivide the wadi into communicating zones. Along the coast, the inward land becomes a river terrace or alluvial plain. It is locally dissected by stream channels filled with cobble and gravel. The number and the depth of channels decrease towards the coast. The wadi plain is used for extensive agricultural activities. Some new industries have commenced in the vicinity of the coastal zone.

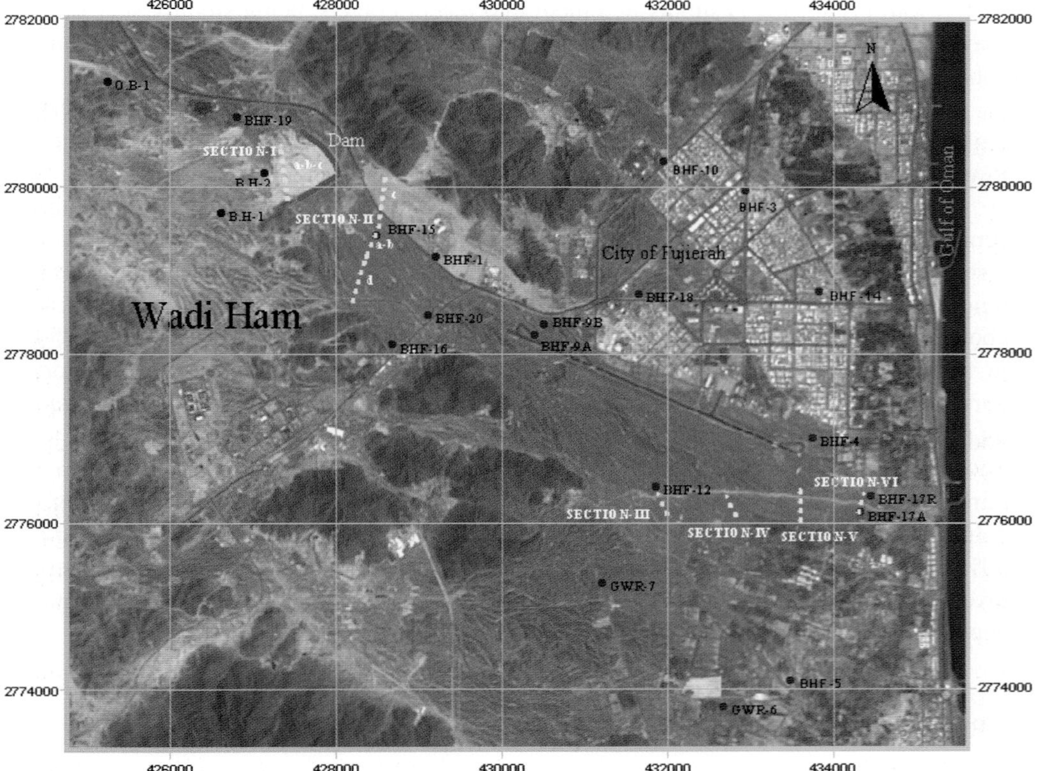

Fig. 2 A remote sensing image for the study area including the locations of existing wells and geophysical sections.

The thickness of wadi gravel in Wadi Ham varies from 18 m at the upstream side of the dam to about 100 m near the coast. The minimum thickness is found in the area of well number BHF-19, at the upstream of Wadi Ham dam and close to the mountain series. The maximum thickness is observed in the area of well number BHF-14 which is very close to the coast of Oman Gulf. The cross-sectional depth of wadi gravels and sand along the wadi course varies from 45 to 64 m.

The cross-sections (longitudinal profiles) in the area near to the Gulf of Oman show that the gravel depth varies from 24 m to 99 m. Its thickness decreases with increasing the distance from the shoreline. For example, within a distance of 3 km it varies from 24 m to 73 m. This is attributed to the regional dipping of the Ophiolite series towards the wadi channel. This information is restricted to the available boreholes and wells.

HYDROGEOLOGICAL PARAMETERS

Two main aquifers can be identified in the area of Wadi Ham, namely the Quaternary aquifer which is composed of wadi gravels and constitutes the main aquifer, and the Fractured Ophiolite which has a low groundwater availability potential. The gravels are highly permeable and of variable hydraulic properties. They tend to be unconsolidated at the ground surface, becoming better cemented and consolidated with depth. Electrowatt (1980, 1981) subdivided them into recent gravels, being slightly silty sand gravel with some cobbles; young gravels, which are silty sandy gravels with many cobbles and boulders; and finally old gravels, which are weathered and cemented.

Values of the hydraulic conductivity of the unconsolidated gravels tend to be very high, typically 6 to 17 m/day and in the range 0.086–0.86 m/day for the cemented lower layers, Electrowatt (1980, 1981).

In the unconsolidated gravels, primary porosity is very high when compared to the cemented gravels. The storage coefficients typically range from 0.1 to 0.3. At a distance of 3.5 km directly downstream of the dam, the saturated aquifer thickness ranges between 10 m and 40 m, with a transmissivity ranging from less than 100 to about 200 m^2/day. In sections where the saturated aquifer thickness varies between 50 m to 100 m the transmissivity may reach more than 1000 m^2/day. Fourteen short duration (8 to 300 minutes) pumping tests performed by ENTEC (1996) were analysed by using both Cooper and Jacob, and Theis methods.

Annual rainfall in the Wadi Ham ranges from 3.7 mm to 505.8 mm. The normal rainfall estimated for 24 years is 151 mm with a standard deviation 126.8 mm, kurtosis 1.36 and coefficient of asymmetry 1.18. The probability of occurrence of 75% and 50% normal rainfall were estimated as 51 and 64 percent, respectively. These numbers reveal that the distribution of rainfall in Wadi Ham is highly scattered and not dependable on annual basis (Sherif *et al.*, 2005, 2006).

GROUNDWATER FLOW MODEL

The study domain of Wadi Ham aquifer comprises an area of 117.81 km^2 with total length of 11.9 km east to west (dam to coast of Oman Gulf) and 9.9 km north to south

(Fujairah to Khalba). The modular three-dimensional finite-difference groundwater model, MODFLOW, (McDonald & Harbaugh, 1988) was employed in this study. It has a modular structure that allows modifying the code for a particular application.

The model area and the aquifer boundaries were delineated by digitizing the remote sensing image of Wadi Ham. The model domain includes the Gulf of Oman and the Ophiolite sequence rock out crops. The Ophiolite outcrops are specified as inactive or no flow areas. The area of separated outcrop is about 6.56 km^2. At the coast, many cells are located in the sea which is considered to be a constant (zero) head boundary. A ponding area was delineated and marked on the study domain. The total area of the pond at flood level was estimated as about 0.40 km^2.

The study area was divided into 119 columns and 99 rows with the size of each cell being 100 by 100 m. The model is comprised of a total of 11 781 equally spaced and square cells. However, the net area of aquifer consisted of only about 64.94 km^2 with 6494 active cells. The area of inactive cells is about 52.87 km^2 with 5287 cells.

The calibration period was selected as five years from January 1989 to December 1993 (1826 days). The length of stress period in this exercise was taken as one real month. However, the period during which recharge of either rainfall or dam storage took place was considered as an extra stress period. The total number of stress periods considered for the calibration period was 146. The model calibration was achieved by changing three parameters, namely, permeability, specific yield and pumping rates. Abstraction and inflow across the boundaries were also simulated by a number of computer runs until a reasonable groundwater level in each observation wells was achieved.

The validation of the flow model was carried out during the period from January 1994 to March 2005, for a total simulation time of 4108 days. The flow vectors were developed as shown in Figs 3 and 4 to identify the direction of flow, particularly at the boundary of the Gulf of Oman. In Fig. 3 the flow is directed towards the sea from the aquifer, a situation which represents the wet conditions with a relatively high rainfall. However, in the case of dry years (Fig. 4) the flow is coming from the aquifer only at the northeast part of the coast. In the southeast part, the flow is from the sea to the aquifer, where the Kalbha groundwater well field is in operation. The flow pattern was also affected by the recharge from the Wadi Han Dam.

TRANSPORT MODEL

A transport model was developed for the identification of seawater intrusion and movement of saline water into the coastal aquifer along the Kalbha and Fujairah coast of UAE. The validated flow model was used to simulate the salinity level of the groundwater aquifer and its variation in time and space from January 1994 to January 2006. The initial concentration grid was developed using the available historical data of some locations in the study domain during 1994 and 1995.

Dispersion is a physical process that tends to disperse the contaminant mass in different directions along and across the advective path of the contaminants and acts to reduce the solute concentration. Dispersion is caused by the tortuosity of the flow-paths of the groundwater as it travels through the interconnected pores of the soil and

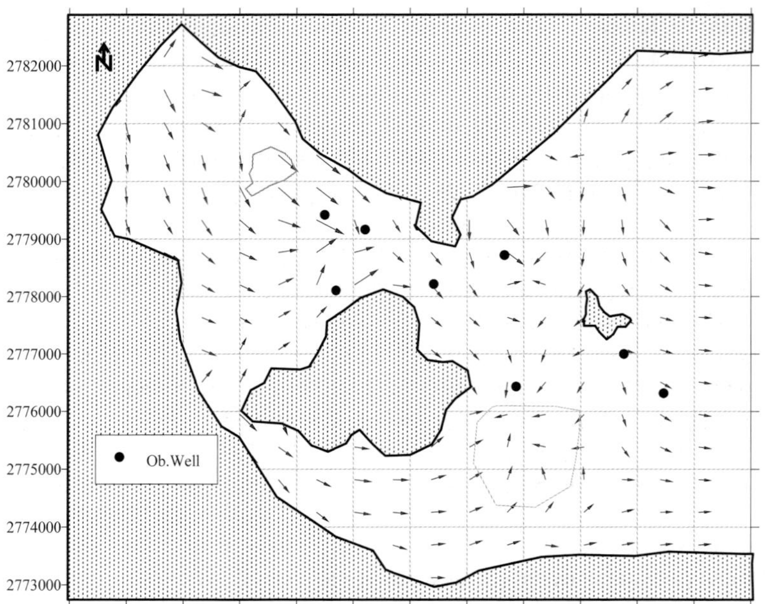

Fig. 3 Flow vector the January 1996 (wet year).

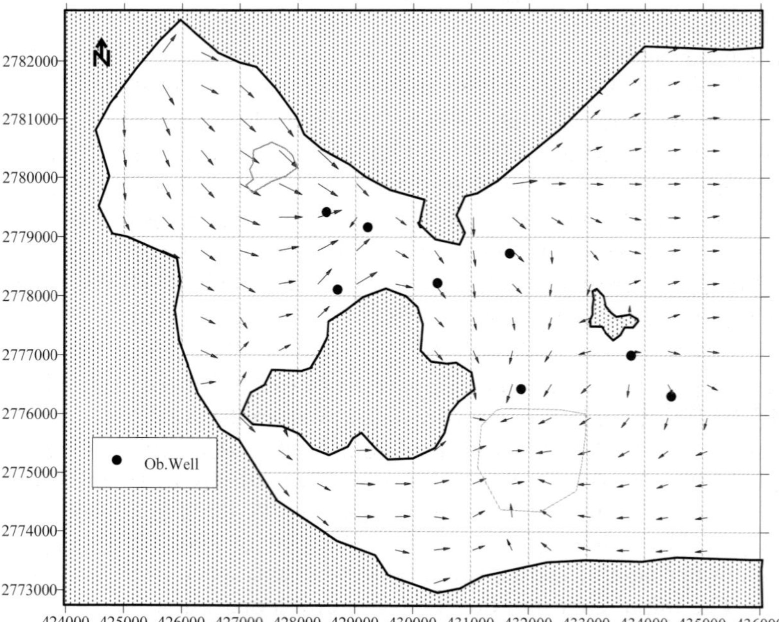

Fig. 4 Flow vector the December 1996 (dry year).

other factors, e.g. undetected heterogeneity of hydraulic properties of the rock. In the present simulation, 20 m of longitudinal dispersivity for each transport grid cell was considered. The ratio of horizontal to longitudinal dispersivity of the layer was 0.1 and the ratio of vertical to longitudinal dispersivity of the layer 0.01. The molecular diffusion coefficient 0.1 UNIT is missing! was considered for the layer.

The constant concentration boundary condition acts as a contaminant source providing solute mass to the model domain in the form of known concentration for a relatively long period of time. Therefore, the area covering the coast of Gulf of Oman in the study domain was taken as a constant concentration boundary with an average salinity (TDS) of 35 000 mg/L. The fluid density was assumed constant. The concentration of the recharge water in the ponding area of the Wadi Ham Dam was set as 500 mg/L and was applied over 40 cells of the model. The evapotranspiration concentration boundary condition specifies the concentration of each species accompanying the evapotranspiration flux specified in the corresponding model. In the present simulation, the extinction depth was considered to be 2 m, hence the evapotranspiration from the water table is negligible. Therefore, concentration accompanying the evapotranspiration flux was neglected.

EFFECT OF ARTIFICIAL RECHARGE

A recharge pond of 100 by 100 m was introduced in the study area using the calibrated and validated model where the experimental pond was constructed. Different scenarios were selected and the simulation was conducted for the flow and concentration levels in the study area, especially in the coastal zone of the Gulf of Oman. Tertiary treated wastewater and excess of desalination water during low-demand periods are the possible sources for the implementation of groundwater artificial recharge projects.

Two different recharge scenarios of 0.5 m/day and 1.0 m/day were applied from January 1994 to January 2006 for 4108 days. The changes of groundwater regime and the quality aspects were also simulated. The impact on the groundwater regime was compared in different locations of the study area. Examples of the water levels in some of the observation wells in the study area are provided in Fig. 5.

Figure 5 shows that there is a clear rise in the groundwater levels due to the recharge from the pond. However, the maximum impact was observed in the wells BHF-4, BHF-12, BHF-17A, BHF-17R and BHF-18 which are relatively closer to the recharge pond. The minimum impact was observed in BHF-1, BHF-9, BHF-15 and BHF-16. These wells are relatively far away from the pond.

The salinity levels were also compared under different recharge conditions. Figure 6 represents a comparison between the salinity distribution under no recharge and 0.5 m/day recharge after a simulation time of 4018 days. The salinity levels were also compared for the case of a recharge level of 1 m/day, as shown in Fig. 7. It is obvious that the recharge of freshwater from the pond will have a significant improvement effect on the groundwater quality. It should be noted, however, this improvement of the groundwater quality is mainly encountered in the vicinity of the recharge pond. A limited improvement is also observed elsewhere.

In order to provide a quantitative assessment for the expected quality improvements under artificial recharge conditions, specific contour lines were compared for the cases of no recharge, 0.5 m/d recharge and 1.0 m/day recharge. To that end,

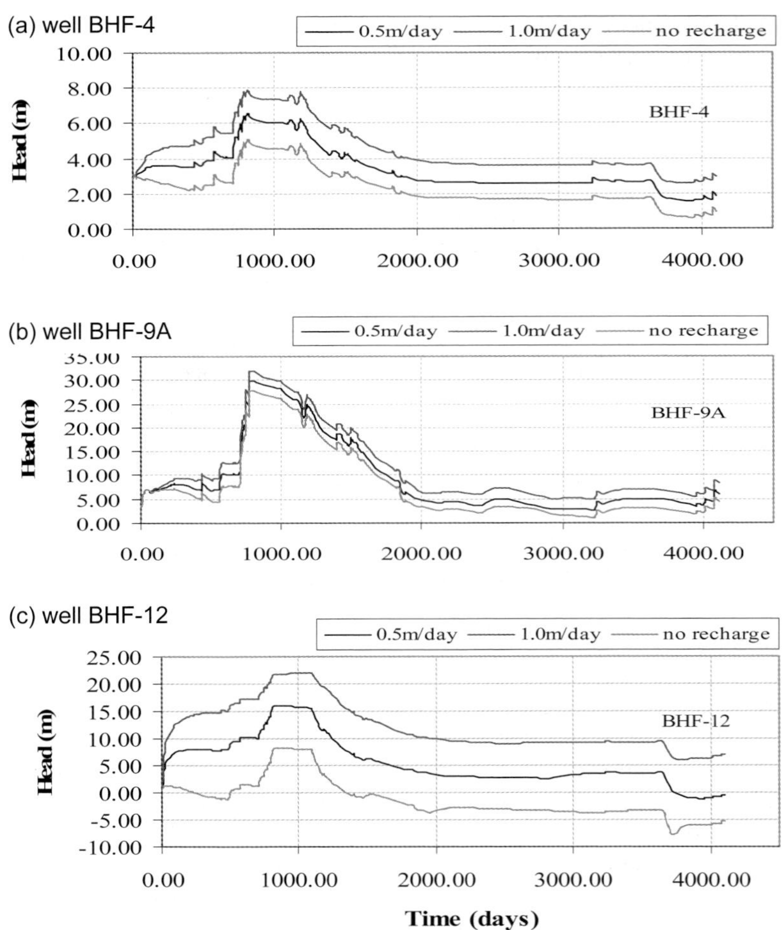

Fig. 5 Comparison of water level with and without recharge.

contour lines of 10 000 mg/L were compared for the above three recharge conditions. An example of the output for simulation period of 4018 days is given in Fig. 8. At the end of the simulation time, the 10 000 mg/L equi-concentration line moved a distance of about 1.25 km toward the Gulf of Oman. Other equi-concentration lines moved in the same direction but with different degrees indicating a significant improvement in the groundwater quality. It should be noted, that, like contamination in porous media, the quality improvement in groundwater systems is a lengthy process and may take a few decades to be recognized.

CONCLUSIONS

The seawater intrusion phenomenon in Wadi Ham is affected by the dry and wet conditions. During the dry years, the velocity vectors are directed from the Gulf of Oman to the aquifer causing severe intrusion problem. During wet years where rainfall

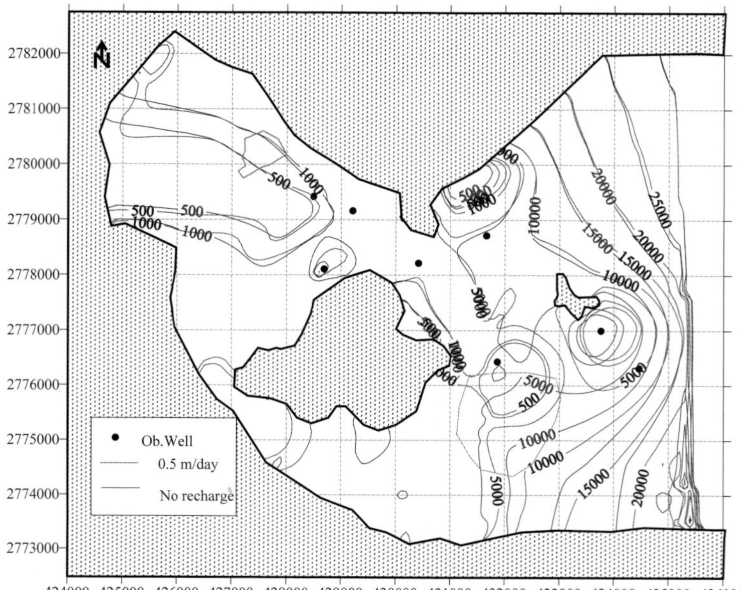

Fig. 6 Comparison of the salinity level after 4018 days with a recharge of 0.5 m/d.

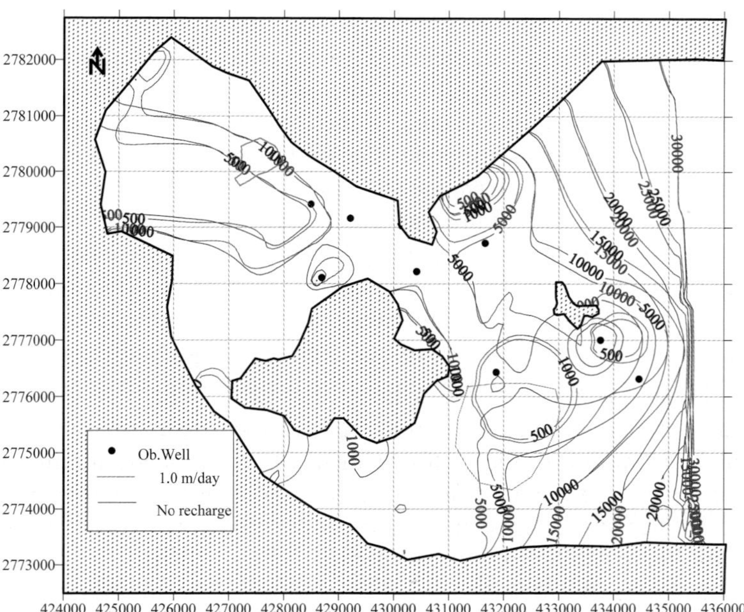

Fig. 7 Comparison of the salinity level after 4018 days with a recharge of 1.0 m/d.

is relatively high and groundwater recharge is encountered from the ponding area of the dam of Wadi Ham, the velocity vectors are reversed and the flow is directed from the aquifer to the Gulf of Oman through out the interface.

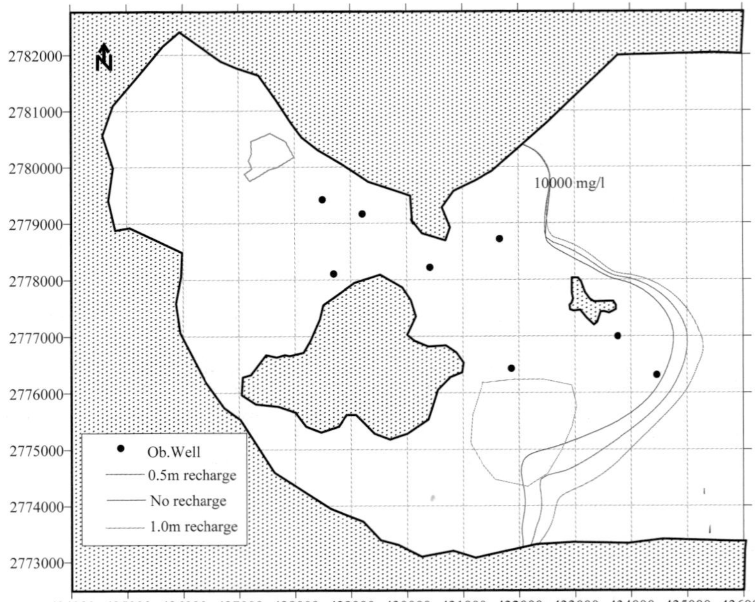

Fig. 8 Variation of TDS (10 000 mg/L) at 4018 days for the different recharge conditions.

The influence of artificial recharge from a pond with dimensions of 100 × 100 m was demonstrated using MODFLOW and MTD. The maximum impact of recharge (significant improvement in the quality of the groundwater) was observed in the wells relatively close to the recharge pond. The minimum impact was observed in the wells which are located far away from the recharge area.

The results of the numerical simulation indicated a backward movement (retardation) towards the Gulf of Oman of about 1.25 km of the equi-concentration line 10000 mg/L under a recharge condition of 1.0 m/day over a period of 4018 days. All parameters, including the pumping rates, were based on the actual data over the last 10 years. In the long term, the artificial recharge of groundwater in coastal aquifers will not only maintain, but will also enhance the groundwater quality in such aquifers.

Acknowledgements The work presented in this paper was carried out within the framework of an interdisciplinary project entitled "Modeling of Groundwater in Two Selected Coastal Aquifers of UAE and Oman as a Precursor for Water Resources Management". The project is funded equally by the research sectors of both UAE University and Sultan Qaboos University. The support of the Ministry of Agriculture and Fisheries (MAF), UAE, Sharjah Electricity and Water Authority (SEWA), UAE and the Ministry of Regional Municipalities, Environment and Water Resources (MRMEWR), Oman is highly appreciated.

REFERENCES

Al Hammadi, M. K. (2003) Assessment of Groundwater Resources using Remote Sensing and GIS. MSc Thesis, Water Resources Master Program, UAE University, UAE.

Baghdady, A. R. (1998) Petrography, mineralogy and environmental impacts of the sand dune fields of the Greater Al Ain area, United Arab Emirates. PhD Thesis, Faculty of Science, Ain Shams University, Egypt.

Electrowatt Engineering Services Ltd (1980) Wadi Ham dam for groundwater recharge and loss prevention. Preliminary Report, Ministry of Agriculture and Fisheries, UAE.

Electrowatt Engineering Services Ltd (1981) Wadi Ham dam and groundwater recharge facilities, vol. I, Design. Ministry of Agriculture and Fisheries, UAE.

Entec Europe Limited, ENTEC (1996) Survey on groundwater recharge and flow in Wadi Ham, vol. 1. Ministry of Agriculture and Fisheries, UAE.

MAF (2001) *Climatological Data,* vol. 4, *1992–93 to 1999–2000.* Water and Soil Department, Ministry of Agriculture and Fisheries, UAE.

McDonald, M. G. & Harbaugh, A. W. (1988) A modular three-dimensional finite-difference ground-water flow model. *US Geological Survey Techniques of Water-Resources Investigations, Book 6, chap.* A1.

Sherif *et al.* (2005) Assessment of the effectiveness of Al Bih, Al Tawiyean and Ham dams in groundwater recharge using numerical models, vol. 4, Final Report. Ministry of Agriculture and Fisheries, Dubai.

Sherif, M. M. & Kacimove, A. (2006) Modeling of groundwater in two selected coastal aquifers of UAE and Oman as a precursor for water resources management. Final Report, Research Sectors of UAEU/SQU, UAE

Waterloo Hydrogeologic Inc. (2000) Visual MODFLOW User's Manual, Waterloo Hydrogeologic Inc., Ontario, Canada.

Zheng, C. (1990) MT3D*,* A modular three-dimensional transport model for simulation of advection, dispersion and chemical reactions of contaminants in groundwater system*s*. Report to the US Environmental Protection Agency.

A simulation of groundwater discharge and nitrate delivery to Chesapeake Bay from the lowermost Delmarva Peninsula, USA

WARD E. SANFORD[1] **& JASON P. POPE**[2]

1 *US Geological Survey, Mail Stop 431, Reston, Virginia 20192, USA*
wsanford@usgs.gov

2 *US Geological Survey, 1730 East Parham Road, Richmond, Virginia 23228, USA*

Abstract A groundwater model has been developed for the lowermost Delmarva Peninsula, USA, that simulates saltwater intrusion into local confined aquifers and nitrate delivery to the Chesapeake Bay from the surficial aquifer. A flow path and groundwater-age analysis was performed using the model to estimate the timing of nitrate delivery to the bay over the next several decades. The simulated mean and median residence times of groundwater in the lowermost peninsula are 30 and 15 years, respectively. Current and future nitrate concentrations in coastal groundwater discharge were simulated based on local well data that include nitrate concentrations and groundwater age. A simulated future-trends analysis indicates that nitrate that has been applied to agricultural regions over the last few decades will continue to discharge into the bay for several decades to come. This study highlights the importance of considering the groundwater lag time that affects the mean transport time from diffuse contamination sources.

Key words groundwater model; submarine groundwater discharge; nitrate; Chesapeake Bay, USA

INTRODUCTION

The Chesapeake Bay in the eastern USA is an impaired water body that has been adversely affected for decades by an oversupply of nutrients that feeds algal blooms and creates regions of low dissolved oxygen. In the mid-1980s, the Chesapeake Bay Program was created as a partnership between the states and the federal government to reduce nutrient loading to the bay. Although criteria have been developed for water-quality levels and remedial actions have begun, it is unknown how long it will take for conditions in the bay to improve. One important factor in estimating the timing of loadings from the land surface to the bay is the influence of groundwater on the transport time. Studies have shown that typical residence times of groundwater in subwatersheds within the bay watershed are ten years or more (Lindsey *et al.*, 2003; Phillips & Lindsey, 2003). A substantial portion of dissolved nitrogen (primarily nitrate) loading to the bay is from agricultural regions, and one of the largest of these regions is on the Delmarva Peninsula, which lies between the bay and the Atlantic Ocean (Fig. 1(a)). If future levels of dissolved nitrogen loading to the bay are to be estimated accurately, a better understanding of the movement of groundwater that discharges to the bay from agricultural regions would be required.

The southernmost part of the Delmarva Peninsula is called the Eastern Shore of Virginia (Fig. 1(a)). The Eastern Shore contains a long topographic high that divides the peninsula into two watersheds (Fig 1(b)), with the western side draining to the

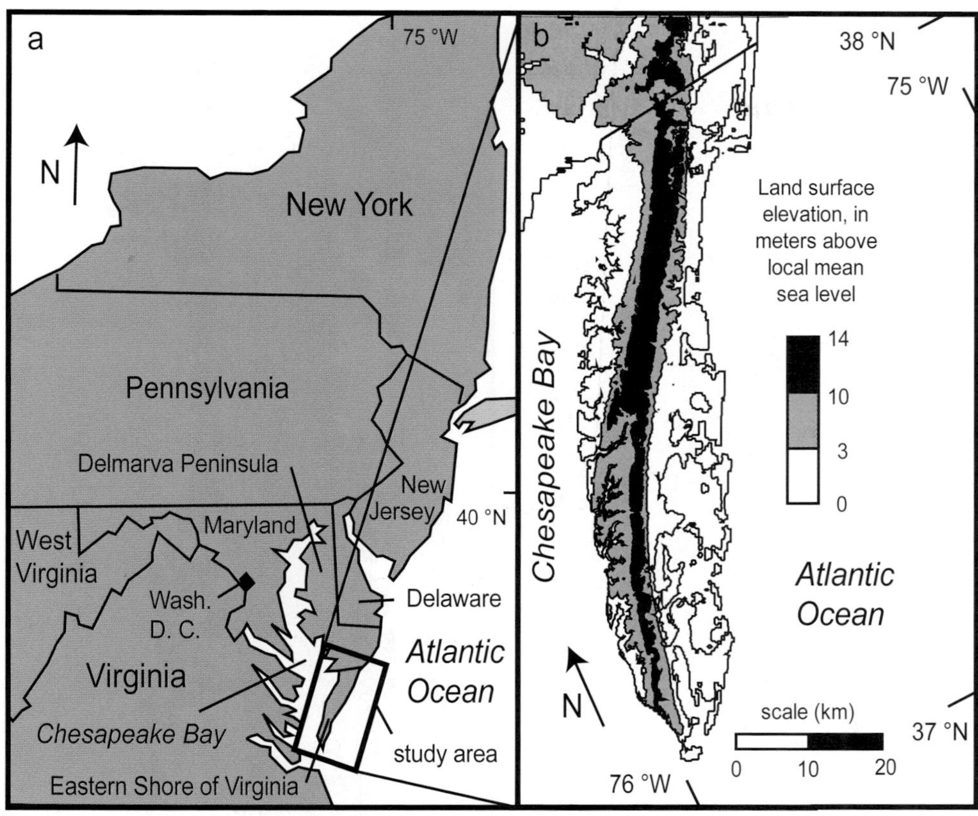

Fig. 1 Maps showing: (a) the location and (b) topography of the Eastern Shore of Virginia.

Chesapeake Bay. Agriculture has long been a substantial portion of the land-use cover on the Eastern Shore (Fig. 2(a)). In the absence of streams on the Eastern Shore, groundwater from a confined aquifer system is the only source of potable water (Richardson, 1994). There is also a shallow water-table aquifer that is used primarily for agricultural irrigation. The water table in this aquifer is close to land surface and thus mimics the topography (Fig. 2(b)). The shallow aquifer is contaminated in many areas by high nitrate levels, and is the primary source of nitrogen loading to the bay from the Eastern Shore of Virginia (Raey *et al.*, 1992; Gallagher *et al.*, 2001). The purpose of this study was to use a recently developed groundwater model of the Eastern Shore to study the movement and discharge of nitrogen-laden groundwater to the bay, and to simulate future changes in nitrogen loadings in response to changes in nitrogen (mostly fertilizer) application rates.

GROUNDWATER FLOW MODEL

A groundwater flow model of the Eastern Shore of Virginia is currently (2007) being developed as a joint project between federal, state, and local government offices. The model will be used primarily by water-resources managers to help plan future

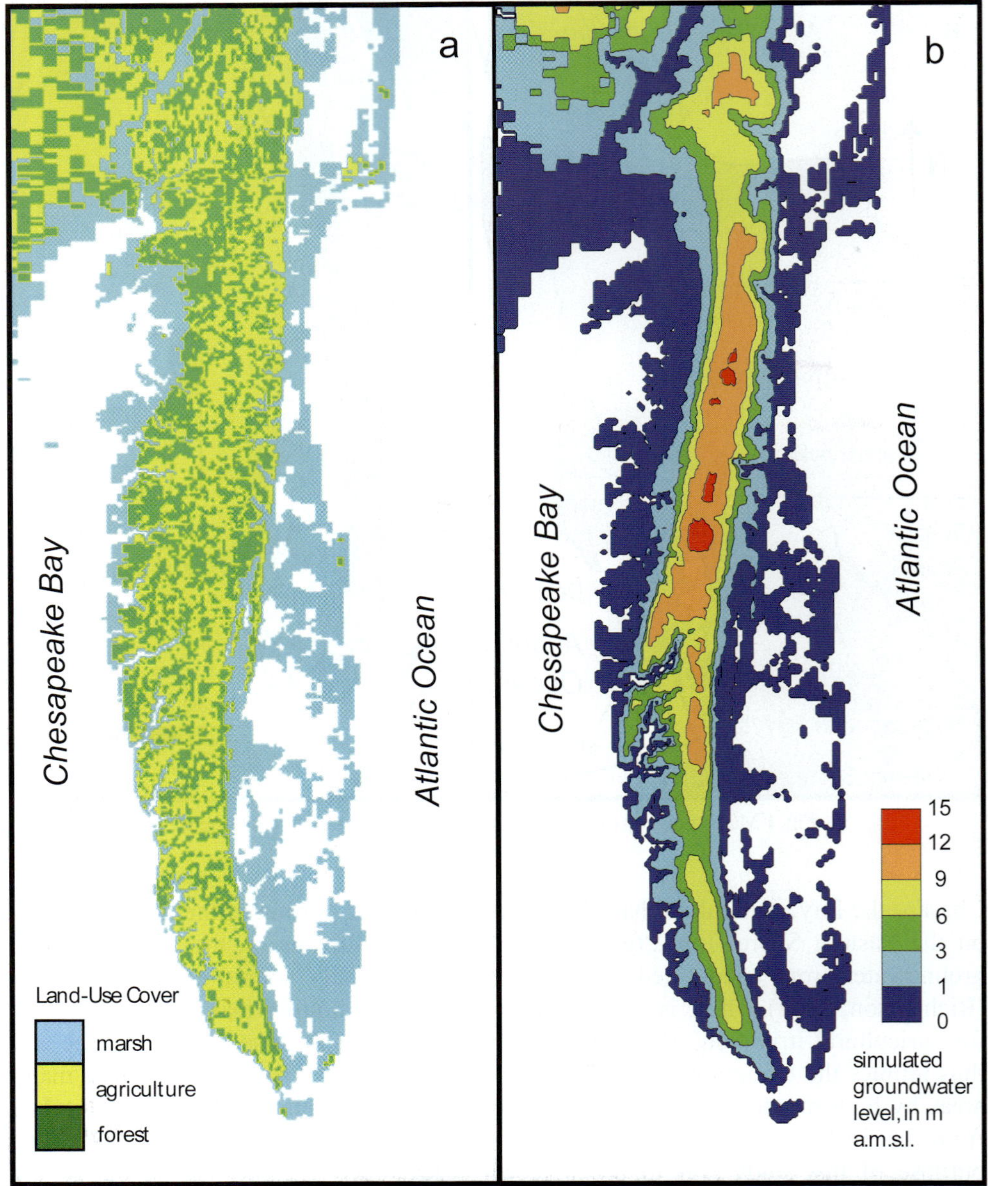

Fig. 2 Maps of the Eastern Shore of Virginia showing: (a) land use from 1992 USGS National Land Cover Data Set (http://landcover.usgs.gov/natllandcover.php), and (b) simulated groundwater levels in the surficial aquifer.

groundwater development within the region, and to assess adverse impacts such as lowering of regional water levels and saltwater encroachment. The US Geological Survey (USGS) code SEAWAT (Langevin *et al.*, 2003) was used to construct the groundwater flow model. SEAWAT can account for the differences in density between groundwater and seawater and so can simulate the encroachment of saltwater into the confined freshwater aquifer system.

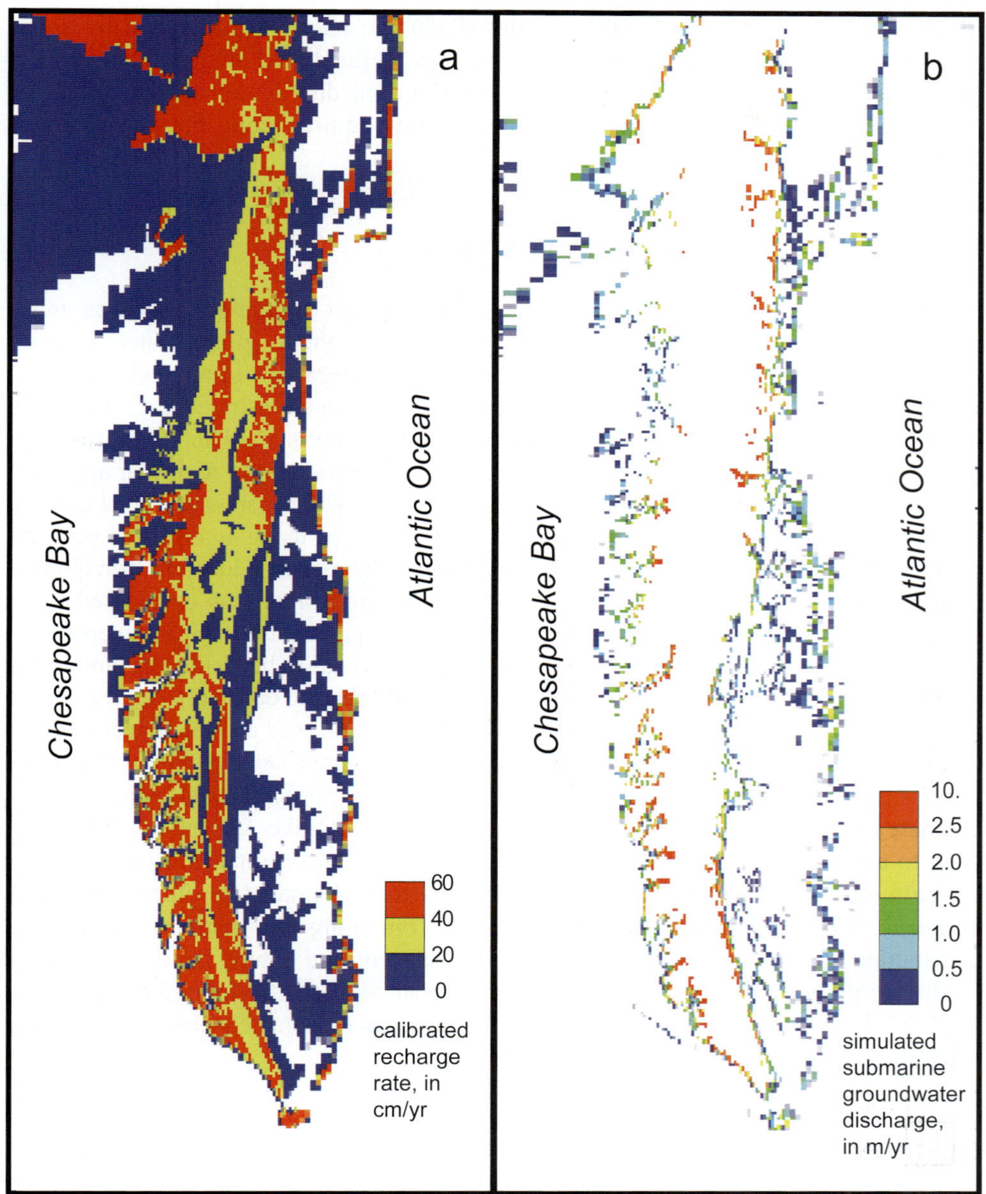

Fig. 3 Maps showing: (a) calibrated recharge rates assigned by zones based on soil texture, and (b) simulated fluxes of submarine groundwater discharge.

The model was calibrated using over 500 groundwater levels in wells spanning over 50 years. In the confined aquifer, these water levels have been declining because of regional pumping stresses. The model also includes the shallow water-table aquifer. In the shallow aquifer, young groundwater ages were measured in wells (Spieran, 1996; USGS, unpublished data), and these were used to help calibrate the shallow part of the model. In the shallow system, evapotranspiration rates were calibrated along with hydraulic conductivities and recharge rates. Recharge rates were assigned by zones (Fig. 3(a)) that coincided with soil textures as defined and mapped by the US Department of Agriculture. The model also simulates the distribution and magnitude of submarine groundwater discharge from the shallow aquifer along the coastal inlets (Fig. 3(b)).

GROUNDWATER TRANSPORT OF NITRATE

To simulate the transport of nitrate through the shallow aquifer to discharge zones along the coastal inlets, an approach was taken that combines groundwater-flowpath simulation with the estimation of historical nitrate concentrations in the recharge zone. Flowpaths throughout the shallow aquifer were calculated using the USGS code MODPATH (Pollock, 1994). One simulated pathline was started in every recharge grid cell of the model and tracked forward to its discharge location. A similar approach has been taken in previous studies (Modica *et al.*, 1998; Kauffman *et al.*, 2001). In the current study, travel times for the pathlines were calculated using an effective porosity of 0.25. The thousands of pathlines were then sorted by recharge location, travel time, land use at the recharge point, and discharge location. Nitrate was considered to be conservative along the flow path, but consumed by denitrification if it intercepted the root zone at the discharge cell. As the faster flow paths tend to be shallower (and in the root zone), denitrification was assigned in the discharge cell only to the faster pathlines. The proportion that was assigned denitrification was pro-rated according to the ratio of evapotranspiration to total discharge in the cell. Only pathlines that were not assigned denitrification were used to tally the total nitrogen load to the bay. An estimate of historical nitrate concentrations over time (Bohlke & Denver, 1995) was used to assign nitrate concentration to those pathlines that originated at cells with an agricultural land-use cover (Fig. 2(a)). Pathlines that originated in other areas were assigned lower, ambient concentrations. By combining the pathline travel-time distribution with nitrate concentration history, current and future trends in nitrogen loading to the bay were calculated for various scenarios of change in nitrate application load at the land surface. A runoff component was estimated from hydrograph separation at a small creek on the peninsula.

RESULTS

The age distribution for groundwater discharging to the bay is highly skewed (Fig. 4). The mean simulated travel time of all groundwater from recharge to discharge on the Eastern Shore is 30 years. The median simulated travel time is 15 years. The mean and median ages of groundwater directly discharging to coastal inlets are 40 and 18 years, respectively. The mean direct-discharge travel time is older than the mean recharge travel time because much of the younger water escapes by evapotranspiration, which is not counted in the directed discharge mean age. Mean and median ages varied by land-use type at the recharge location. Water recharged in agricultural areas that directly discharged had a mean and median age of 31 and 17 years, respectively, whereas water from forest land had means and median ages of 28 and 18 years, respectively, and

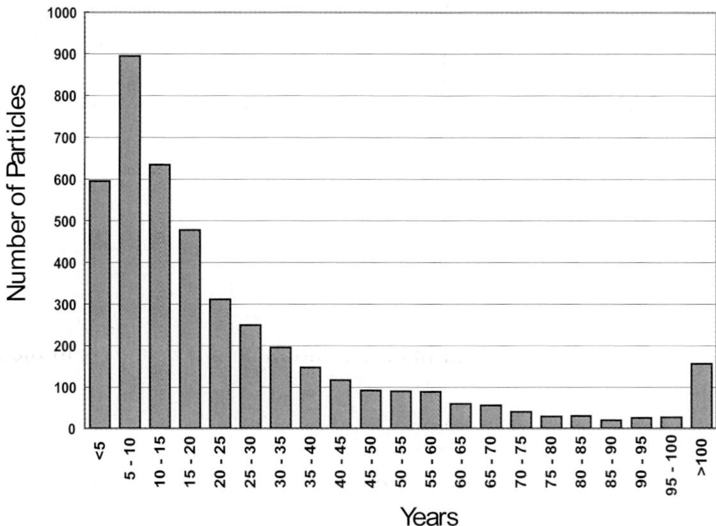

Fig. 4 The distribution of groundwater age on the Eastern Shore of Virginia according to simulated pathlines in the flow model.

marsh lands had 200 and 37 years respectively. Low recharge rates in the marsh soils resulted in very slow flow through those regions.

Even with a median age of 18 years for groundwater discharge, a substantial fraction of the water has an age of many decades (Fig. 4). Nitrate concentrations began rising substantially only in the last 30 years or so, when fertilizer application rates began increasing greatly (Bohlke & Denver, 1995). Therefore, much of the groundwater with elevated concentrations of nitrate is not yet discharging to the bay. Thus we expect that with no change in current nitrate application at the land surface, nitrogen load to the bay will continue to increase until the discharge load equals the recharge load, which will take many decades to occur. The path-line age distribution was used to calculate future trends in the nitrogen load to the bay, given possible changes in future nitrogen application rates (Fig. 5). The load calculation is made by multiplying the flux associated with each pathline by the concentration during the year of recharge, and the year is calculated from the travel time. The load estimates include a runoff component as well as the groundwater discharge component. As is expected, a "no change" in application load scenario led to increasing nitrogen discharge over the next several decades. Cutting the nitrogen application by half resulted in loads staying close to those currently observed. Eliminating the nitrogen input altogether resulted in an initial small decline caused by the loss of the runoff component, but several decades were required before the loads were reduced to substantially lower levels.

IMPLICATIONS FOR ESTIMATING NUTRIENT DISCHARGE

For the Eastern Shore of Virginia, our simulated estimates of nitrogen loading were similar to the cap load allocations (Fig. 5) for this section of the bay suggested by the

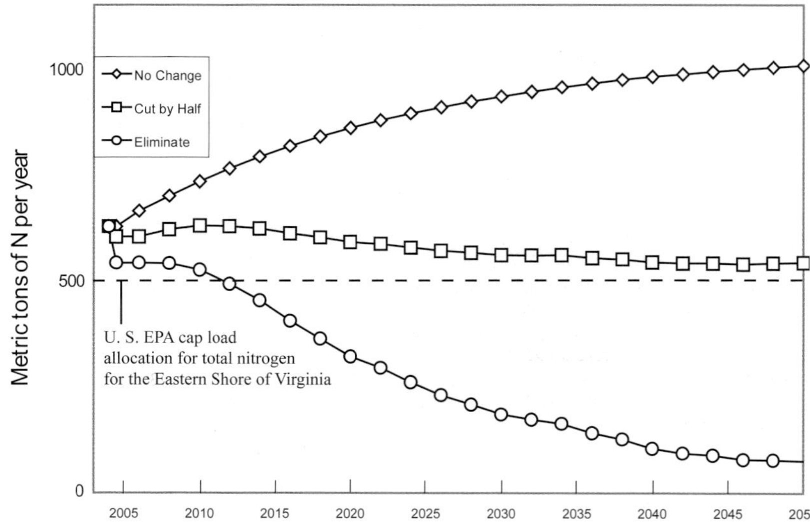

Fig. 5 Future trends in nitrogen loading to the Chesapeake Bay from the Eastern Shore of Virginia calculated with the model groundwater age distribution and three different scenarios for future nitrogen application.

US Environmental Protection Agency (Koroncai *et al.*, 2003). Assessments of contaminant loading to coastal water bodies, such as the Chesapeake Bay, are sometimes performed using models that primarily account for the surface-water transport components of the hydrological system (Linker *et al.*, 1996). Although such models may be appropriate for constituents, such as phosphorus or sediment, that are transported predominantly by surface water, our study has shown that for dissolved constituents that are transported by groundwater, such as nitrate, the lag time introduced by the flow system can be the dominant factor affecting loading trends for many decades. Models of nitrogen loading that neglect the groundwater-flow component will likely perform poorly if used to estimate future trends, and will underestimate the time required for total loads to respond to changes in management practices.

Acknowledgements This work has been funded in part by the USGS Office of Ground Water and the Virginia Department of Environmental Quality.

REFERENCES

Bohlke, J. K. & Denver, J. M. (1995) Groundwater dating, chemical, and isotopic analyses to resolve the history and fate of nitrate contamination in two agricultural watersheds, Atlantic Coastal Plain, Maryland. *Water Resour. Res.* **31**(9), 2319–2339.

Gallagher, D. L., Wynn, J. W., Reay, W. G. & Robinson, M. (2001) A geographic information system analysis of submarine groundwater discharge on the Eastern Shore of Virginia. In: *Proc. First International Conference on Saltwater Intrusion and Coastal Aquifers—Monitoring, Modeling, and Management* (ed. by D. Ouazar & A. H.-D. Cheng), 1–13. SWICA, Essaouira, Morocco.

Kauffman, L. J., Baehr, A. L., Ayers, M. A. & Stackelberg, P. E. (2001) Effects of land use and travel time on the distribution of nitrate in the Kirkwood-Cohansey aquifer system in southern New Jersey. *US Geol. Survey Water-Resour. Invest. Report 01-4117.*

Koroncai, R., Linker, L., Sweeney, J. & Batiuk, R. (2003) Setting and allocating the Chesapeake Bay basin nutrient and sediment loads. *US Environmental Protection Agency Report 93-R-03-007.*

Langevin, C. D., Shoemaker, W. B. & Guo, W. (2003) MODFLOW-2000, the U. S. Geological Survey modular ground-water model—documentation of the SEAWAT-2000 version with the variable-density flow process (VDF) and the integrated MT3DMS transport process (IMT). *US Geol. Survey Open-File Report 03-426.*

Lindsey, B. D., Phillips, S. W., Donnelly, C. A., Speiran, G. K, Plummer, N. L., Bohlke, J. K., Focazio, M. J. & Burton, W. C. (2003) Residence time and nitrate transport in ground water discharging to streams in the Chesapeake Bay watershed. *US Geol. Survey Water-Resour. Invest. Report 03-4035.*

Linker, L. C., Stigall, C. H., Chang, C. H. & Donigan, A. S., Jr (1996) Aquatic accounting: Chesapeake Bay watershed model quantifies nutrient loads. *Water Environ. Technol.* **8**(1), 48–52.

Modica, E., Buxton, H. T. & Plummer, L. N. (1998) Evaluating the source and residence times of groundwater seepage to streams, New Jersey Coastal Plain. *Water Resour. Res.* **34**(11), 2797–2810.

Richardson, D. L. (1994) Hydrogeology and analysis of the ground-water-flow system of the Eastern Shore, Virginia. *US Geol. Survey Water-Supply Paper 2410.*

Phillips, S. W. & Lindsey, B. D. (2003) The influence of ground water on nitrogen delivery to the Chesapeake Bay. *US Geol. Survey Fact Sheet FS-091-03.*

Pollock, D. W. (1994) User's guide for MODPATH/MODPATH-PLOT, version 3: A particle tracking post-processing package for MODFLOW, the US Geological Survey finite-difference ground-water flow model. *US Geol. Survey Open-File Report 94-464.*

Raey, W. G., Gallagher, D. L. & Simmons, G. M., Jr. (1992) Groundwater discharge and its impact on surface water quality in a Chesapeake Bay inlet. *Water Resour. Bull.* **28**(6), 1121–1134.

Speiran, G. K. (1996) Geoydrology and geochemistry near coastal ground-water discharge areas of the Eastern Shore, Virginia. *US Geol. Survey Water-Supply Paper 2479.*

A New Focus on Groundwater–Seawater Interactions
(Proceedings of Symposium HS1001 at IUGG2007, Perugia, July 2007). IAHS Publ. 312, 2007.

Effect of tidal fluctuations on contaminant transfer to the ocean

IVANA LA LICATA[1], CHRISTIAN D. LANGEVIN[2] & ALYSSA M. DAUSMAN[2]

1 *Dipartimento di Ingegneria Idraulica Ambientale e del Rilevamento (D.I.I.A.R.), Politecnico di Milano, I-20133 Milan, Italy*
ivana.lalicata@polimi.it

2 *US Geological Survey, Florida Integrated Science Center, Fort Lauderdale, Florida 33315, USA*

Abstract Variable-density groundwater flow was simulated to examine the effects that tide has on the coastward migration of a contaminant through a freshwater/saltwater interface and toward a coastal ocean boundary. Simulated ocean tides did not significantly affect the total contaminant mass input to the ocean; however, the difference in tidal and non-tidal simulated concentrations could be as much as 15%. It may be possible to numerically approximate the tidal-driven hydraulic transients in transport models that do not explicitly include tides by locally increasing dispersivity.

Key words tidal variation; contaminant transport; density

INTRODUCTION

Simulation of contaminant transport in coastal aquifers is intrinsically complex and computationally expensive because of the complex flow patterns that develop when freshwater mixes with saline groundwater. Because of its density, seawater intrudes landward below freshwater and a diffuse transition zone occurs between these two different fluids (Henry, 1959). The extent of saltwater intrusion is affected by a large number of physical and hydraulic parameters, such as recharge and aquifer dispersivity. The location and shape of the transition zone influences groundwater flow direction and its velocity. Furthermore, contaminant migration in groundwater is also affected by the characteristics of the transition zone between freshwater and saltwater. An additional complication in simulating contaminant transport in coastal aquifers is the confounding effect of tides which necessitates the use of a short time step, resulting in substantial computational effort (Volker *et al.*, 1998). Several studies have been completed on these topics: Schincariol & Schwartz (1990) and Oostrom *et al.* (1992a,b) studied the behaviour of dense plumes in a horizontal and uniform flow field in porous media; Knoch & Zhang (1992) considered also the effects of contaminant density on its movement in a steady horizontal flow field. Li *et al.* (1999) showed, by a theoretical model, an important contribution of local tide variations on groundwater discharge into the ocean. In order to investigate the consequence of simplifying the seaward boundary condition by neglecting the seawater density and tidal variations in numerical prediction of contaminant transport, Zhang *et al.* (2001) presented a comparison of numerical prediction with experimental results. The comparison indicates that neglecting seawater intrusion and tidal variations does not markedly affect the migration rate of the plume before it reaches the saltwater interface.

Shoemaker (2004) demonstrated that the dispersivity parameter is necessary to reproduce an exact distribution of hydraulic head, salinity and flow in the transition zone between freshwater and saltwater in a coastal aquifer system. Recently, Robinson *et al.* (2006) showed that tidal variations significantly affect the transport pathway of contaminants discharging to coastal waters in a study that examined mixing mechanisms in the subterranean estuary driven by tidal forcing.

The purpose of this paper is to investigate the influence of tidal variation on solute transport where contaminants reach the freshwater–saltwater transition zone and discharge into the ocean. This is accomplished using simulations based on a simple two-dimensional cross-sectional model that explicitly represents coastal groundwater flow within the freshwater and saltwater transition zone. Simulations that neglected (*No Tide*) and included (*Tide*) a tidally fluctuating ocean boundary were compared with the aim to analyse the influence of tidal variation on contaminant behaviour and mass flux toward the sea. Similar simulations were conducted using three different dispersivity values. These simulations also considered the effects of a pumping well on capturing a contaminant moving toward the sea. Simulations with spatially varying dispersivity values were also run to determine if the mechanical dispersion attributed to hydraulic transients could be approximated in a steady-flow model with increases in dispersivity.

Coastal groundwater flow problems require careful representation of fluid density and its effect on groundwater flow. For this reason, the SEAWAT computer program (Langevin *et al.*, 2003; Langevin & Guo, 2006) was used for the analyses. SEAWAT is a combination of MODFLOW (McDonald & Harbaugh, 1996) and MT3DMS (Zheng, 1990), designed to simulate variable-density groundwater flow coupled with solute transport. SEAWAT has been tested with many of the commonly used benchmark problems (Guo & Langevin, 2002; Langevin *et al.*, 2003; Bakker *et al.*, 2004; Langevin & Guo, 2006). The program has been applied to issues related to submarine groundwater discharge (Langevin, 2001, 2003), saltwater intrusion (Shoemaker & Edwards, 2003; Rao *et al.*, 2004; Shoemaker, 2004; Masterson, 2004; Dausman & Langevin, 2005), coastal wetland hydrology (Langevin *et al.*, 2004, 2005), and island hydrology (Schneider & Kruse, 2003).

PROBLEM DESCRIPTION

Groundwater flow and transport simulations were run using a two-dimensional cross-section model (Fig. 1) in which active cells were assigned properties similar to those of a coastal aquifer located near a coastal refinery in eastern Italy. The parameters assigned to the model are shown in Table 1 and are based on general knowledge, field data and results from prior studies (Alberti *et al.*, 2006).

A constant head equal to 1.25 m was assigned on the western boundary. Constant heads and salinities were assigned to the eastern sea boundary. In the simulation without tides (*No Tide*) the constant-head boundary representing the sea was set to mean sea level (0 m) and salinity to 35 kg/m^3. For *Tide*, the sea boundary was assigned temporally fluctuating heads to represent tides (Fig. 2). A diurnal tidal cycle is simulated in accordance with the relation $h_T = A\sin(\varpi t)$, where h_T is the time-varying

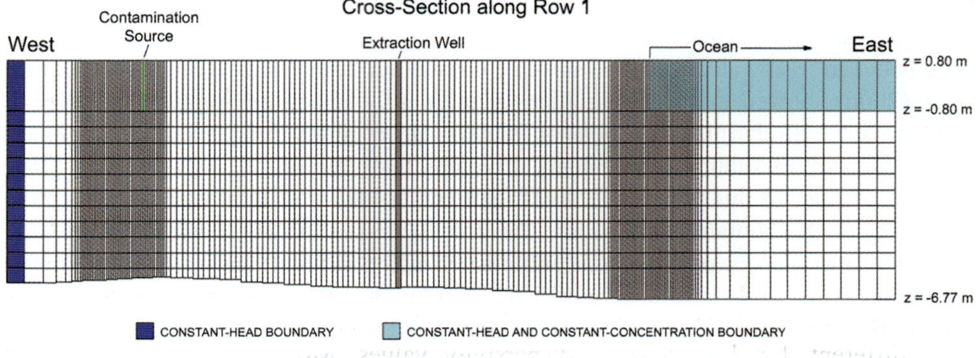

Fig. 1 Cross-section showing model grid and boundary conditions.

Table 1 Parameters assigned to the model.

Height (m)	7.5	Number of layers	12
Length (m)	500	Hydraulic conductivity (m/s)	5×10^{-3}; 1×10^{-5}
Number of rows	1	Recharge (m/d)	4.8×10^{-3}
Number of columns	201	Porosity	0.25

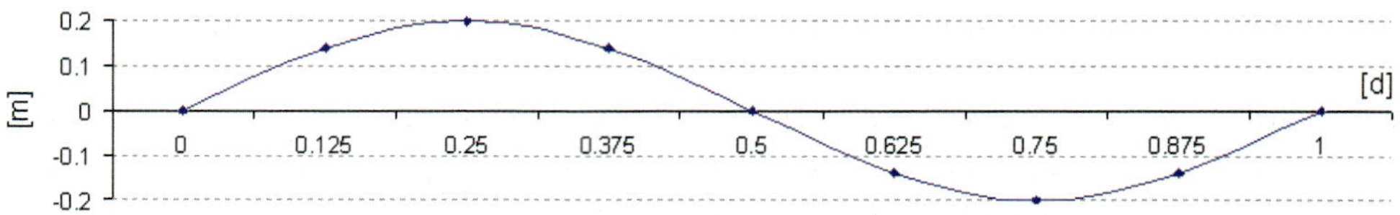

Fig. 2 Head variations in one tidal cycle.

head (relative to the level 0 m), A is the tidal amplitude, and ϖ the tidal frequency (6.283 d^{-1}; period = 1 day). The resulting tidal signal roughly approximates the tidal signal observed in the northern Adriatic Sea. *Tide* represents a 2-year period using 3-hour stress periods (a total of 5760 stress periods).

Simulations were performed using a variety of dispersivity values. For several simulations, a heterogeneous dispersivity distribution was used. For all representations, the ratio of longitudinal to transverse to vertical transverse dispersivity was held constant at 1:0.1:0.01 (Table 2). A contamination source, releasing a pollutant similar to MTBE, was located at a distance of 75 m from the western border of the model domain. The constant-concentration cell representing the source was assigned an arbitrary concentration value of 100 kg/m^3. The contaminant is simulated as being conservative.

Table 2 Dispersivity values used.

	α_L	α_T	α_V
Case1	16 m	1.6 m	0.16 m
Case2	10 m	1.0 m	0.10 m
Case3	5 m	0.5 m	0.05 m

α_L: longitudinal dispersivity; α_T: transverse dispersivity, α_V: vertical dispersivity.

RESULTS AND DISCUSSION

At the end of the 2-year simulation, the contaminant is at steady state, and the leading edge of the plume has reached the ocean (Fig. 3). As expected, the plume moves toward the freshwater–seawater interface and rises to a shallower part of the section as it reaches the denser water of the transition zone. The contaminant discharges into the ocean through a narrow outflow face near the shoreline.

Cumulative contaminant transfers were compared for simulations that neglected and included a tidally fluctuating ocean boundary to analyse the influence of tidal variation on contaminant transfer from the aquifer to the ocean. These comparisons were performed for three separate cases (Case 1, Case 2, and Case 3) characterized by three different, but homogenous, dispersivity values. Moreover, the concentration distribution was analysed for Case 2 and the differences in calculated concentration for simulations with and without tide were compared. Then, simulations with spatially varying dispersivity values were performed (case2a, case2b, case2c) to determine if the mechanical dispersion attributed to tidally driven hydraulic transients could be approximated in the model without tides by adjusting the dispersivity value near the transition zone (i.e. where the hydraulic transients are the largest).

Analysis 1 – Estimation of tidal effects on contaminant transfer into the ocean

In the first analysis, the results from the three simulations were compared to examine the effect of dispersivity on the contaminant flux to the ocean. A plot of the cumulative mass transfer from the aquifer to the ocean is shown in Fig. 4. This figure shows that tidal variations do not affect the cumulative transfer of the contaminant into the ocean. There essentially is no difference between simulations with and without tidal variations at the end of the 2-year simulation period. Close inspection of Fig. 4 reveals minor differences in the contaminant transfer between the tide and no-tide simulations from about 250 to 550 days. This slight difference is due to slightly higher concentrations reaching the ocean (for the simulation without tides) compared to the tidal simulation where tidally-driven mixing disperses the contaminant as it reaches the ocean. After this analysis, the effects of an extraction well were included in the simulations. The extraction well (representing a line sink in three dimensions), pumping at a rate of 1.44 m^3/d (approximately 60% of recharge), was assigned to column 100. Results from these simulations also show that the difference in contaminant transfer to the ocean is unimportant with and without tides.

The modelling results also demonstrate that the aquifer dispersivity affects the rate and cumulative contaminant transfer to the ocean. More dispersive mixing will result in a higher contaminant flux and higher cumulative contaminant transfer to the ocean. This difference is due to the constant-concentration boundary type used to represent the contaminant source. In fact, because of the constant-concentration boundary, the dispersive flux from the boundary into the aquifer is larger for higher dispersivity values than for lower values. Application of another type of concentration boundary (e.g. constant flux) would probably result in the same cumulative contaminant transfer regardless of dispersivity value.

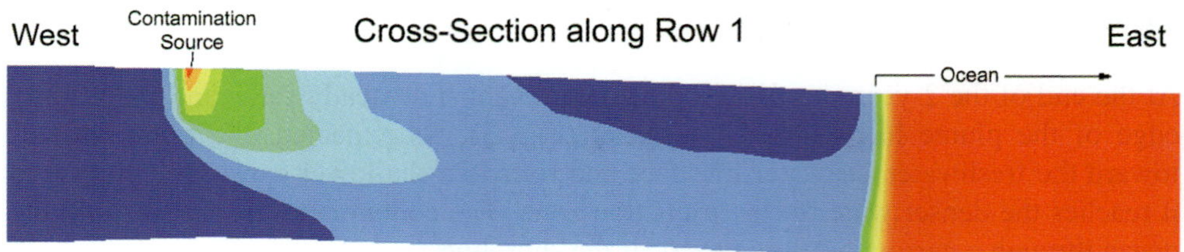

Fig. 3 Simulation of contaminant path toward the ocean.

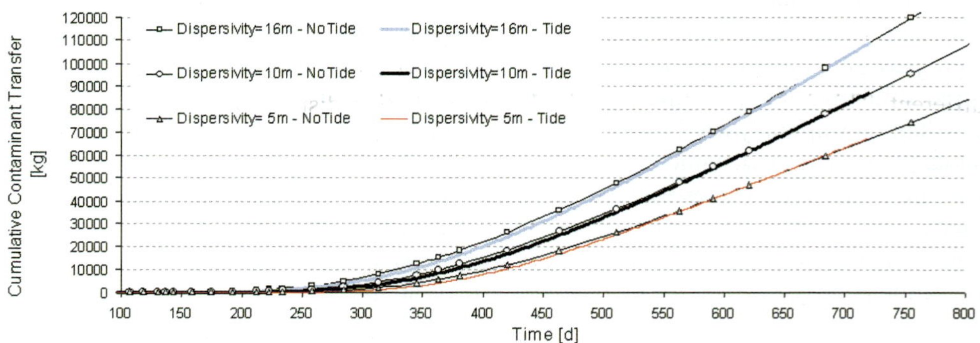

Fig. 4 Cumulative contaminant transfer from the aquifer to the ocean for the 2-year simulation period, without extraction well.

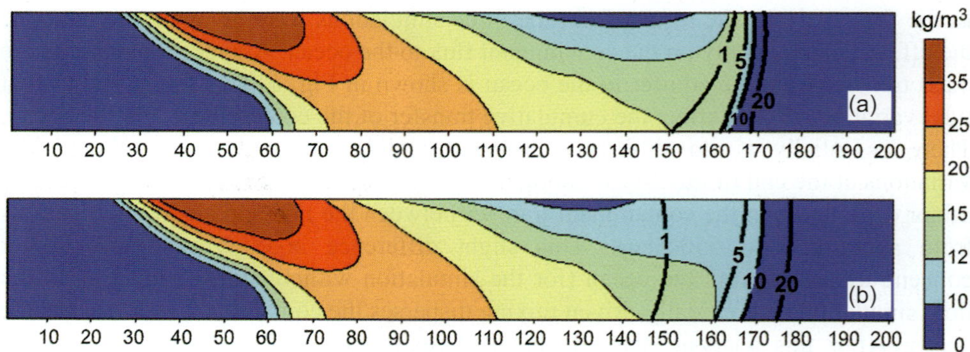

Fig. 5 Overlap of calculated contaminant concentration on freshwater–saltwater interface in simulation without (a) and with (b) tidal variations. Dark black contours represent salinity isosurfaces in kg/m^3.

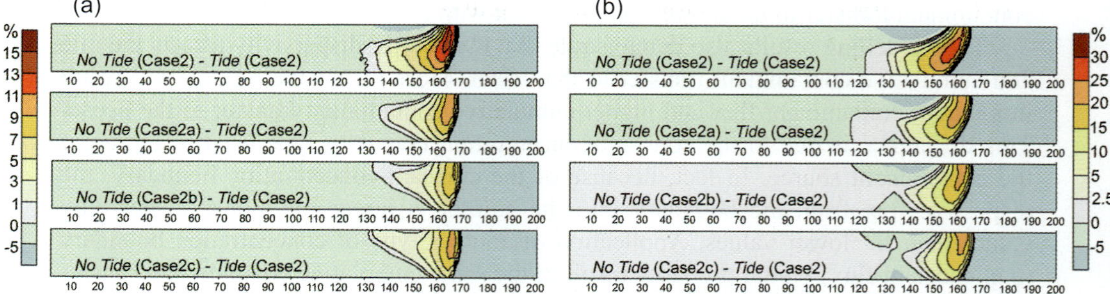

Fig. 6 Percentage difference in contaminant concentration without (a) and with (b) the extraction well.

Analysis 2 – Estimation of apparent dispersivity

Although tides do not appear to have a substantial affect on the contaminant transfer into the ocean, results show substantial differences in contaminant and salinity concentrations between *No Tide* and *Tide*. These concentration differences are due to the tidally-driven hydraulic transients in *Tide* that increase mixing. Simulations with tidal effects show a larger transition zone than simulations where the tide is not represented (Fig. 5). Contaminant concentrations near the ocean are generally higher for *No Tide* than for *Tide* (Fig. 5). This difference is due to the fact that the contaminant flux is the same (the contaminant flux from the constant-concentration boundary is identical between the two models); however, the mixing patterns are different. This induces a difference in the concentration distribution. In the following analysis, the results from three simulations were compared to examine the effect of dispersivity on the distribution of contamination concentrations.

Goode & Konikow (1990) define apparent dispersivities as "those values that yield the best match or calibration of the solute transport model under steady state flow conditions to a plume that developed under transient flow conditions". The concept of transient dispersion is used here to test the hypothesis that the effects of tidal mixing can be included in a model with a constant ocean stage boundary by increasing the aquifer dispersivity value. In this analysis the dispersivity values for a model without tides are adjusted to try and match the concentrations from the Case 2 simulation with tides. For the model without tides, the dispersivity (apparent) was increased only near the freshwater–seawater interface (from column 135 to column 190). Increasing dispersivity over the whole section creates large differences in calculated concentrations far from the zone influenced by the tide, and causes a larger contaminant flux into the aquifer from the constant-concentration boundary.

Three simulations with spatially varying dispersivity values were run to determine if the mechanical dispersion attributed to hydraulic transients could be approximated in a steady-flow model with increases to dispersivity. In these simulations, a larger dispersivity value was used in the tidal zone. These cases are referred to as cases 2a, 2b, and 2c. The dispersivity values used in the tidal zone are listed in Table 3.

Table 3 Used dispersivity values across the interface in steady-state simulations and percentage difference in contaminant concentration without pumping well.

	α_L	α_T	α_V	%
Case 2	10 m	1.0 m	0.10 m	15
Case 2a	16 m	1.6 m	0.16 m	10
Case 2b	20 m	2.0 m	0.20 m	8
Case 2c	26 m	2.6 m	0.26 m	10

α_L: longitudinal dispersivity; α_T: transverse dispersivity, α_V: vertical dispersivity.

Figure 6 shows the contaminant concentration difference (as a percentage) between Case 2 (with tides) and cases 2a, 2b, and 2c (tides represented by an increase in dispersivity). For Case 2 (without the extraction well) the maximum difference is between 13% and 15% and decreases when the longitudinal dispersivity in the coastal zone increases to 20 m. In this case (Case 2b), the largest difference is at its minimum (equal to 8%). By doubling the dispersivity value, the maximum concentration difference is halved. When the dispersivity is larger than 20 m, the maximum difference increases. This occurs with and without the extraction well so it is not necessary to calibrate the dispersivity value and the extension of the area again when adding the well. These results indicate that, in order to achieve the calibration, it may be possible to approximate tidal effects in a model with a constant ocean boundary by optimizing the dispersivity value near the boundary.

CONCLUSION

Contaminant transport models require extensive computational resources that can result in lengthy runtimes, especially for those simulations that deal with seawater intrusion and tidal fluctuations. Several simulations were performed to investigate the influence of tidal variation on contaminant transport patterns for the situation in which a contaminant migrates through a coastal aquifer through the transition zone and into the ocean. This paper presents a comparison of numerical results between contaminant transport simulations with and without tidal effects. The comparison of simulations indicates that tidal variations do not affect the cumulative flux of a conservative contaminant from the aquifer to the ocean. In general, the cumulative mass transfer to the ocean does not differ between simulations. Model results also show that the aquifer dispersivity affects the rate and cumulative contaminant transfer to the ocean. This difference is due to the boundary type used to represent the contaminant source.

Simulations also show that a difference in the contaminant and salinity concentration distributions occurs when tides are represented. This is because of the larger transition zone that develops when tidal effects are included in the simulation. Thus, even if the contaminant transfer to the ocean is not affected by the ocean boundary conditions, the concentration distribution is different when tide is accounted for because of the different mixing zone. This conclusion has potentially severe implications for model calibration where erroneous adjustments may be required to match concentrations for models that do not explicitly represent tides. Results indicate that the dispersive mixing effects of tides can be represented in a model that does not explicitly represent tides by increasing the dispersivity value near the ocean. Thus, when tidal effects are neglected, the error that occurs in the estimation of concentration near the ocean is a function of the apparent dispersivity in correspondence to the transition zone. Therefore, it may be possible to replace tidal effects with a change in the dispersivity value in this zone when calibrating a model.

Acknowledgements I.L.L. thanks Prof. V. Francani, Politecnico di Milano (D.I.I.A.R), Italy, for allowing her to study abroad with the US Geological Survey in Fort Lauderdale, Florida. She also extends special thanks to her USGS colleagues for their guidance and expertise. The authors thank Mike Deacon, USGS, for his thorough review of an earlier draft of this manuscript.

REFERENCES

Alberti, L., Francani, V. & La Licata, I., (2006) Hydrogeologic parameters and human activities influence on sea water intrusion at a refinery site. In: International FEFLOW User Conference (Berlin, Germany, September 2006).

Bakker, M., Oude Essink, G. H. P. & Langevin, C. D. (2004) The rotating movement of three immiscible fluids–a benchmark problem. *J. Hydrol.* **278**, 270–278.

Dausman, A. M. & Langevin, C. D. (2005) Movement of the saltwater interface in the superficial aquifer system in response to hydrologic stresses and water-management practices, Broward County, Florida. *US Geol. Survey Scientific Investigation Report 2004-5256.*

Goode, D. J. & Konikow, L. F. (1990) Apparent dispersion in transient groundwater flow. *Water Resour. Res.* **26**(10), 2339–2351.

Guo, W. & Langevin, C. D. (2002) User guide to SEAWAT: a computer program for simulation of three-dimensional variable-density groundwater flow. *US Geol. Survey Open File Report 01-434.*

Henry, H. R. (1959) Salt intrusion into fresh-water aquifers. *J. Geophys. Res.* **64**, 1911–1019.

Knoch, M. & Zhang, G., (1992) Numerical simulation of the effects of a variable density in a contaminant plume. *Ground Water* **30**(5), 731–742.

Langevin, C. D. (2001) Simulation of ground-water discharge to Biscayne Bay, southeastern Florida. *US Geol. Survey Water Resources Investigation Report* 00-4251.

Langevin, C. D. (2003) Simulation of submarine ground water discharge to a marine estuary: Biscayne Bay, Florida. *Ground Water* **41**(6), 758–771.

Langevin, C. D. & Guo, W. (2006) MODFLOW/MT3DMS-based simulation of variable density ground water flow and transport. *Ground Water* **44**(3), 339–351.

Langevin, C. D., Shoemaker, W. B. & Guo, W. (2003) MODFLOW-2000, the US Geological Survey Modular Ground-Water Model: Documentation of the SEAWAT-2000 Version with the Variable-Density Flow Process (VDF) and the Integrated MT3DMS Transport Process (IMT). *US Geol. Survey Open File Report 2003-426.*

Langevin, C. D., Oude Essink, G. H. P., Panday, S., Bakker, M., Prommer, H., Swain, E. D., Jones, W., Beach, M. & Barcelo, M. (2004) MODFLOW-based tools for simulation of variable-density groundwater flow. In: *Coastal Aquifer Management: Monitoring, Modeling, and Case Studies* (ed. by A. Cheng & D. Ouazar), 49–76. Lewis Publishers, Boca Raton, Florida, USA.

Langevin, C. D., Swain, E. D. & Wolfert, M. A. (2005) Simulation of integrated surface-water/ground-water flow and salinity for a coastal wetland and adjacent estuary. *J. Hydrol.* **314**, 212–234.

Li, L., Barry, D. A., Stagnitti, F. & Parlange, J.-Y. (1999) Submarine groundwater discharge and associated chemical input to a coastal sea. *Water Resour. Res.* **35**(11), 3253–3259.

Masterson, J. P. (2004) Simulated interaction between freshwater and saltwater and effects of ground-water pumping and sea-level change, lower Cape Cod aquifer system, Massachusetts. *US Geol. Survey Scientific Investigation Report 2004-5014.*

McDonald, M. G. & Harbaugh, A. W. (1996) Programmer's documentation for Mudflow-96, an update to the US Geological Survey modular finite-difference groundwater flow model. *US Geol. Survey Open File Rep. 96-486.*

Oostrom, M., Hayworth, J. S., Dane, J. H. & Guven, O. (1992a) Behaviour of dense aqueous phase leachate plumes in homogeneous porous media. *Water Resour. Res.* **28**(8), 2123–2134.

Oostrom, M., Dane, J. H., Guven, O. & Hayworth, J. S. (1992b) Experimental investigation of dense solute plumes in an unconfined aquifer model. *Water Resour. Res.* **28**(9), 2315–2326.

Rao, S. V. N., Sreenivasulu, V., Bhallamudi, S. M., Thandaveswara, B. S. & Sudheer, K. P. (2004) Planning groundwater development in coastal aquifers. *Hydrol. Sci. J.* **49**(1), 155–170.

Robinson, C., Li, L. & Barry, D. A., (2006) Effect of tidal forcing on a subterranean estuary. *Adv. Water Resour.* doi: 10.1016/j.advwatres.2006.07.006.

Schincariol, R. A. & Schwartz, F. W. (1990) An experimental investigation of variable density flow and mixing in homogeneous and heterogeneous media. *Water Resour. Res.* **26**(10), 2317–2329.

Schneider, J. C. & Kruse, S. E. (2003) Assessing selected natural and anthropogenic impacts on freshwater lens morphology on small barrier islands: Dog Island and St George Island, Florida, USA. *J. Hydrogeol.* **14**(1-2), 131–145.

Shoemaker, B. W. & Edwards, M. K., (2003) Potential for saltwater intrusion into the lower Tamiami aquifer near Bonita Springs, southwestern Florida. *US Geol. Survey Water Resources Investigation Report 2003-4262.*

Shoemaker, B. W. (2004) Important observations and parameters for a salt water intrusion model. *Ground Water* **42**(6), 829–840.

Volker, R. E., Ataie-Ashtiani, B. & Lockington, D. A. (1998) Unconfined coastal aquifer response to sea boundary condition. In: Proc. Int. Groundwater Conf. Melbourne, Australia (ed. by T. R. Weaver & C. R. Lawrence), 771–776.

Zhang, Q., Volker, R. E. & Lockington, D. A. (2001) Influence of seaward boundary condition on contaminant transport in unconfined coastal aquifer. *J. Contam. Hydrol.* **49**(2001), 201–215.

Zheng, C. (1990) MT3D: A modular three-dimensional transport model for simulation of advection, dispersion and chemical reactions of contaminants in groundwater systems. Report to the US Environmental Protection Agency, Ada, Oklahoma, USA.

Key word index

Hydrology 2020

An Integrating Science to Meet World Water Challenges

Edited by Taikan Oki, Caterina Valeo & Kate Heal

H 2020 Hydrology
IAHS Working Group

Taikan Oki Japan (Chair)
Jeanna Balonishnikova Russia
Wolfgang Diernhofer Austria
Pierre Etchevers France
Stewart W. Franks Australia
Guobin Fu Australia
Kate Heal Scotland, UK
Susan S. Hubbard USA
Harouna Karambiri Burkina Faso
Johan Kuylenstierna Sweden
Stefan Uhlenbrook The Netherlands
Caterina Valeo Canada

A milestone capturing the state of the art in hydrological science at the beginning of the 21st century, a chart for hydrologists exploring the new frontiers in hydrology, and a guide for those involved with developing and implementing water policies.

A group of 12 younger hydrologists, with experience across the full spectrum of hydrology, was commissioned by the International Association of Hydrological Sciences (IAHS) in 2001 to look to the future and explore how hydrological sciences can evolve to meet the world water challenges that are expected to prevail by 2020. This book reports their deliberations. It considers the capability that hydrological sciences will, and should have by 2020, and what needs doing now in order to achieve this. There is an emphasis on societal issues and interdisciplinary work pertinent to hydrology as hydrologists cannot and should not work in isolation from society and other scientific disciplines.

The Executive Summary is available in English, Arabic, Chinese, French, German, Japanese, Russian and Spanish at:

www.iahs.info

IAHS Publ. 300
(2006)

ISBN 978-1-901502-33-6
190 + xxxii pp
Price £45.00

Published by IAHS Press

Please send book orders and enquiries to:

Mrs Jill Gash
IAHS Press, Centre for Ecology and Hydrology
Wallingford, Oxfordshire OX10 8BB, UK

jilly@iahs.demon.co.uk
tel.: + 44 1491 692442
fax: + 44 1491 692448/692424